VOLUME FIVE HUNDRED AND FORTY FIVE

METHODS IN ENZYMOLOGY

Regulated Cell Death Part B: Necroptotic, Autophagic and other Non-apoptotic Mechanisms

METHODS IN ENZYMOLOGY

Editors-in-Chief

JOHN N. ABELSON and MELVIN I. SIMON
Division of Biology
California Institute of Technology
Pasadena, California

ANNA MARIE PYLE
Departments of Molecular, Cellular and Developmental Biology and Department of Chemistry Investigator
Howard Hughes Medical Institute
Yale University

Founding Editors

SIDNEY P. COLOWICK and NATHAN O. KAPLAN

VOLUME FIVE HUNDRED AND FORTY FIVE

METHODS IN ENZYMOLOGY

Regulated Cell Death Part B: Necroptotic, Autophagic and other Non-apoptotic Mechanisms

Edited by

AVI ASHKENAZI
Cancer Immunology, Genentech, Inc., San Francisco, CA, USA

JAMES A. WELLS
Departments of Pharmaceutical Chemistry and Cellular & Molecular Pharmacology, University of California – San Francisco, CA, USA

JUNYING YUAN
Department of Cell Biology, Harvard Medical School, Boston, MA, USA

AMSTERDAM • BOSTON • HEIDELBERG • LONDON
NEW YORK • OXFORD • PARIS • SAN DIEGO
SAN FRANCISCO • SINGAPORE • SYDNEY • TOKYO
Academic Press is an imprint of Elsevier

Academic Press is an imprint of Elsevier
525 B Street, Suite 1800, San Diego, CA 92101-4495, USA
225 Wyman Street, Waltham, MA 02451, USA
The Boulevard, Langford Lane, Kidlington, Oxford, OX5 1GB, UK
32 Jamestown Road, London NW1 7BY, UK

First edition 2014

ISBN: 978-0-12-801430-1
ISSN: 0076-6879

For information on all Academic Press publications
visit our website at store.elsevier.com

CONTENTS

CONTRIBUTORS

Eric H. Baehrecke
Department of Cancer Biology, University of Massachusetts Medical School, Worcester, Massachusetts, USA

Rhesa Budhidarmo
Department of Biochemistry, Otago School of Medical Sciences, University of Otago, Dunedin, New Zealand

Catherine L. Day
Department of Biochemistry, Otago School of Medical Sciences, University of Otago, Dunedin, New Zealand

Alexei Degterev
Department of Developmental, Molecular & Chemical Biology, Tufts University School of Medicine, Boston, Massachusetts, USA

Paul C. Driscoll
Division of Molecular Structure, Medical Research Council, National Institute for Medical Research, London, United Kingdom

Peter Geserick
Section of Molecular Dermatology, Department of Dermatology, Venereology, and Allergology, Medical Faculty Mannheim, University Heidelberg, Heidelberg, Germany

Tae-Bong Kang
Department of Biological Chemistry, The Weizmann Institute of Science, Rehovot, Israel, and Department of Biotechnology, College of Biomedical and Health Science, Konkuk University, Chung-Ju, Republic of Korea

Maxime J. Kinet
Laboratory of Developmental Genetics, The Rockefeller University, New York, USA

Andrew Kovalenko
Department of Biological Chemistry, The Weizmann Institute of Science, Rehovot, Israel

Martin Leverkus
Section of Molecular Dermatology, Department of Dermatology, Venereology, and Allergology, Medical Faculty Mannheim, University Heidelberg, Heidelberg, Germany

Jenny L. Maki
Department of Developmental, Molecular & Chemical Biology, Tufts University School of Medicine, Boston, Massachusetts, USA

Adam J. Middleton
Department of Biochemistry, Otago School of Medical Sciences, University of Otago, Dunedin, New Zealand

Charles Nelson
Department of Cancer Biology, University of Massachusetts Medical School, Worcester, Massachusetts, USA

Vassiliki Nikoletopoulou
Institute of Molecular Biology and Biotechnology, Foundation for Research and Technology—Hellas, Heraklion, Greece

Ramon Schilling
Section of Molecular Dermatology, Department of Dermatology, Venereology, and Allergology, Medical Faculty Mannheim, University Heidelberg, Heidelberg, Germany

Pascal Schneider
Department of Biochemistry, University of Lausanne, Epalinges, Switzerland

Shai Shaham
Laboratory of Developmental Genetics, The Rockefeller University, New York, USA

John Silke
The Walter and Eliza Hall Institute of Medical Research, and Department of Medical Biology, University of Melbourne, Parkville, Victoria, Australia

Cristian R. Smulski
Department of Biochemistry, University of Lausanne, Epalinges, Switzerland

Brent R. Stockwell
Department of Biological Sciences; Department of Chemistry, and Howard Hughes Medical Institute, Columbia University, New York, USA

Nektarios Tavernarakis
Institute of Molecular Biology and Biotechnology, Foundation for Research and Technology—Hellas, Heraklion, Greece

Beata Toth
Department of Biological Chemistry, The Weizmann Institute of Science, Rehovot, Israel

Domagoj Vucic
Department of Early Discovery Biochemistry, Genentech, Inc., South San Francisco, California, USA

David Wallach
Department of Biological Chemistry, The Weizmann Institute of Science, Rehovot, Israel

Laure Willen
Department of Biochemistry, University of Lausanne, Epalinges, Switzerland

Adam J. Wolpaw
Residency Program in Pediatrics, The Children's Hospital of Philadelphia, Philadelphia, Pennsylvania, USA

Seung-Hoon Yang
Department of Biological Chemistry, The Weizmann Institute of Science, Rehovot, Israel

Junying Yuan
Department of Cell Biology, Harvard Medical School, Boston, Massachusetts, USA

Wen Zhou
Department of Cell Biology, Harvard Medical School, Boston, Massachusetts, USA

PREFACE

Cell turnover is a fundamental feature of metazoan biology. Severe damage to cellular integrity usually causes passive, nonregulated cell death. In contrast, more confined disruption can lead to more deliberate cell elimination, through specific mechanisms of Regulated Cell Death. In these two volumes of *Methods in Enzymology*, we aim to highlight the current molecular understanding of the major processes of Regulated Cell Death and to illustrate basic and advanced methodologies to study them. Volume A focuses on the most extensively studied mode of cell death—apoptosis. Volume B covers several nonapoptotic mechanisms. These include necroptosis, which shares certain signal transduction aspects with apoptosis but is unique in its execution phase, and autophagic cell death, which is an offshoot of autophagy—a more basic prosurvival metabolic adaptation mechanism. Chapters 1–4 cover how to measure necroptosis and various molecular components and complexes that signal this process. Chapter 5 discusses approaches to interrogating interactions between tumor necrosis factor superfamily ligands and receptors. Chapters 6–8 highlight nonapoptotic cell death mechanisms in the model organisms, *C. elegans* and *D. melanogaster*. Chapters 9 and 10 discuss structural aspects of death receptor complexes and strategies to study posttranslational modification of downstream signaling components by RING E3 ubiquitin ligases. Finally, Chapter 11 describes a multidimensional profiling approach to studying small-molecule-induced cell death. We hope these chapters will be both conceptually informative and practically useful for readers interested in the current understanding and the key open questions in each area, as well as in experimental strategies and techniques to interrogate nonapoptotic regulated cell death mechanisms.

AVI ASHKENAZI
JAMES A. WELLS
JUNYING YUAN

CHAPTER ONE

Assays for Necroptosis and Activity of RIP Kinases

Alexei Degterev*, Wen Zhou†, Jenny L. Maki*, Junying Yuan†,1

*Department of Developmental, Molecular & Chemical Biology, Tufts University School of Medicine, Boston, Massachusetts, USA

†Department of Cell Biology, Harvard Medical School, Boston, Massachusetts, USA

[1]Corresponding author: e-mail address: junying_yuan@hms.harvard.edu

Contents

Methods in Enzymology, Volume 545
ISSN 0076-6879
http://dx.doi.org/10.1016/B978-0-12-801430-1.00001-9

Abstract

Necrosis is a primary form of cell death in a variety of human pathologies. The deleterious nature of necrosis, including its propensity to promote inflammation, and the relative lack of the cells displaying necrotic morphology under physiologic settings, such as during development, have contributed to the notion that necrosis represents a form of pathologic stress-induced nonspecific cell lysis. However, this notion has been challenged in recent years by the discovery of a highly regulated form of necrosis, termed regulated necrosis or necroptosis. Necroptosis is now recognized by the work of multiple labs, as an important, drug-targetable contributor to necrotic injury in many pathologies, including ischemia–reperfusion injuries (heart, brain, kidney, liver), brain trauma, eye diseases, and acute inflammatory conditions. In this review, we describe the methods to analyze cellular necroptosis and activity of its key mediator, RIP1 kinase.

1. INTRODUCTION

1.1. Distinguishing features of necroptotic cell death

Discovery of regulated necrosis originates from the observations that "canonical" inducers of apoptosis, such as agonists TNFα family of death domain receptors (DRs), can trigger cell death morphologically resembling necrosis in cells either intrinsically deficient in caspase activation (e.g., mouse fibrosarcoma L929 cells) or under conditions when caspase activation is inhibited (e.g., caspase-8-deficient Jurkat cells or cells treated with pan-caspase inhibitor zVAD.fmk) (Holler et al., 2000; Matsumura et al., 2000; Vercammen, Vandenabeele, Beyaert, Declercq, & Fiers, 1997). The lack of caspase activation as well as the absence of other typical features of apoptosis, such as cytochrome *c* release, membrane blebbing, phosphatidylserine (PS) exposure, and intranucleosomal DNA cleavage, served as important initial differentiators between necroptosis and apoptosis (Tait & Green, 2008).

Electron microscopy has also proved very useful in distinguishing necroptosis from apoptosis in morphology. Necroptotic cells are characterized by the lack of typical nuclear fragmentation, swelling of cellular organelles especially mitochondria, and the loss of plasma membrane integrity, whereas apoptotic cells exhibit shrinkage, blebbing, nuclear fragmentation, and chromatin condensation (Degterev et al., 2005). Robust activation of autophagy is another feature of necroptosis which provides useful means to distinguish this form of cell death *in vitro* and *in vivo* both morphologically (e.g., by EM) and at the molecular level (e.g., by measuring of LC3II formation) (Degterev et al., 2005; Yu et al., 2004). This leads to necroptosis in some cases being referred to as "autophagic cell death," such as zVAD-induced death of L929 cells (Yu et al., 2004). It should be noted, however, that functional role of autophagy varies greatly depending on the specifics of necroptosis activation, with instances where this process promotes, inhibits, or does not affect cell death (Degterev et al., 2005; Shen & Codogno, 2012; Yu et al., 2004). Furthermore, activation of necroptosis-inducing necrosome complex (discussed below) can also happen downstream from autophagosome formation (Basit, Cristofanon & Fulda, 2013).

A detailed comparison of TNF-induced necroptosis and H_2O_2-induced necrosis was performed by Vanden Berghe et al. (2010). Despite the different kinetics of cellular events including ROS production, mitochondrial polarization changes, and lysosomal membrane permeabilization, the major hallmarks of necroptosis and oxidant-induced necrosis were remarkably similar, leading to an important conclusion that necroptosis is a subtype of necrosis, morphologically indistinguishable from other types of necrosis but defined by a specific mode of activation (discussed below).

Generation of DAMPs as a result of cell lysis is an important consequence of necroptotic death both *in vitro* and *in vivo* (Duprez et al., 2011; Murakami et al., 2013). In addition, recent evidence suggests that synthesis of TNFα occurs independently of cell death as a result of specific signaling by key necroptosis initiator RIP1 kinases (RIPK1) (Christofferson et al., 2012; Kaiser et al., 2013; McNamara et al., 2013). Autocrine TNFα can promote cell death dependent on a cytosolic complex "ripoptosome" consisting of RIPK1, FADD, and caspase-8 (Biton & Ashkenazi, 2011; Hitomi et al., 2008; Kaiser et al., 2013; Tenev et al., 2011). Several instances have also been reported where RIPK1 and RIPK3 promote inflammatory signaling through the production of IL-1α and IL-1β/IL-18 in the absence of cell death (Kang, Yang, Toth, Kovalenko, & Wallach, 2013; Lukens et al., 2013). These data highlight complex interrelationship between necroptosis and inflammation.

1.2. Pathways and mediators of necroptosis

We refer the readers to a number of in-depth reviews on the subject (Christofferson et al., 2012; Christofferson, Li, & Yuan, 2014;

Christofferson & Yuan, 2010b; Fulda, 2013; Zhou, Han, & Han, 2012). We will just briefly summarize some of the key findings. Initiation of necroptosis is best understood in the context of TNFα signaling. Engagement of TNFR1 leads to the formation of a membrane-bound complex named Complex I, containing RIPK1, TRADD, and TRAF2 as key components (Micheau & Tschopp, 2003). Ubiquitination of Lys377 of RIPK1 within this complex leads to the assembly of NF-kB-activating complexes involving TAK1 and IKK kinases (Ea, Deng, Xia, Pineda, & Chen, 2006). Dissociation of the components from TNFR1 is followed by the assembly of cytosolic signaling complexes: either Complex IIa/DISC including RIPK1, FADD, and caspase-8 which leads to apoptosis (Micheau & Tschopp, 2003), or Complex IIb/necrosome including FADD, RIPK1, and RIPK3 which leads to necroptosis in the absence of caspase activity (summarized in Galluzzi, Kepp, & Kroemer, 2009). Activation of necroptosis requires cross-phosphorylation of RIPK1 and RIPK3, utilizing Ser/Thr kinase domains of both proteins (Cho et al., 2009). RIPK1 and RIPK3 kinases further form amyloid-like fibers (Li et al., 2012), and RIPK3 recruits and phosphorylates pseudokinase MLKL on Thr357/Ser358, which serves as a critical gateway to necroptosis execution (Murphy et al., 2013; Sun et al., 2012; Wu et al., 2013). Downstream events are currently less well understood. As discussed above, oxidative stress mediated by mitochondrial Complex I and NADPH oxidase was found to play a role in some cell types. Other factors, such as Ca^{2+}, ceramide, activation of autophagy, and HtrA2 and UCH-L1 proteases (Sosna et al., 2013), have also been proposed to play a role. However, connections between these factors and necrosome remain unknown.

Other signals were also shown to promote necrosome activation, but the mechanisms may differ. For example, multiple Toll-like receptors (TLRs) were found to induce necroptosis (He, Liang, Shao, & Wang, 2011; Kaiser et al., 2013). The mechanisms differ depending on the specific signals and cell types. TLR3 and TLR4 act through adaptor TRIF to directly recruit RIPK1 and RIPK3 through their RHIM domains, while other TLRs signaling through MyD88 adaptor trigger necroptosis through an autocrine TNFα loop. Furthermore, while RIPK1 is required for TRIF-mediated necroptosis in macrophages, it is dispensable in epithelial and fibroblast cells. Additional signals directly triggering RIPK3, such as activation of viral DNA sensor DAI (Upton, Kaiser, & Mocarski, 2012), have also been described, and overexpression of RIPK3 was shown to reduce the requirement for RIPK1 in necroptosis initiation (Moujalled et al., 2013). Interferons were also found to be efficient inducers of necroptosis, utilizing kinase PKR to initiate necrosome formation (Thapa et al., 2013).

While RIPK3 clearly plays an indispensable role in necroptosis, RIPK1 appears to serve a critical role as a master regulator controlling multiple cell fate decisions, including cell survival, apoptosis, and necroptosis. RIPK1 is a multidomain protein, which contains N-terminal Ser/Thr kinase, followed by intermediate domain including K377 ubiquitination site and RHIM motif, and C-terminal death domain mediating binding to DRs. E3 ubiquitin ligases cIAP1/2 in concert with TRAF2 ubiquitinates RIPK1 in Complex I, providing conditions for TAK1 and IKK kinase complex binding, activating the downstream proinflammatory and prosurvival pathways (Arslan & Scheidereit, 2011). RIP1 deubiquitinase CYLD, operating in Complex II (Moquin, McQuade, & Chan, 2013), is critical for necrosome formation and activation of necroptosis (Hitomi et al., 2008). Notably, CYLD is cleaved by caspase-8/c-FLIP$_L$ heterodimer (Oberst et al., 2011), explaining reciprocal regulation of apoptosis and necroptosis. RIPK1 kinase activity is required for necrosome formation and necroptosis. Finally, inhibition of cIAP1/2 and TAK1 can also promote another function of RIP1 kinase activity, that is, activation of caspase-8 and apoptosis (Dondelinger et al., 2013; Feoktistova et al., 2011; Tenev et al., 2011). RIPK1-dependent apoptosis is activated by DRs, TLRs, DNA-damaging agents, and other anticancer drugs *in vitro* (Abhari et al., 2013; Feoktistova et al., 2011; Loder et al., 2012; Tenev et al., 2011; Wagner et al., 2013). However, the physiologic role of this pathway and details of its activation are currently unknown. The role of RIPK3 in RIPK1-dependent apoptosis is also not entirely clear.

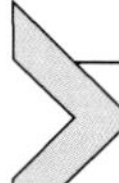

2. CELLULAR MODELS OF NECROPTOSIS

2.1. Cell types (Table 1.1)

A number of different cell types (both cell lines and primary cells) have been reported to undergo necroptosis *in vitro* in response to different stimuli, which have provided convenient systems to study this pathway. Some of the widely used cellular models are listed in Table 1.1. Conversely, a number of commonly used epithelial cancer cell lines, such as HEK293, HeLa, and

Table 1.1 Several widely used cellular models of necroptosis

Cell type	ATCC number	Typical necroptosis conditions
Jurkat A3 cells	CRL-2570	FasL (5 ng/mL), cycloheximide (CHX) (1 μg/mL), zVAD (100 μ*M*) (Holler et al., 2000)
FADD-deficient Jurkat cells	CRL-2572	Human TNFα (10 ng/mL) (Degterev et al., 2008)
U-937 cells	CRL-1593.2	Human TNFα (40 ng/mL), zVAD (100 μ*M*) (Degterev et al., 2005)
MEFs		Mouse TNFα (1–100 ng/mL), zVAD (50–100 μ*M*), CHX (1 μg/mL) (Degterev et al., 2005; Thapa et al., 2013)
FADD-deficient MEFs		IFNα,β,γ (5 ng/mL) (Thapa et al., 2013)
HT-29 cells	HTB-38	Human TNFα (20 ng/mL), zVAD (20 μ*M*), SMAC mimetic (100 n*M*) (Sun et al., 2012)
L929 cells	CRL-2148	Mouse TNFα (1–10 ng/mL), zVAD (20–100 μ*M*), or combination (McNamara et al., 2013); poly(I:C) (25 μg/mL), IFNγ (1000 U/mL) (Hitomi et al., 2008)
Primary bone marrow derived or peritoneal macrophages, macrophage/monocyte cell lines (THP-1, RAW264.7, J77.4)	TIB-202 (THP-1), TIB-71 (RAW264.7)	LPS (5–500 ng/mL), zVAD (25 μ*M*) (Kaiser et al., 2013)

MCF-7, are resistant to necroptosis. In some cases, this was linked to the lack of RIPK3 expression (He et al., 2009). It should be noted that different clones of the same cell line, for example, NIH3T3, were shown to display widely different sensitivity to necroptosis (Zhang et al., 2009). Therefore, caution is recommended in ensuring the sensitivity to necroptosis or lack thereof is not a result of genetic variability, for example, in different clones of MEF cells. Activation of necroptosis *in vitro* typically requires the presence of caspase inhibitors, such as zVAD.fmk or Q-VD-OPh, or inhibition of upstream apoptotic signaling, especially FADD or caspase-8, through knockout or siRNA knockdown. In some cases, the presence of additional sensitizing agents (discussed in Section 2.2) may also be required.

Among cell lines frequently used for analysis are chemically mutagenized FADD-deficient Jurkat cells (Juo et al., 1999) (ATCC CRL-2572; control ATCC CRL-2570), which undergo necroptosis in response to TNFα (Degterev et al., 2005), but are resistant to Fas-induced death. FADD$^{-/-}$ MEFs were shown to undergo necroptosis in response to interferon stimulation, but are resistant to TNFα-induced necroptosis (Thapa et al., 2013). Caspase-8$^{-/-}$ Jurkat T cells and primary mouse T cells are also sensitive to necroptosis (Bell et al., 2008; O'Donnell et al., 2011). In L929 cells, necroptosis can be directly induced by TNFα alone (Vercammen et al., 1997), zVAD.fmk, or poly(I:C) + IFN-γ (Hitomi et al., 2008). Conversely, RIP1-deficient Jurkat cells are resistant to necroptosis (Ting, Pimentel-Muinos, & Seed, 1996), providing a useful tool to study mutations in RIPK1 (Degterev et al., 2008). Similarly, RIP3$^{-/-}$ and MLKL$^{-/-}$ cells provide convenient means to study these two important mediators of necroptosis (Cho et al., 2009; He et al., 2009; Murphy et al., 2013; Wu et al., 2013).

2.2. Inducers of necroptosis

There is a rapidly growing repertoire of extracellular and intracellular inducers of necroptosis, summarized in detail in a recent review by Vanlangenakker, Vanden Berghe, and Vandenabeele (2012). Several of these have been used extensively, including members of the TNFα family: Fas ligand (FasL, such as *Super*FasLigand; Axxora, cat no. ALX-522-020 (we will indicate the sources of the reagents that we use, other sources exist as well), typically: 5–50 ng/mL), TNFα (human or mouse depending on cell type; Peprotech, cat no. 300-01A (human) and 315-01A (mouse), typically: 10–100 ng/mL), TRAIL (Super*Killer*TRAIL; Axxora, cat no. ALX-201-115, typically: 5–10 ng/mL). Various TLR agonists were also found to induce necroptosis in epithelial, fibroblast, and macrophage cells (He et al., 2011; Hitomi et al., 2008; Kaiser et al., 2013). TLR3 (poly(I:C); Sigma, cat no. P9582, typically: 50 ng/mL to 50 μg/mL) and TLR4 (LPS; Invivogen, cat no. tlrl-3pelps, typically: 10–1000 ng/mL) agonists trigger necroptosis through TRIF, while agonists of other TLRs act through MyD88-dependent autocrine TNFα loop (Kaiser et al., 2013). Interferons (IFN-α, PBL Assay Science, cat no. 12100-1; IFN-β, PBL Assay Science, cat no. 12405-1; IFN-γ, Peprotech, cat no. 315-05, typically: 5–10 ng/mL for all IFNs) also induce necroptosis, especially in MEF and macrophage cells. Activation of necroptosis *in vitro* under most circumstances requires the presence of caspase inhibitors, such as zVAD.fmk (Bachem, N-1510, typically: 20–100 μ*M*). Depending on the cell type, addition of protein synthesis inhibitor cycloheximide (CHX) (Sigma, cat no. C1988, typically: 1–10 μg/mL) can promote necroptosis (Holler et al., 2000). As discussed above, cIAP1/2 inhibitors (SM164 (Christofferson et al., 2012), BV6 (Wagner et al., 2013), Compound A (Dondelinger et al., 2013), none currently commercially available to our knowledge) or TAK1 inhibitor ((5*Z*)-7-oxozeaenol; AnalytiCon Discovery, cat no. NP-009245, 1 μ*M*) can also promote RIPK1-dependent necroptosis or apoptosis, depending on the presence of caspase inhibitors.

2.3. Inhibitors of necroptosis

Inhibitors of necroptosis provide useful tools to explore activation of this pathway *in vitro* and *in vivo*. Three major classes of inhibitors have been described to date. First, specific inhibitors of RIPK1, necrostatins, have been developed (Degterev et al., 2008). These molecules, termed Necrostatin-1, Necrostatin-3, and Necrostatin-4, are structurally dissimilar, but bind the same DLG-out pocket on RIP1 kinase, stabilizing its inactive conformation (Xie, Peng, Liu, et al., 2013). Of these molecules, optimized Nec-1, 7-Cl-O-Nec-1 (BioVision, cat no. 2263-1, typically: 1–30 μ*M*) displays superior activity and stability *in vitro* and *in vivo* and is exclusively selective toward RIPK1 (Christofferson et al., 2012; Degterev, Maki, & Yuan, 2013). Hsp90 inhibitor, geldanamycin (Sigma, cat no. 3381, typically: 0.25–1 μg/mL), causes degradation of RIPK1, providing an additional, albeit not selective tool to inhibit necroptosis (Holler et al., 2000). Second, RIPK3 inhibitors (GSK-843 and GSK-872; GlaxoSmithKline) have been recently described and shown to efficiently inhibit necroptosis (Kaiser et al., 2013). Third, an irreversible inhibitor of human MLKL, necrosulfonamide (Millipore, cat no. 432531-71-0, 0.5 μ*M*) has been reported (Sun et al., 2012). This molecule forms a covalent bond with Cys86 of human MLKL, but lacks activity against mouse protein due to the absence of the orthologous Cys.

3. MEASUREMENT OF NECROPTOTIC CELL DEATH

3.1. Analysis of viability of FADD-deficient Jurkat cells treated with TNFα using CellTiter-Glo assay (Fig. 1.1)

1. Cells are routinely cultured in the media containing RPMI1640 (Invitrogen, cat no. 11875-093) supplemented with 10% FetalPlex serum (Gemini, cat no. 100-602) and 1% antibiotic–antimycotic mix (Invitrogen, cat no. 15240062). Cell density should be maintained in the range from 1×10^5 to 1×10^6 cells/mL.
2. On the day of the experiment, cells are diluted in fresh media at the density of 5×10^5 cells/mL. 100 μL is plated into each well of a white clear bottom 96-well plate (Corning, cat no. 3903) to allow subsequent analysis as well as microscopic observation of the cells.
3. Human TNFα (Peprotech, cat no. 300-01A) is dissolved in sterile water to the concentration of 100 μg/mL and further diluted to 1 μg/mL in sterile PBS. 1 μL of TNFα is added to the wells to induce necroptosis, and plate is returned into 37 °C incubator for 24 h.
4. 25 μL of reconstituted CellTiter-Glo assay reagent (Promega, cat no. G7570) is added into each well and plate is incubated at room temperature on a rocking platform for 10 min.
5. Luminescence (integration time 0.3–1 s) is measured using a platereader, such as Victor^{3}V (Perkin Elmer) or similar.
6. Viability is calculated according to the formula: Viability (%) = (RLU TNFα well/RLU control well) × 100%.

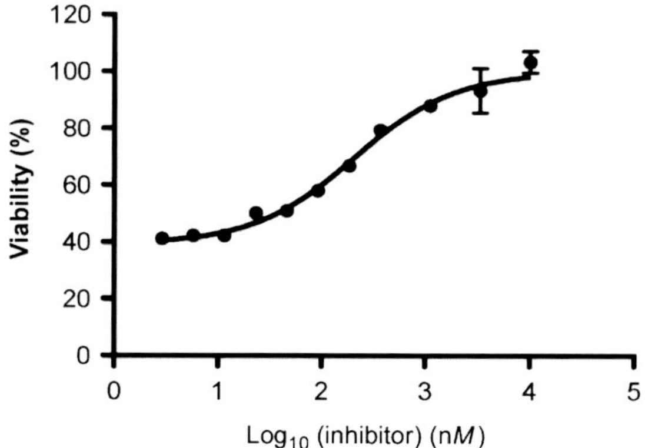

Figure 1.1 Titration of 7-Cl-O-Nec-1 (3 n*M* to 10 μ*M*) in TNF-treated FADD-deficient cells. Viability was determined using CellTiter-Glo assay.

3.2. Determination of specific cell death using SYTOX Green assay (Fig. 1.2)

CellTiter-Glo assay is based on a luciferase reaction and measures cellular ATP levels. It provides a robust and sensitive measurement of cell viability. However, decreased proliferation or cellular stress can also lead to the decrease in values. Therefore, assays specifically detecting dead cells are also useful. There are a number of approaches that can be used, including LDH release assay (Promega, cat no. G1780), MultiTox-Glo (Promega, cat no. G9270), FACS-based assays (see below). We prefer SYTOX Green-based assay due to its good signal-to-noise ratio and relatively low cost. SYTOX Green is a cell-impermeable dye, which increases fluorescence upon DNA binding. This provides a convenient readout for cell lysis during necroptosis.

1. Cells are cultured as described in Section 3.1, except cells are seeded into black clear bottom plates (Corning, cat no. 3904) in phenol-red-free RPMI1640 media (Invitrogen, cat no. 11835-030), supplemented with 10% FetalPlex serum and 1% antibiotic–antimycotic mix.
2. At a selected time point (typically 24–48 h), SYTOX Green (Invitrogen, cat no. S7020) is added to the wells at the final concentration of 1 μ*M*. Cells are incubated at 37 °C for 30 min, and fluorescence (green channel, ex. 488 nm, em. 523 nm) is measured using a platereader (1-s integration time).

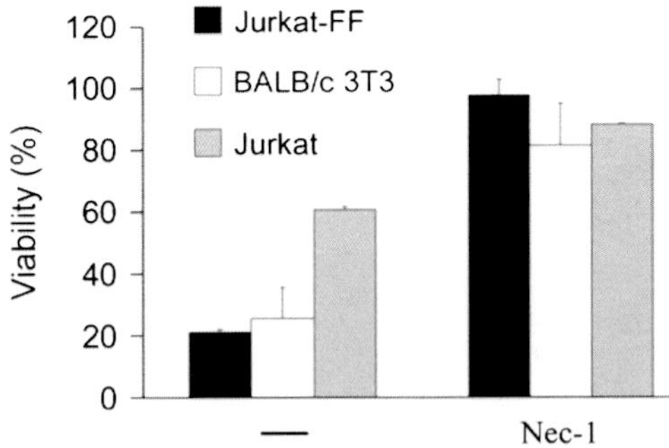

Figure 1.2 Analysis of cell death using SYTOX Green assay. BALB/c 3T3 cells were treated with TNFα/zVAD.fmk and Nec-1 for 24 h, or Jurkat cells were treated with FasL/CHX/zVAD.fmk and Nec-1 for 48 h. Jurkat-FF (Jurkat cells stably expressing chemically dimerizable FADD) were treated with dimerizer AP20187/zVAD.fmk and Nec-1 for 48 h. Data presented as: Viability (%) = 100% − dead cells (%). *Reproduced with permission from Degterev et al. (2005).*

3. To produce maximal cell lysis, 5 μL of 20% Triton X-100 is subsequently added into each well for 1 h at 37 °C (or 4–24 h at room temperature), followed by second fluorescence measurement.
4. Values obtained in the negative control wells containing media without cells are subtracted from corresponding sample values.
5. Percentage of dead cells is calculated as a ratio: Dead (%) = ((RFU TNF well/total RFU TNF well) − (RFU control well/total RFU control well)) × 100%.

3.3. Annexin V/PI assay (Fig. 1.3)

While assays in Sections 3.1 and 3.2 provide simple and robust methods for measuring necroptotic death, these assays cannot distinguish between necroptosis and other forms of death, for example, apoptosis. Thus, more specific necroptosis assays are needed in establishing that this mode of cell death is activated. A number of apoptosis-specific assays, such as mitochondrial cytochrome *c* release, DNA fragmentation, and caspase activation, are useful in excluding the activation of this mechanism of cell death. Early release of Cyclophilin A has been found to potentially represent a specific marker of necroptosis (Christofferson & Yuan, 2010a). Release of HMGB1 protein has also been observed, but it may represent a more downstream event indicative of cell lysis (Christofferson & Yuan, 2010a). Annexin V/PI assay provides another simple approach to differentiate apoptosis and necroptosis. Annexin V protein binds to PS exposed in the outer leaflet

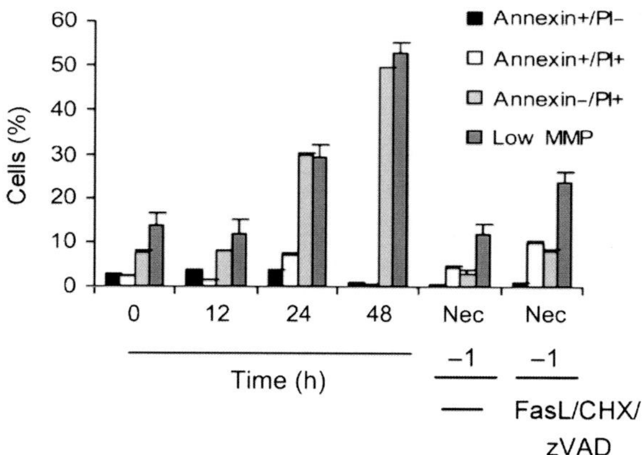

Figure 1.3 Annexin V/PI and mitochondrial membrane potential assays of Jurkat cells treated with FasL/CHX/zVAD at different time points. *Reproduced with permission from Degterev et al. (2005).*

of plasma membrane of apoptotic cells in a caspase-dependent fashion. This precedes the loss of plasma membrane integrity (Rimon, Bazenet, Philpott, & Rubin, 1997; Vanags, Porn-Ares, Coppola, Burgess, & Orrenius, 1996). Propidium iodide (PI) is a cell-impermeable DNA dye. Thus, the appearance of Annexin V^+/PI^- cells is characteristic for apoptosis. These cells progress to become Annexin V^+/PI^+ due to secondary necrosis. Activation of necroptosis in Jurkat cells results in the appearance of Annexin V^-/PI^+ cells (Degterev et al., 2005). Other cell types, such as MEFs, proceed to become Annexin V^+/PI^+ as a result of necroptosis (Wu et al., 2013). Overall, this assay provides convenient means to determine the numbers of dead cells and establishes the lack of apoptotic Annexin V^+/PI^- cells in the sample.

1. FADD-deficient Jurkat cells are seeded into a 12-well plate (Costar, cat no. 3513) at the density of 5×10^5 cells/mL (2 mL/well, 1×10^6 cells). Necroptosis is induced as described in Section 3.1.
2. Cells are collected by centrifugation for 5 min at $400 \times g$ at room temperature. Cell pellet is resuspended in 500 μL of 1 × binding buffer (ApoAlert Annexin V kit; Clontech, cat no. 630109), followed by centrifugation.
3. Cells are resuspended in 200 μL of 1 × binding buffer supplemented with 5 μL of Annexin V-GFP and 10 μL of PI.
4. After 15 min incubation in the dark, cells are further diluted to 500 μL with 1 × binding buffer and analyzed by FACS using FL1 (green, Annexin V-FITC) and FL3 (red, PI) channels.

3.4. Analysis of ROS increase (Fig. 1.4)

Increase in ROS is one of the important features of necroptotic cell death in a number of cell types, such as MEFs and L929 cells (Shindo, Kakehashi, Okumura, Kumagai, & Nakano, 2013; Vanden Berghe et al., 2010). Two sources of increased ROS have been reported: mitochondrial Complex I and NADPH oxidase (Kim, Beg, & Haura, 2013; Vanden Berghe et al., 2010). It should be noted that ROS may not be a universal feature of necroptosis as no increase in ROS accompanies necroptosis in Jurkat cells (Degterev et al., 2005). A number of ROS sensors can be used to measure ROS increase, including CM-H_2DCFDA (Invitrogen, cat no. C6827), CellROX sensors (Invitrogen, cat no. C10444), dihydrorhodamine 123 (Invitrogen, cat no. D632), and others. Sensors differ in fluorescence spectra, sensitivity, and repertoire of ROS species detected. In our case, MitoSOX Red (Invitrogen, cat no. M36008), measuring mitochondrial superoxide, provides an excellent tool for measuring necroptosis-associated ROS by FACS.

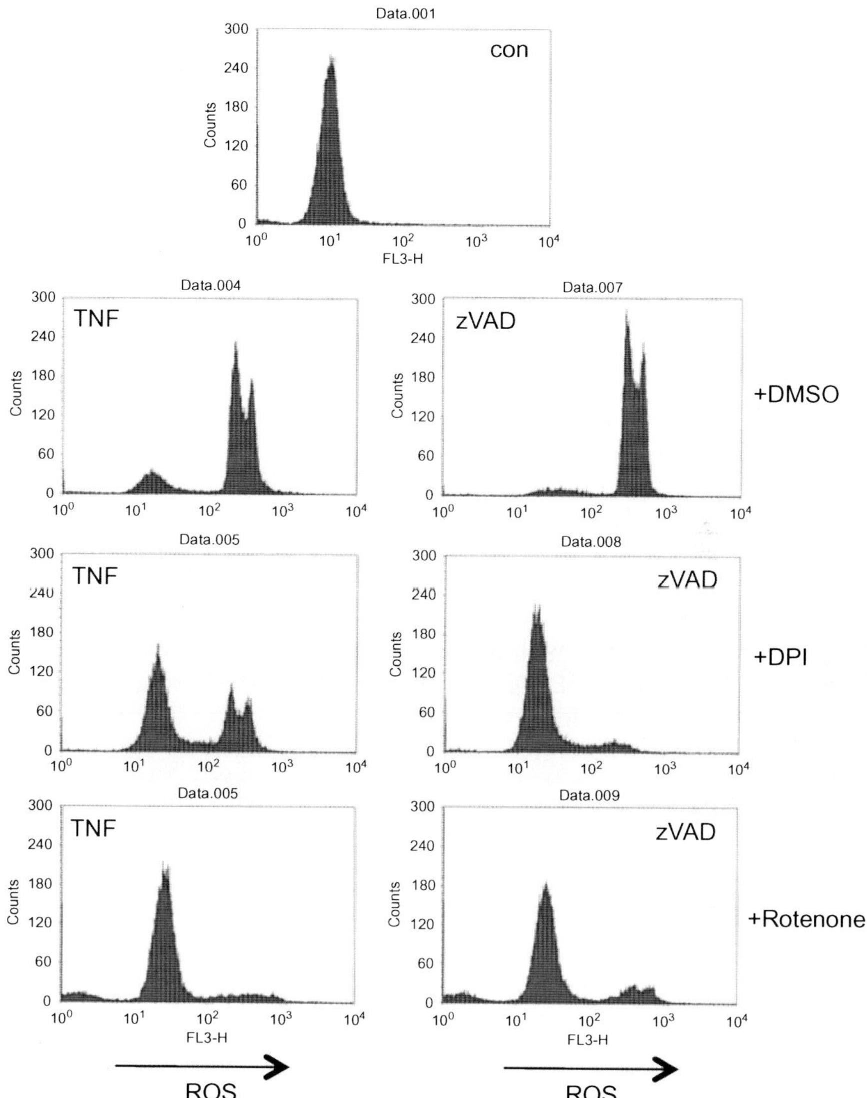

Figure 1.4 Attenuation of ROS increase in L929 cells by Complex I and NADPH oxidase inhibitors. L929 cells were treated with TNFα or zVAD.fmk and 50 μ*M* rotenone (Complex I inhibitor) or 25 μ*M* diphenylene iodonium (DPI, NADPH inhibitor) for 12 h.

1. Cells are cultured as described in Section 3.3.
2. MitoSOX reagent (5 m*M* stock in DMSO) is added to the cells to a final concentration of 5 μ*M*, and cells are returned into the 37 °C incubator for additional 15 min.

3. Cells can be directly analyzed by FACS using FL3 (red) channel or washed several times with culture media and observed using fluorescent microscope.

3.5. Mitochondrial membrane depolarization (Fig. 1.3)

Change in mitochondrial transmembrane potential is another hallmark of necrosis, in general, and necroptosis, in particular (Vanden Berghe et al., 2010). Early transient hyperpolarization (Vanden Berghe et al., 2010) is followed by the loss of membrane potential, concomitant with cell death (Degterev et al., 2005; Temkin, Huang, Liu, Osada, & Pope, 2006). Fluorescent probes, such as tetramethylrhodamine (TMRM; Invitrogen, cat no. T668), JC-1 (Invitrogen, cat no. T3168), and 3,3′-dihexyloxacarbocyanine iodide (DiOC6(3); Invitrogen, cat no. D273), can be used, although JC-1 could be more specific (Salvioli, Ardizzoni, Franceschi, & Cossarizza, 1997).

1. Cells are cultured as described in Section 3.3.
2. DiOC6(3) reagent is added to the cells to a final concentration of 40 μ*M*, and cells are returned into the 37 °C incubator for additional 30 min.
3. Cells are washed once with prewarmed media and can be directly analyzed by FACS using FL1 (green) channel or observed using fluorescent microscope.

3.6. Analysis of TNFα gene expression changes by qPCR (Fig. 1.5)

In addition to activation of cell death, RIPK1 activation has been shown to promote TNFα synthesis (Christofferson et al., 2012; Hitomi et al., 2008; McNamara et al., 2013), further highlighting connections between necrotic cell death and inflammation. In some cases, such as L929 cells treated with zVAD.fmk (Hitomi et al., 2008) and cells stimulated with antagonists of MyD88-dependent TLRs (Kaiser et al., 2013), autocrine TNF signaling is critical for necroptosis activation.

1. MEFs are seeded into a 12-well plate (Costar, cat no. 3513) in 1 mL of media at the density of $1.5–2 \times 10^5$ cells/well.
2. On the following day, cells are stimulated with 10 ng/mL mouse TNFα, 50 μ*M* zVAD.fmk, and 1 μg/mL CHX for 6–8 h. Specific concentrations may differ and CHX may not be necessary, depending on the strain of MEFs.
3. Total RNA is isolated using one of the commercial kits, for example, Quick-RNA MiniPrep kit (Zymo Research, cat no. R1054). RNA

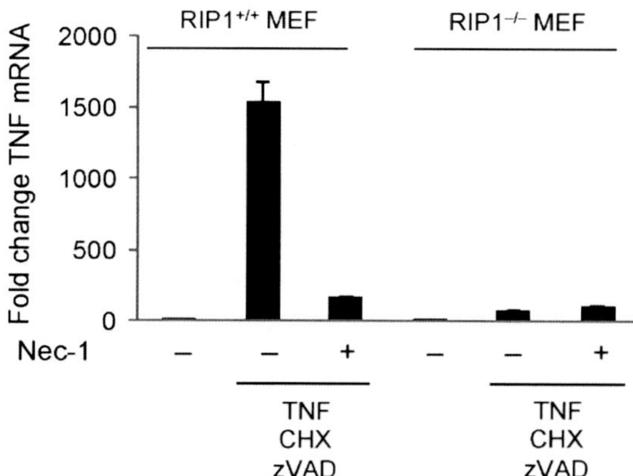

Figure 1.5 RIPK1-dependent upregulation of TNF mRNA. RIP1$^{+/+}$ and RIP1$^{-/-}$ MEFs (gift of Dr. Michelle Kelliher, UMass Medical School) were treated with TNF/CHX/zVAD.fmk for 6 h followed by qPCR analysis of TNF mRNA. Data are normalized to 18S RNA levels.

concentration is determined based on OD_{260}. Typical RNA yields are 5–15 μg.

4. cDNA is synthesized using one of the commercial cDNA kits using random primers, for example, iScript cDNA synthesis kit (BioRad, cat no. 170-8891). 1 μg of total RNA is diluted to 15 μL with RNase-free water and combined with 4 μL of 5 × reaction buffer and 1 μL of enzyme mix. Reactions are incubated in a standard PCR machine: 25 °C—5 min, 42 °C—30 min, 85 °C—5 min. After completion, reactions are diluted with 30–80 μL of water. qPCRs are set up in duplicate or triplicate for TNFα and 18S (or another housekeeping gene such as GAPDH or β-actin). Sequences of qPCR primers are: mouse TNFα—forward 5′-CCCTCACACTCAGATCATCTTCT-3′, reverse 5′-GCTACGAC GTGGGCTACAG-3′; mouse 18S—forward 5′-ATAACAGGTCTG TGATGCCCTTAG-3′, reverse 5′-CTAAACCATCCAATCGGTA GTAGC-3′. Primers are dissolved in water at 100 μ*M* and primer mix combining 10 μ*M* forward and reverse primers is prepared.
5. qPCRs are set up in white 96-well PCR plates (Geneseesci, cat no. 27-409) to include: 2 μL of cDNA, 1 μL of primer mix, 7 μL of water, and 10 μL of 2 × VeriQuest SYBR Green master mix (Affymetrix, cat no. 75665).
6. Plate is sealed using ThermalSeal RTS film (Geneseesci, cat no. 12-537) and loaded into LightCycler 480 qPCR machine (Roche). Cycling parameters are: 50 °C—2 min, 95 °C—10 min, 45 cycles: 95 °C—15 s, 60 °C—30 s (detection).

7. Relative expression of TNFα in TZ versus control samples is calculated according to the formula: Foldchange $= 2^{-((C_t(\mathrm{TNF, TZ}) - (C_t(\mathrm{GADPH, TZ}) - C_t(\mathrm{GADPH, con}))) - C_t(\mathrm{TNF, con}))}$.

4. RECAPITULATION OF RIP1 KINASE EXPRESSION IN RIP1-DEFICIENT JURKAT CELLS

Reexpression of RIPK1 and RIPK3 mutants in corresponding deficient cells provides excellent means to perform structure–activity relationship analysis of RIPK signaling, for example, by expressing kinase and RHIM domain mutants. However, (a) many of the cell types typically used to study necroptosis, especially Jurkat, L929, and macrophage cells, are difficult to transfect, and (b) we found that stable expression of RIPK1 in lentivirally or retrovirally transduced cells is either readily lost or is difficult to achieve.

4.1. Transient transfection (Fig. 1.6)

1. Transfections are performed using pcDNA3.1-based expression vectors for human or mouse RIPK1 (Degterev et al., 2008). Transfection mix is prepared by diluting 4 μg of RIPK1 DNA and 1 μg of pEGF-N1 vector (Addgene, cat no. 6085-1) with 500 μL of Opti-MEM I media (Invitrogen, cat no. 31985070). Next, 15 μL of X-tremeGENE HP

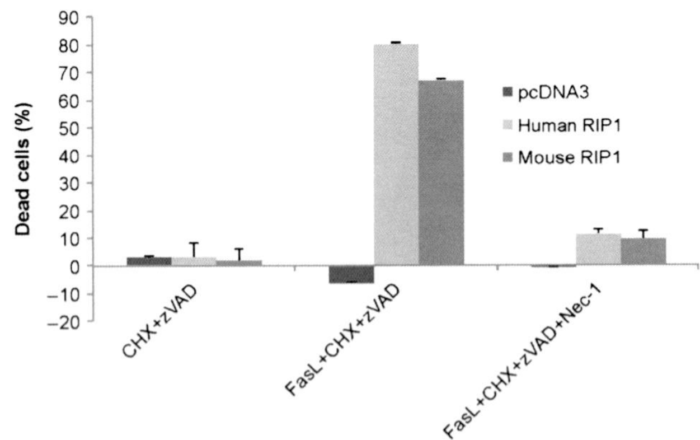

Figure 1.6 Transient reexpression of human or mouse RIPK1 restores necroptosis in RIP1-deficient Jurkat cells. Necroptosis was induced by treatment with FasL/CHX/zVAD for 24 h followed by FACS analysis of GFP+/PI+ cells as described in Section 4.1.

(Roche, cat no. 06366244001) transfection reagent is added, and transfection complexes are allowed to form for 30 min at room temperature.

2. 5×10^5 RIP1-deficient Jurkat cells are resuspended in 5 mL of RPMI1640 media supplemented with 10% FetalPlex and 1% antibiotic–antimycotic. Transfection mix is added to the cells for 48 h.
3. Cells are collected by centrifugation (5 min, $400 \times g$) and divided into two samples (1 mL media each). One sample is control treated with 1 μg/mL CHX and 100 μ*M* zVAD.fmk; second sample (necroptosis) is additionally treated with 10 ng/mL KillerFas ligand (Axxora, see above).
4. After incubation for 24 h, cells are placed in FACS tubes (Falcon, cat no. 352054), supplemented with 1 μg/mL PI (Sigma, cat no. P4864), and analyzed by FACS using FL1 (green, GFP) and FL3 (red, PI) gates.
5. Percentage of cell death is calculated as a combination of % decrease in GFP^+ cells due to cell lysis and % increase combining in PI^+/GFP^+ (dead/transfected) cells.

4.2. Generation of stable-inducible cell lines (Fig. 1.7)

In this case, RIP1-deficient Jurkat cells are consequently infected with retroviruses encoding reverse tetracycline-regulated transactivator (rtTA, pMA2641; Addgene, cat no. 25435, blasticidin and GFP markers) and RIPK1 (pRetroX-Tight-Pur; Clontech, cat no. 632104, puromycin). Standard procedures to generate VSV-G pseudotyped viruses can be used. We

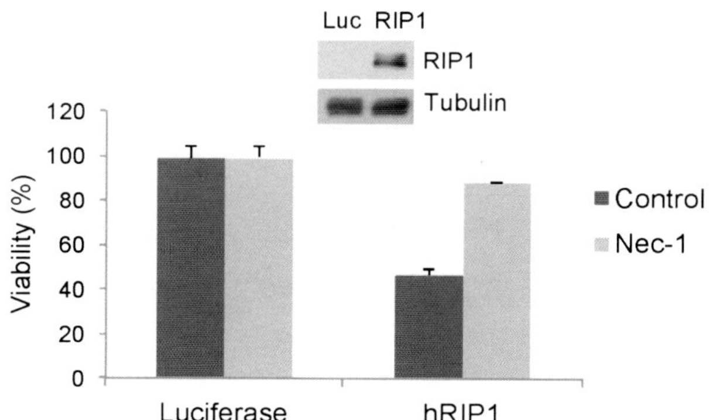

Figure 1.7 Stable reexpression of RIPK1 in RIP1-deficient Jurkat cells restores necroptosis in response to FasL/CHX/zVAD.fmk. As a control, cells were infected with the virus encoding luciferase gene. Viability was determined using CellTiter-Glo assay and normalized to corresponding CHX/zVAD.fmk-treated controls (set as 100% viability). Western blot indicating expression of RIP1 is also shown.

typically use Lenti-X 293T cells (Clontech, cat no. 632180) to generate 2 mL of viral supernatant by transfecting 2.5×10^5 cells with 500 μL Opti-MEM transfection mix, containing 2 μg of viral DNA and 1 μg of each VSV-G and gag/pol plasmids, and 10 μL of Lipofectamine 2000 reagent (Invitrogen, cat no. 12566014). Virus-containing supernatants are collected 48–72 h after transfection and filtered through 0.45-μ*M* syringe filter (Millipore, cat no. SLHV033RS). Virus-producing Lenti-X cells are supplied with 2 mL of fresh DMEM (Invitrogen, cat no. 11965-092) supplemented with 10% FBS (Tet system approved FBS; Clontech, cat no. 631106) and 1% antibiotic–antimycotic, and viruses are collected against after additional 48 h.

1. To perform infections with rtTA virus, 1×10^6 RIP1-deficient Jurkat cells are resuspended in 1 mL complete RPMI media (containing Tet-approved FBS), combined with 2 mL of viral supernatant and 8 μg/mL polybrene (Sigma, cat no. 107689). Cells are infected by spinning at $1000 \times g$ for 90 min and returned to 37 °C in the virus-containing media. After 24 h, cells are collected by centrifugation and resuspended in 5 mL of fresh media for two additional infections.
2. After three infections, cells are selected in 5 μg/mL blasticidin (Invivogen, cat no. ant-bl-1). Selection is usually complete in 3–5 days. We find that further FACS of GFP-positive cells is helpful in ensuring subsequent adequate expression of RIPK1.
3. Selected cells are infected with RIPK1 virus using the same procedure, followed by selection with 1 μg/mL puromycin (Sigma, cat no. P8833). Selection is typically complete in 3–4 days.
4. Expression of RIPK1 is induced by addition of 2 μg/mL doxycycline (Sigma, D9891) for 24 h, after which the cells are ready for downstream analyses.

5. ANALYSIS OF NECROSOME COMPLEX FORMATION

5.1. Immunoprecipitation of necrosome complex (Fig. 1.8)

Formation of RIPK1/RIPK3-containing necrosome complex is unique for necroptosis and provides a useful method for analyzing the initiation of this pathway. Variations of this method using antibodies to multiple proteins, present in Complex II (FADD, RIPK1, RIPK3, caspase-8) (Cho et al., 2009; He et al., 2009; Thapa et al., 2013; Vince et al., 2012), have been described. Immunoprecipitation of necrosome using RIPK3 antibody has been found reliable in our lab. The protocol is as follows:

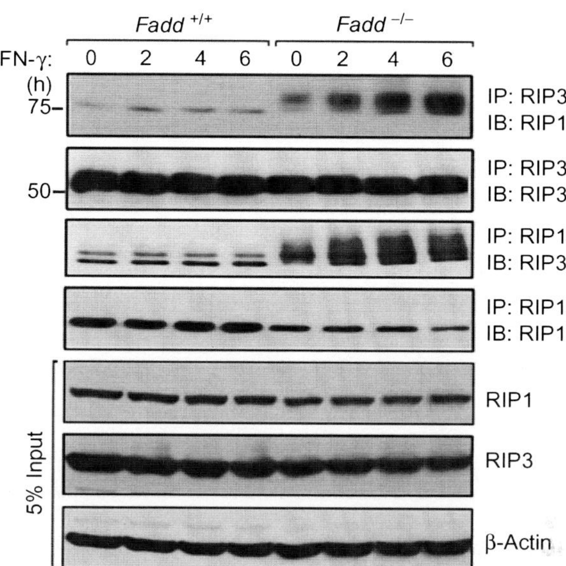

Figure 1.8 Coimmunoprecipitation of RIP1 and RIP3 from FADD$^{-/-}$ MEFs induced to undergo necroptosis by treatment with IFNγ. *Reproduced with permission from Thapa et al. (2013).*

1. Seed FADD-deficient Jurkat cells 1 day before treatment. A total of $1–5 \times 10^7$ cells are used for each condition. Necroptosis is induced with 20 ng/mL recombinant human TNFα. The necrosome is detectable 4–5 h after induction. Necrosome can also be detected in a variety of other cells, such as MEF and HT-29, in the presence of 20 μ*M* zVAD to prevent caspase activation. The concentration of TNFα and the duration of treatment need to be adjusted depending on the sensitivity of the cells to the stimuli, but necrosome can also be typically detected 2–5 h after TNFα treatment.
2. Cells are washed twice with ice-cold PBS and lysed in 0.5–1 mL lysis buffer containing 0.2% (vol/vol) Triton X-100, 150 m*M* NaCl, 20 m*M* Tris–HCl (pH 7.4), 1 m*M* EDTA, 5 m*M* NaF, 1 m*M* $NaVO_3$ (*ortho*), 1 m*M* PMSF, and Complete protease inhibitor cocktail (Roche). Incubate the cells on ice for 30 min to 1 h with periodic mixing.
3. Lysates are cleared by centrifugation at 12,000–14,000 rpm in a tabletop 4 °C microcentrifuge for 10–15 min.
4. Protein concentrations are normalized based on one of the standard protein assays (e.g., Pierce 660 nm Protein Assay kit, cat no. 22662).

5. The lysates are precleared by incubating with 5–10 μL Protein A/G UltraLink Resin (Thermo Scientific, product no. 53133) at 4 °C for 1 h with gentle rocking.
6. 2 μg of rabbit anti-RIP3 antibody (ProSci, mouse specific, cat no. 2283) is incubated with each sample overnight at 4 °C.
7. Add 5–10 μL of Protein A/G UltraLink Resin to the lysate and incubate at 4 °C for 2 h with gentle rocking.
8. Beads are washed four times with lysis buffer, and proteins are eluted by boiling in 1 × SDS-PAGE loading buffer.
9. Analyze the whole cell extract samples and immunoprecipitation samples by Western blotting. RIPK1 and FADD (unless in FADD-deficient cells) should be detected in the TNFα-treated immunoprecipitation samples.

5.2. Immunoprecipitation of TNFR1 complex

The formation of necrosome is preceded by the formation of TNFR1 complex, also named complex I, at the plasma membrane. When assessing the assembly of necrosome, or the lack thereof, it is important to examine whether TRADD, TRAF2, cIAP1/2, RIP1, and IKKγ/NEMO are recruited to TNFR1, and whether this complex dissociates and transforms to the cytosolic complexes. Notably, two TNFα receptors have been identified in vertebrates, named TNFR1 and TNFR2. While TNFR1 mediates the death signaling, TNFR2 lacks the death domain and was originally proposed to activate survival pathways. However, more recent evidence suggest that coligation of TNFR1 and TNFR2 results in more efficient prodeath signaling, for example, in the absence of CHX that is required for induction of cell death by TNFR1 alone, possibly through the regulation of TRAF2 and cIAPs (Chan & Lenardo, 2000; Chan et al., 2003; Pimentel-Muinos & Seed, 1999). Human TNFα has good affinity to murine TNFR1 but low affinity to murine TNFR2 (Lewis et al., 1991), providing a platform to assess TNFR1 complex specifically. When using human cells for the immunoprecipitation, human TNFα should be used but the results need to be interpreted accordingly as both TNFR1 and TNFR2 will be present. The protocol adapted from Micheau and Tschopp (2003) is as follows:

1. A variety of cell lines are capable of forming TNFR1 complex, regardless of whether they proceed to apoptosis or necroptosis. Adherent cells are recommended because of the requirement of accurate timing control.

2. Treat 1–5 × 10^7 cells with 1.5–2 μg/mL Flag-hTNFα (for murine cells; Enzo Life Sciences, cat no. ALX-522-008-C050) or 50–100 ng/mL Flag-hTNFα (for human cells) for 0, 2, 5, 15 min.
3. Wash the cells with cold PBS twice. The cells are lysed in 0.5–1 mL lysis buffer containing 0.2% (vol/vol) NP-40, 150 m*M* NaCl, 20 m*M* Tris–HCl (pH 7.4), 5 m*M* NaF, 1 m*M* $NaVO_3$ (*ortho*), 1 m*M* PMSF, and Complete protease inhibitor cocktail (Roche). Incubate the cells on ice for 30 min to 1 h with periodic mixing.
4. Lysates are cleared by centrifugation at 12,000–14,000 rpm in a tabletop 4 °C microcentrifuge for 10–15 min.
5. Incubate the lysate with 5–10 μL anti-Flag M2 affinity gel (Sigma, cat no. A2220) at 4 °C for 2–4 h.
6. Wash the beads four times with lysis buffer. Elute the immunoprecipitated proteins with 200 μg/mL Flag peptide at 4 °C for 1 h.
7. Analyze the whole cell extract samples and immunoprecipitation samples by Western blotting. Nemo and ubiquitinated RIPK1 in the immunoprecipitation samples should appear 2 min after TNFα treatment. Disassembly of the complex is typically observed 15 min after treatment.

5.3. Assessment of necrosome formation by fluorescence microscopy

Another useful method in assessing necrosome formation is immunofluorescence-based detection of RIPK3 aggregation. RIPK3 is present as a diffuse cytosolic signal in the control cells. Activation of necroptosis leads to initial formation of distinct punctae, which continuously enlarges as necroptosis progresses (Sun et al., 2012). Punctae formation is blocked by RIPK1 inhibitor Necrostatin-1, while MLKL inhibitor, necrosulfonamide, inhibits its enlargement (Sun et al., 2012). We refer readers to the Methods sections of the manuscripts by He et al. (2009) and Li et al. (2012) for the discussion of the protocols to analyze FLAG-RIPK3 and mCherry-RIPK3 aggregation.

6. ENDOGENOUS RIPK AUTOPHOSPHORYLATION ASSAYS (FIG. 1.9)

Analysis of catalytic activity of RIPK1 and RIPK3 kinases using autophosphorylation or phosphorylation of myelin basic protein (MBP) and histone H1 substrates provides useful tool to study inhibitors of these kinases and their activation in the cells, respectively (Cho et al., 2009;

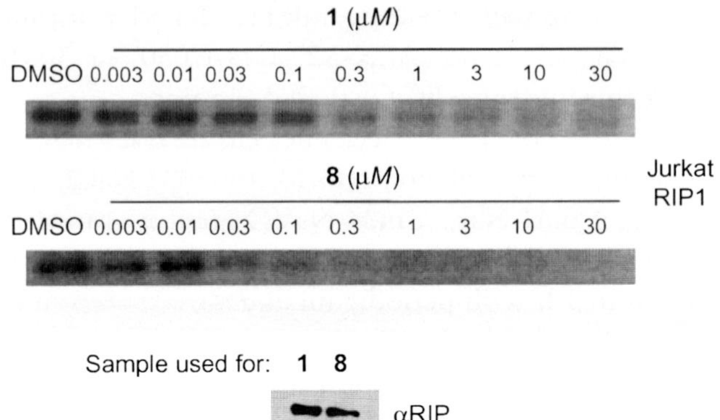

Figure 1.9 ^{32}P autophosphorylation assay of RIPK1 immunoprecipitated from Jurkat cells. Inhibition by original Nec-1 (**1**) and optimized 7-Cl-O-Nec-1 (**8**, active *R*-isomer) is shown. Western blot indicating the amount of RIP1 protein in representative reactions is also shown. *Reproduced with permission from Degterev et al. (2008).*

Degterev et al., 2008; Li et al., 2012). As an extension of this approach, Cho et al. (2009) described a sequential FADD/RIP3 IP approach, which allowed detection of increased RIPK1-dependent catalytic activity toward MBP in Complex II. Because RIPK1 and RIPK3 form NP40-insoluble aggregates upon activation, analysis of catalytic activity in post-NP40 lysates was also shown to reveal activation using H1 as a substrate (Li et al., 2012). The modified protocol based on Degterev et al. (2008) is as follows:

1. Approximately, 1×10^7 Jurkat cells are used per reaction. Cells are lysed in 1 mL of the buffer containing 1% Triton X-100, 150 m*M* NaCl, 20 m*M* HEPES, pH 7.3, 5 m*M* EDTA, 5 m*M* NaF, 0.2 m*M* $NaVO_3$ (*ortho*), and Complete protease inhibitor cocktail (Roche) for 20 min on ice (occasionally mixing side-to-side) and spun down at 14,000 rpm for 10 min at 4 °C.
2. Immunoprecipitation is carried out for 16 h at 4 °C using 1–2 μg of mouse anti-RIP1 antibody (BD Transduction Labs, cat no. 610458) per sample. It is recommended that a pilot experiment is performed to ensure efficient IP as amounts of RIPK1 can vary between cell types. 10 μL of Protein A + Protein G magnetic Dynabeads (Invitrogen, cat no. 10001D and 10003D) are added the following day for 2 h, followed by three washes with lysis buffer and two washes with 20 m*M* HEPES/0.025% NP-40, pH 7.3. Cross-linking of antibody to the beads using DMP reagent (Pierce, cat no. 21666) according to the protocol: http://www.neb.com/nebecomm/products/protocol52.asp

is also recommended to increase consistency and facilitate analysis. We typically incubate 100 μg of antibody with 400 μL of beads, collect the beads, and incubate residual antibody with fresh 400 μL of beads, followed by coupling of the resulting 800 μL of beads. It is recommended that titration of the beads is performed for each new cross-linking.

3. Beads are subsequently resuspended in 9.5 μL of the kinase reaction buffer containing 20 m*M* HEPES, pH 7.3, 5 m*M* $MnCl_2$, 5 m*M* $MgCl_2$, and 0.025% NP-40 and incubated with 0.5 μL of inhibitors in DMSO for 10–15 min at room temperature. Reactions are initiated by the addition of 5 μL of 30 μ*M* ATP (Sigma, cat no. A7699) and 3 μCi of γ-^{32}P-ATP (Perkin Elmer, cat no. BLU002Z250UC), diluted in kinase reaction buffer. Reactions are performed for 30 min at 30 °C with shaking (we use Eppendorf Thermomixer set at 650 rpm).
4. Reactions are stopped by the addition of 5 μL of standard 4 × SDS-PAGE loading buffer and heating at 95 °C for 5 min. 15 μL of the supernatant is loaded on an 8% SDS-PAGE gel, and RIPK1 band is visualized by autoradiography.

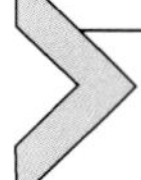

7. ANALYSIS OF RECOMBINANT RIPK1 KINASE ACTIVITY AND INHIBITION BY NECROSTATINS

7.1. Expression and purification of recombinant RIP1 and RIP3

Recombinant catalytically active RIPK1 kinase domain can be efficiently expressed using baculoviral infection of Sf9 cells. We found that RIP1 (8–322)-His_6 and GST-RIP1(1–327) proteins can be expressed as catalytically active proteins that are inhibited by necrostatins. Coinfection of Sf9 cells with both RIPK1 and Cdc37 co-chaperone viruses typically increases yield and solubility of recombinant RIPK1 especially of RIP1(8–322)-His_6. In general, the GST-fusion protein displays better kinase activity. We refer the readers to Maki and Degterev (2013) and Maki, Tres Brazell, Teng, Cuny, and Degterev (2013) for detailed protocols regarding virus production, infection, and protein purification. In addition, methods for expression and purification of recombinant RIPK3 have also been recently described (Xie, Peng, Yan, et al., 2013).

7.2. Kinase-Glo assay (Fig. 1.10)

The Kinase-Glo Luminescent Kinase Assay (Promega, cat no. V6711) is a homogeneous assay format, based on quantification of ATP depletion

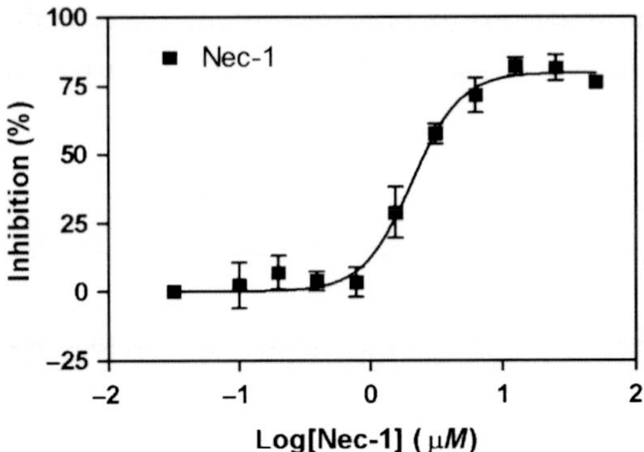

Figure 1.10 Inhibition of recombinant GST-RIP1 kinase activity by 7-Cl-O-Nec-1 using Kinase-Glo assay. Experiment was performed as described in Section 7.2 using 11-point dose range of 7-Cl-O-Nec-1 (50 μ*M* to 100 n*M*).

during kinase reaction. The assay generates a luminescent signal produced by the patented Ultra-Glo Luciferase in a proprietary buffer system. Kinase activity is inversely related to the luminescent signal. We discovered that autophosphorylation and intrinsic ATPase activities of recombinant RIP1K are measureable in this assay system and are inhibited by necrostatins. At the same time, addition of exogenous substrate, MBP, does not lead to a significant increase in activity, suggesting that identification of more efficient RIPK1 substrate is necessary for further development of this assay.

1. 2 μL of 5 × kinase buffer (100 m*M* HEPES, pH 7.3, 5 m*M* $MgCl_2$, 5 m*M* $MnCl_2$, 750 m*M* NaCl, 0.5% BSA) is added to each well of a 384-well white polystyrene plate (Labsystems, cat no. 95040230). Subsequently, 4 μL of 3.1 μ*M* GST-RIP1 8–327 and 2 μL of inhibitor (both diluted in 1 × kinase buffer, final concentration of DMSO = 2%) are added into each well. The plate is covered with the lid and incubated at room temperature for 10 min.
2. 2 μL of 100 μ*M* ATP (Sigma, cat no. A7699), diluted in 1 × kinase buffer, is added to the reaction. The plate is covered with the lid and incubated at room temperature for 90 min.
3. Kinase reaction is terminated by adding 10 μL of room temperature Kinase-Glo reagent. The plate is covered and incubated at room temperature for 10 min.
4. The luminescence signal is measured using a Victor^{3}V (Perkin Elmer) platereader (integration time 0.3–1 s).
5. The degree of inhibition is calculated by subtracting value of the control protein sample without inhibitor from values in all other samples,

followed by using the formula: Inhibition (%) = (luminescent signal (inhibitor)/luminescent signal (no protein and no inhibitor)) × 100%.

7.3. HTRF KinEASE assay (Fig. 1.11)

Homogeneous time-resolved fluorescence (HTRF) KinEASE assay (CIsBio, cat no. 62ST3PEB) utilizes a homogenous kinase assay format based on phosphorylation of one of the proprietary biotinylated kinase substrates, provided in the universal ST kinase kit. Time-resolved energy transfer between streptavidin-XL665 and Eu(K)-labeled phospho-specific antibody results in fluorescence signal. We found that peptide substrate 3 is an efficient substrate for recombinant RIP1K (Maki & Degterev, 2013).

1. To set up a reaction, GST-RIP1 8–327 is diluted to 500 n*M* in 1 × kinase buffer provided in the kit supplemented with 12.5 μ*M* $MgCl_2$, 12.5 μ*M* $MnCl_2$, and 0.1% BSA. 2 μL of the protein is mixed with 4 μL of inhibitor (final concentration of DMSO = 2%) in a well of 384-well black polystyrene plate (Corning, cat no. 3677). The plate is sealed with Thermowell sealing tape (Costar, cat no. 6570) and incubated at room temperature for 10 min.
2. 2 μL of 5 μ*M* substrate 3 is added to the reaction, followed by 2 μL of 250 μ*M* ATP (both diluted in 1 × kinase buffer). The plate is resealed and incubated at room temperature for 3 h.

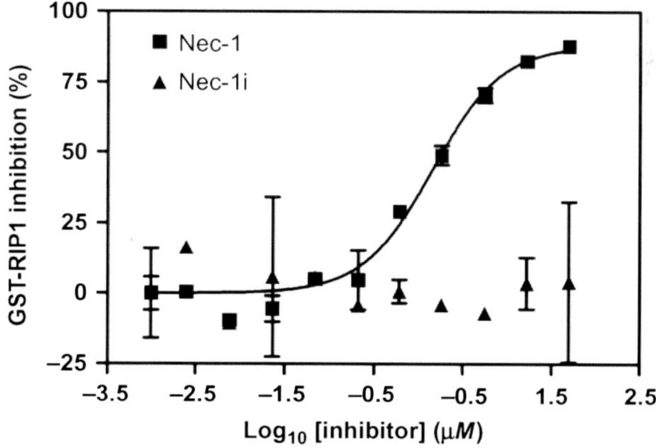

Figure 1.11 Inhibition of recombinant GST-RIP1 kinase activity by *R*-7-Cl-O-Nec-1 using KinEASE HTRF assay. Experiment was performed as described in Section 7.2 using 11-point dose range of *R*-7-Cl-O-Nec-1 (100 μ*M* to 2 n*M*). As a negative control, inactive analog of original Nec-1 (Nec-1i) was tested as well. *Reproduced with permission from Maki and Degterev (2013).*

3. Kinase reaction is terminated by the addition of 5 μL of 250 n*M* XL-665, diluted in HTRF detection buffer provided in the kit, followed by 5 μL of cryptate reagent. The plate is resealed and incubated at room temperature for 1 h.
4. Sealing tape is removed and fluorescence at 620 nm (cryptate) and 665 nm (XL-665) is measured in Victor^{3}V platereader (Perkin Elmer) using standard DELFIA program.
5. The data are analyzed by first calculating the ratio for each reaction: Ratio = (665 nm/620 nm) × 10^4. Next, the specific signal for each reaction is determined as SS = ratio (sample) − ratio (empty 1 × kinase buffer control). To convert specific signal to percent inhibition, the following formula is used: Inhibition (%) = ((ratio (DMSO sample) − SS (sample))/ ratio (DMSO sample)) × 100.

7.4. Fluorescence polarization assay (Fig. 1.12)

Previously described ^{32}P, Kinase-Glo, and HTRF assays provide useful tools to assess catalytic activity of RIPK1. In addition, simple and robust homogenous binding assays can be used to directly measure interactions of RIPK1 with small-molecule inhibitors. The first assay is based on fluorescence polarization (FP), that is, differential ability of a fluorophore to change polarization plane of the fluorescent light based on its size (Lea & Simeonov, 2011). For the purpose of this assay, we described development of fluorescent FITC analogs of Nec-1 and Nec-3 (Maki et al., 2012).

1. Assays are set up in 384-well black polystyrene plates. For the reactions, 12.5 μL of dilution series (0–100 μ*M*) of recombinant RIPK1 in 50 m*M* Tris, pH 8.0, 150 m*M* NaCl, 20% glycerol are added into the wells.

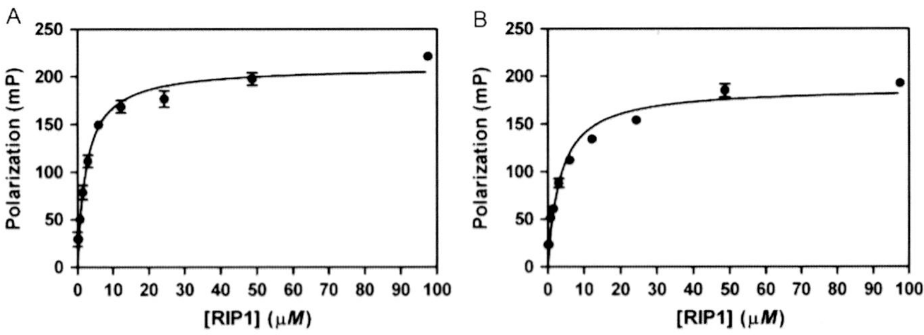

Figure 1.12 Binding of GST-RIP1 to fluorescent Nec-1 (A) and Nec-3 (B) analogs measured by FP assay. *Reproduced with permission from Maki et al. (2012).*

2. Fluorescent compounds are diluted to 750 n*M* in 50 m*M* Tris, pH 8.0, 20% glycerol, 5 m*M* $MgCl_2$, and 5 m*M* $MnCl_2$ and 1 μL is added into each well along with 1.35 μL of reaction buffer (100 m*M* HEPES, pH 7.3, 50 m*M* $MgCl_2$, 50 m*M* $MnCl_2$).
3. Plates are sealed with Thermowell seals and incubated at room temperature for 10 min. FP values are determined in Victor^{3}V platereader using an excitation wavelength of 485 nm (15-nm bandwidth) and an emission wavelength of 535 nm (25-nm bandwidth) with a count time of 1 s and a *G*-factor of 1.4.

7.5. Thermomelt assay (Fig. 1.13)

Fluorescence thermal shift analysis utilized Sypro Orange dye, which increases its fluorescence upon binding to exposed hydrophobic areas of proteins. The assay is based on measuring increased thermal stability (ΔT_m value) of the proteins in the presence of small-molecule inhibitors.

1. The reaction mix, containing 9 μ*M* RIP1-His_6, 180 μ*M* necrostatins (final concentration of DMSO = 3%), and 5× Sypro Orange (Invitrogen, cat no. S6650) in thermal shift buffer (50 m*M* Tris, pH 8.0, 150 m*M* NaCl, 1 m*M* $MgCl_2$), is added into the wells (20 μL/well) of 96-well white PCR plates (Geneseesci, cat no. 27-409). Compounds without RIP1 are also run to determine if compounds interfere with fluorescence readout.

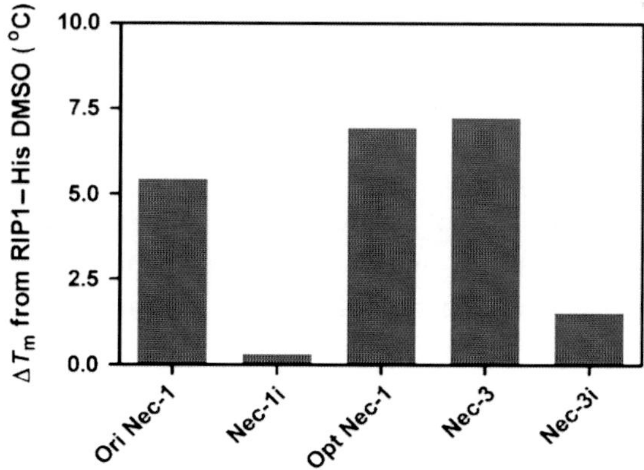

Figure 1.13 Thermomelt analysis of RIP1–His interaction with necrostatins. Original Nec-1 and Nec-3, optimized Nec-1 (7-Cl-O-Nec-1), and inactive Nec-1 and Nec-3 analogs were used. ΔT_m value reflects ability of analogs in each series to inhibit RIPK1. *Reproduced with permission from Maki et al. (2013).*

2. Plate is sealed with ThermalSeal RTS film (Geneseesci, cat no. 12-537) and incubated at room temperature for 5 min.
3. The fluorescence signal is measured using LightCycler 480 (Roche) with excitation at 465 nm and emission at 580 nm. The protein denaturation fluorescence data are collected using melting curve analysis from 25 to 85 °C in continuous acquisition mode with 10 acquisitions per 1 °C and a ramp rate of 0.06 °C/s.
4. The data are analyzed to calculate T_m values using nonlinear regression with Boltzmann sigmoidal equation in GraphPad 5 using approximately 2° of pre- and posttransition baseline data.

8. CONCLUSIONS

Since the original description of necroptosis as a distinct form of regulated cell death (Degterev et al., 2005), major advances have been made in understanding mechanisms and pathophysiologic roles of necroptosis. Availability of inhibitors of all three major regulators of necroptosis (RIPK1, RIPK3, MLKL) and emerging understanding of upstream molecular events (such as necrosome formation and RIPK activation) provide a useful toolset for identification and analysis of necroptosis. At the same time, more work is needed to further define specific molecular events universally contributing to execution of necroptosis. Phosphorylation of MLKL on residues Thr357/Ser358 potentially represents the first specific marker of necroptosis, which could be of broad value once phospho-specific antibodies similar to the one described by Chen et al. (2013) are widely available. Sites of cross-phosphorylation between RIPK1 and RIPK3 also hold promise, especially since none of the other specific substrates of RIPK1 are currently known, but are yet to be identified. Given that RIPK3 and MLKL-deficient cells are highly resistant to necroptosis, it is unclear whether necrosome components are exclusive substrates of RIPKs or additional substrates of RIPK1 and RIPK3 may exist and what role they may play in this pathway. A growing number of studies revealed that RIPKs play important roles in pathological injuries in mouse models. It will be important to extend these observations to corresponding human conditions, which, in turn, necessitates further identification of necroptosis-specific markers.

Availability of RIPK1 and RIPK3 structures (Xie, Peng, Liu, et al., 2013; Xie, Peng, Yan, et al., 2013) in conjunction with identification of assays to measure their activity will undoubtedly further facilitate the development of small-molecule inhibitors of these kinases. One important outstanding question relates to the surprisingly exclusive selectivity of existing RIPK1 and RIPK3 inhibitors, despite very close similarity of both kinases. The molecular basis for this selectivity is currently unknown. Furthermore, it remains to be determined whether high degree of homology may allow development of efficient and selective dual RIPK1/RIPK3 inhibitors and whether this approach may result in improved activity compared to separate targeting of RIPKs.

ACKNOWLEDGMENTS

This work was supported in part by grants to J. Y. from the National Institute on Aging (R37AG012859), the National Institute of Neurological Disorders and Stroke (1R01NS082257), and a Senior Fellowship from the Ellison Foundation and to A. D. from the National Institute of General Medical Sciences (R01GM080356, R01GM084205).

REFERENCES

Abhari, B. A., Cristofanon, S., Kappler, R., von Schweinitz, D., Humphreys, R., & Fulda, S. (2013). RIP1 is required for IAP inhibitor-mediated sensitization for TRAIL-induced apoptosis via a RIP1/FADD/caspase-8 cell death complex. *Oncogene*, *32*, 3263–3273.

Arslan, S. C., & Scheidereit, C. (2011). The prevalence of TNFalpha-induced necrosis over apoptosis is determined by TAK1-RIP1 interplay. *PLoS One*, *6*, e26069.

Basit, F., Cristofanon, S., & Fulda, S. (2013). Obatoclax (GX15-070) triggers necroptosis by promoting the assembly of the necrosome on autophagosomal membranes. *Cell Death and Differentiation*, *20*, 1161–1173.

Bell, B. D., Leverrier, S., Weist, B. M., Newton, R. H., Arechiga, A. F., Luhrs, K. A., et al. (2008). FADD and caspase-8 control the outcome of autophagic signaling in proliferating T cells. *Proceedings of the National Academy of Sciences of the United States of America*, *105*, 16677–16682.

Biton, S., & Ashkenazi, A. (2011). NEMO and RIP1 control cell fate in response to extensive DNA damage via TNF-alpha feedforward signaling. *Cell*, *145*, 92–103.

Chan, F. K., & Lenardo, M. J. (2000). A crucial role for p80 TNF-R2 in amplifying p60 TNF-R1 apoptosis signals in T lymphocytes. *European Journal of Immunology*, *30*, 652–660.

Chan, F. K., Shisler, J., Bixby, J. G., Felices, M., Zheng, L., Appel, M., et al. (2003). A role for tumor necrosis factor receptor-2 and receptor-interacting protein in programmed necrosis and antiviral responses. *Journal of Biological Chemistry*, *278*, 51613–51621.

Chen, W., Zhou, Z., Li, L., Zhong, C. Q., Zheng, X., Wu, X., et al. (2013). Diverse sequence determinants control human and mouse receptor interacting protein 3 (RIP3) and mixed lineage kinase domain-like (MLKL) interaction in necroptotic signaling. *Journal of Biological Chemistry*, *288*, 16247–16261.

Cho, Y. S., Challa, S., Moquin, D., Genga, R., Ray, T. D., Guildford, M., et al. (2009). Phosphorylation-driven assembly of the RIP1-RIP3 complex regulates programmed necrosis and virus-induced inflammation. *Cell*, *137*, 1112–1123.

Christofferson, D. E., Li, Y., Hitomi, J., Zhou, W., Upperman, C., Zhu, H., et al. (2012). A novel role for RIP1 kinase in mediating TNFalpha production. *Cell Death Disease*, *3*, e320.

Christofferson, D. E., Li, Y., & Yuan, J. (2014). Control of life-or-death decisions by RIP1 kinase. *Annual Review of Physiology*, *76*, 129–150.

Christofferson, D. E., & Yuan, J. (2010a). Cyclophilin A release as a biomarker of necrotic cell death. *Cell Death and Differentiation, 17*, 1942–1943.

Christofferson, D. E., & Yuan, J. (2010b). Necroptosis as an alternative form of programmed cell death. *Current Opinion in Cell Biology, 22*, 263–268.

Degterev, A., Hitomi, J., Germscheid, M., Ch'en, I. L., Korkina, O., Teng, X., et al. (2008). Identification of RIP1 kinase as a specific cellular target of necrostatins. *Nature Chemical Biology, 4*, 313–321.

Degterev, A., Huang, Z., Boyce, M., Li, Y., Jagtap, P., Mizushima, N., et al. (2005). Chemical inhibitor of nonapoptotic cell death with therapeutic potential for ischemic brain injury. *Nature Chemical Biology, 1*, 112–119.

Degterev, A., Maki, J. L., & Yuan, J. (2013). Activity and specificity of necrostatin-1, small-molecule inhibitor of RIP1 kinase. *Cell Death and Differentiation, 20*, 366.

Dondelinger, Y., Aguileta, M. A., Goossens, V., Dubuisson, C., Grootjans, S., Dejardin, E., et al. (2013). RIPK3 contributes to TNFR1-mediated RIPK1 kinase-dependent apoptosis in conditions of cIAP1/2 depletion or TAK1 kinase inhibition. *Cell Death and Differentiation, 20*, 1381–1392.

Duprez, L., Takahashi, N., Van Hauwermeiren, F., Vandendriessche, B., Goossens, V., Vanden Berghe, T., et al. (2011). RIP kinase-dependent necrosis drives lethal systemic inflammatory response syndrome. *Immunity, 35*, 908–918.

Ea, C. K., Deng, L., Xia, Z. P., Pineda, G., & Chen, Z. J. (2006). Activation of IKK by TNFalpha requires site-specific ubiquitination of RIP1 and polyubiquitin binding by NEMO. *Molecular Cell, 22*, 245–257.

Feoktistova, M., Geserick, P., Kellert, B., Dimitrova, D. P., Langlais, C., Hupe, M., et al. (2011). cIAPs block ripoptosome formation, a RIP1/caspase-8 containing intracellular cell death complex differentially regulated by cFLIP isoforms. *Molecular Cell, 43*, 449–463.

Fulda, S. (2013). The mechanism of necroptosis in normal and cancer cells. *Cancer Biology and Therapy, 14*, 999–1004.

Galluzzi, L., Kepp, O., & Kroemer, G. (2009). RIP kinases initiate programmed necrosis. *Journal of Molecular Cell Biology, 1*, 8–10.

He, S., Liang, Y., Shao, F., & Wang, X. (2011). Toll-like receptors activate programmed necrosis in macrophages through a receptor-interacting kinase-3-mediated pathway. *Proceedings of the National Academy of Sciences of the United States of America, 108*, 20054–20059.

He, S., Wang, L., Miao, L., Wang, T., Du, F., Zhao, L., et al. (2009). Receptor interacting protein kinase-3 determines cellular necrotic response to TNF-alpha. *Cell, 137*, 1100–1111.

Hitomi, J., Christofferson, D. E., Ng, A., Yao, J., Degterev, A., Xavier, R. J., et al. (2008). Identification of a molecular signaling network that regulates a cellular necrotic cell death pathway. *Cell, 135*, 1311–1323.

Holler, N., Zaru, R., Micheau, O., Thome, M., Attinger, A., Valitutti, S., et al. (2000). Fas triggers an alternative, caspase-8-independent cell death pathway using the kinase RIP as effector molecule. *Nature Immunology, 1*, 489–495.

Juo, P., Woo, M. S., Kuo, C. J., Signorelli, P., Biemann, H. P., Hannun, Y. A., et al. (1999). FADD is required for multiple signaling events downstream of the receptor Fas. *Cell Growth and Differentiation, 10*, 797–804.

Kaiser, W. J., Sridharan, H., Huang, C., Mandal, P., Upton, J. W., Gough, P. J., et al. (2013). Toll-like receptor 3-mediated necrosis via TRIF, RIP3 and MLKL. *Journal of Biological Chemistry, 288*, 31268–31279.

Kang, T. B., Yang, S. H., Toth, B., Kovalenko, A., & Wallach, D. (2013). Caspase-8 blocks kinase RIPK3-mediated activation of the NLRP3 inflammasome. *Immunity, 38*, 27–40.

Kim, J. Y., Beg, A. A., & Haura, E. B. (2013). Non-canonical IKKs, IKK and TBK1, as novel therapeutic targets in the treatment of non-small cell lung cancer. *Expert Opinion on Therapeutic Targets*, *17*, 1109–1112.

Lea, W. A., & Simeonov, A. (2011). Fluorescence polarization assays in small molecule screening. *Expert Opinion on Drug Discovery*, *6*, 17–32.

Lewis, M., Tartaglia, L. A., Lee, A., Bennett, G. L., Rice, G. C., Wong, G. H., et al. (1991). Cloning and expression of cDNAs for two distinct murine tumor necrosis factor receptors demonstrate one receptor is species specific. *Proceedings of the National Academy of Sciences of the United States of America*, *88*, 2830–2834.

Li, J., McQuade, T., Siemer, A. B., Napetschnig, J., Moriwaki, K., Hsiao, Y. S., et al. (2012). The RIP1/RIP3 necrosome forms a functional amyloid signaling complex required for programmed necrosis. *Cell*, *150*, 339–350.

Loder, S., Fakler, M., Schoeneberger, H., Cristofanon, S., Leibacher, J., Vanlangenakker, N., et al. (2012). RIP1 is required for IAP inhibitor-mediated sensitization of childhood acute leukemia cells to chemotherapy-induced apoptosis. *Leukemia*, *26*, 1020–1029.

Lukens, J. R., Vogel, P., Johnson, G. R., Kelliher, M. A., Iwakura, Y., Lamkanfi, M., et al. (2013). RIP1-driven autoinflammation targets IL-1alpha independently of inflammasomes and RIP3. *Nature*, *498*, 224–227.

Maki, J. L., & Degterev, A. (2013). Activity assays for receptor-interacting protein kinase 1: A key regulator of necroptosis. *Methods in Molecular Biology*, *1004*, 31–42.

Maki, J. L., Smith, E. E., Teng, X., Ray, S. S., Cuny, G. D., & Degterev, A. (2012). Fluorescence polarization assay for inhibitors of the kinase domain of receptor interacting protein 1. *Analytical Biochemistry*, *427*, 164–174.

Maki, J. L., Tres Brazell, J., Teng, X., Cuny, G. D., & Degterev, A. (2013). Expression and purification of active receptor interacting protein 1 kinase using a baculovirus system. *Protein Expression and Purification*, *89*, 156–161.

Matsumura, H., Shimizu, Y., Ohsawa, Y., Kawahara, A., Uchiyama, Y., & Nagata, S. (2000). Necrotic death pathway in Fas receptor signaling. *Journal of Cell Biology*, *151*, 1247–1256.

McNamara, C. R., Ahuja, R., Osafo-Addo, A. D., Barrows, D., Kettenbach, A., Skidan, I., et al. (2013). Akt regulates TNFalpha synthesis downstream of RIP1 kinase activation during necroptosis. *PLoS One*, *8*, e56576.

Micheau, O., & Tschopp, J. (2003). Induction of TNF receptor I-mediated apoptosis via two sequential signaling complexes. *Cell*, *114*, 181–190.

Moquin, D. M., McQuade, T., & Chan, F. K. (2013). CYLD deubiquitinates RIP1 in the TNFalpha-induced necrosome to facilitate kinase activation and programmed necrosis. *PLoS One*, *8*, e76841.

Moujalled, D. M., Cook, W. D., Okamoto, T., Murphy, J., Lawlor, K. E., Vince, J. E., et al. (2013). TNF can activate RIPK3 and cause programmed necrosis in the absence of RIPK1. *Cell Death Disease*, *4*, e465.

Murakami, Y., Matsumoto, H., Roh, M., Giani, A., Kataoka, K., Morizane, Y., et al. (2013). Programmed necrosis, not apoptosis, is a key mediator of cell loss and DAMP-mediated inflammation in dsRNA-induced retinal degeneration. *Cell Death and Differentiation*, *21*, 270–277.

Murphy, J. M., Czabotar, P. E., Hildebrand, J. M., Lucet, I. S., Zhang, J. G., Alvarez-Diaz, S., et al. (2013). The pseudokinase MLKL mediates necroptosis via a molecular switch mechanism. *Immunity*, *39*, 443–453.

Oberst, A., Dillon, C. P., Weinlich, R., McCormick, L. L., Fitzgerald, P., Pop, C., et al. (2011). Catalytic activity of the caspase-8-FLIP(L) complex inhibits RIPK3-dependent necrosis. *Nature*, *471*, 363–367.

O'Donnell, M. A., Perez-Jimenez, E., Oberst, A., Ng, A., Massoumi, R., Xavier, R., et al. (2011). Caspase 8 inhibits programmed necrosis by processing CYLD. *Nature Cell Biology*, *13*, 1437–1442.

Pimentel-Muinos, F. X., & Seed, B. (1999). Regulated commitment of TNF receptor signaling: A molecular switch for death or activation. *Immunity*, *11*, 783–793.

Rimon, G., Bazenet, C. E., Philpott, K. L., & Rubin, L. L. (1997). Increased surface phosphatidylserine is an early marker of neuronal apoptosis. *Journal of Neuroscience Research*, *48*, 563–570.

Salvioli, S., Ardizzoni, A., Franceschi, C., & Cossarizza, A. (1997). JC-1, but not DiOC6(3) or rhodamine 123, is a reliable fluorescent probe to assess delta psi changes in intact cells: Implications for studies on mitochondrial functionality during apoptosis. *FEBS Letters*, *411*, 77–82.

Shen, H. M., & Codogno, P. (2012). Autophagy is a survival force via suppression of necrotic cell death. *Experimental Cell Research*, *318*, 1304–1308.

Shindo, R., Kakehashi, H., Okumura, K., Kumagai, Y., & Nakano, H. (2013). Critical contribution of oxidative stress to TNFalpha-induced necroptosis downstream of RIPK1 activation. *Biochemical and Biophysical Research Communications*, *436*, 212–216.

Sosna, J., Voigt, S., Mathieu, S., Kabelitz, D., Trad, A., Janssen, O., et al. (2013). The proteases HtrA2/Omi and UCH-L1 regulate TNF-induced necroptosis. *Cell Communication and Signalling*, *11*, 76.

Sun, L., Wang, H., Wang, Z., He, S., Chen, S., Liao, D., et al. (2012). Mixed lineage kinase domain-like protein mediates necrosis signaling downstream of RIP3 kinase. *Cell*, *148*, 213–227.

Tait, S. W., & Green, D. R. (2008). Caspase-independent cell death: Leaving the set without the final cut. *Oncogene*, *27*, 6452–6461.

Temkin, V., Huang, Q., Liu, H., Osada, H., & Pope, R. M. (2006). Inhibition of ADP/ATP exchange in receptor-interacting protein-mediated necrosis. *Molecular and Cellular Biology*, *26*, 2215–2225.

Tenev, T., Bianchi, K., Darding, M., Broemer, M., Langlais, C., Wallberg, F., et al. (2011). The ripoptosome, a signaling platform that assembles in response to genotoxic stress and loss of IAPs. *Molecular Cell*, *43*, 432–448.

Thapa, R. J., Nogusa, S., Chen, P., Maki, J. L., Lerro, A., Andrake, M., et al. (2013). Interferon-induced RIP1/RIP3-mediated necrosis requires PKR and is licensed by FADD and caspases. *Proceedings of the National Academy of Sciences of the United States of America*, *110*, E3109–E3118.

Ting, A. T., Pimentel-Muinos, F. X., & Seed, B. (1996). RIP mediates tumor necrosis factor receptor 1 activation of NF-kappaB but not Fas/APO-1-initiated apoptosis. *EMBO Journal*, *15*, 6189–6196.

Upton, J. W., Kaiser, W. J., & Mocarski, E. S. (2012). DAI/ZBP1/DLM-1 complexes with RIP3 to mediate virus-induced programmed necrosis that is targeted by murine cytomegalovirus vIRA. *Cell Host & Microbe*, *11*, 290–297.

Vanags, D. M., Porn-Ares, M. I., Coppola, S., Burgess, D. H., & Orrenius, S. (1996). Protease involvement in fodrin cleavage and phosphatidylserine exposure in apoptosis. *Journal of Biological Chemistry*, *271*, 31075–31085.

Vanden Berghe, T., Vanlangenakker, N., Parthoens, E., Deckers, W., Devos, M., Festjens, N., et al. (2010). Necroptosis, necrosis and secondary necrosis converge on similar cellular disintegration features. *Cell Death and Differentiation*, *17*, 922–930.

Vanlangenakker, N., Vanden Berghe, T., & Vandenabeele, P. (2012). Many stimuli pull the necrotic trigger, an overview. *Cell Death and Differentiation*, *19*, 75–86.

Vercammen, D., Vandenabeele, P., Beyaert, R., Declercq, W., & Fiers, W. (1997). Tumour necrosis factor-induced necrosis versus anti-Fas-induced apoptosis in L929 cells. *Cytokine*, *9*, 801–808.

Vince, J. E., Wong, W. W., Gentle, I., Lawlor, K. E., Allam, R., O'Reilly, L., et al. (2012). Inhibitor of apoptosis proteins limit RIP3 kinase-dependent interleukin-1 activation. *Immunity*, *36*, 215–227.

Wagner, L., Marschall, V., Karl, S., Cristofanon, S., Zobel, K., Deshayes, K., et al. (2013). Smac mimetic sensitizes glioblastoma cells to Temozolomide-induced apoptosis in a RIP1- and NF-kappaB-dependent manner. *Oncogene*, *32*, 988–997.

Wu, J., Huang, Z., Ren, J., Zhang, Z., He, P., Li, Y., et al. (2013). Mlkl knockout mice demonstrate the indispensable role of Mlkl in necroptosis. *Cell Research*, *23*, 994–1006.

Xie, T., Peng, W., Liu, Y., Yan, C., Maki, J., Degterev, A., et al. (2013). Structural basis of RIP1 inhibition by necrostatins. *Structure*, *21*, 493–499.

Xie, T., Peng, W., Yan, C., Wu, J., Gong, X., & Shi, Y. (2013). Structural insights into RIP3-mediated necroptotic signaling. *Cell Reports*, *5*, 70–78.

Yu, L., Alva, A., Su, H., Dutt, P., Freundt, E., Welsh, S., et al. (2004). Regulation of an ATG7-beclin 1 program of autophagic cell death by caspase-8. *Science*, *304*, 1500–1502.

Zhang, D. W., Shao, J., Lin, J., Zhang, N., Lu, B. J., Lin, S. C., et al. (2009). RIP3, an energy metabolism regulator that switches TNF-induced cell death from apoptosis to necrosis. *Science*, *325*, 332–336.

Zhou, Z., Han, V., & Han, J. (2012). New components of the necroptotic pathway. *Protein & Cell*, *3*, 811–817.

CHAPTER TWO

IAP Family of Cell Death and Signaling Regulators

John Silke[*,†,1], **Domagoj Vucic**[‡,1]

[*]The Walter and Eliza Hall Institute of Medical Research, Parkville, Victoria, Australia
[†]Department of Medical Biology, University of Melbourne, Parkville, Victoria, Australia
[‡]Department of Early Discovery Biochemistry, Genentech, Inc., South San Francisco, California, USA
[1]Corresponding authors: e-mail address: silke@wehi.edu.au; domagoj@gene.com

Contents

Abstract

Inhibitor of apoptosis (IAP) proteins interface with, and regulate a large number of, cell signaling pathways. If there is a common theme to these pathways, it is that they are involved in the development of the immune system, immune responses, and unsurprisingly, given their name, cell death. Beyond that it is difficult to discover an underlying logic because sometimes IAPs are required to inhibit or prevent signaling, whereas in other cases they are required for signaling to take place. In whatever role they play, they are recruited into signaling complexes and function as ubiquitin E3 ligases, via their RING domains. This review discusses IAP regulation of signaling pathways and focuses on the mammalian IAPs, XIAP, c-IAP1, and c-IAP2, with a particular emphasis on techniques and methods that were used to uncover their roles. We also provide a perspective on targeting IAP proteins for therapeutic intervention and methods used to define the clinical relevance of IAP proteins.

Methods in Enzymology, Volume 545
ISSN 0076-6879
http://dx.doi.org/10.1016/B978-0-12-801430-1.00002-0

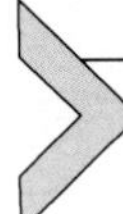

1. IDENTIFICATION OF IAPs, STRUCTURE, AND DOMAIN FUNCTION

1.1. Discovery

Inhibitors of apoptosis (IAPs) were originally found in baculoviruses in genetic screens designed to identify viral factor(s) capable of substituting for baculovirus-encoded caspase inhibitor P35 (Birnbaum, Clem, & Miller, 1994). These screens relied on the absolute dependence of baculovirus propagation on antiapoptotic proteins that can block massive cell death induced by the infection of insect cells (Birnbaum et al., 1994). The *p35*-deficient baculovirus strain AcMNPV is a particularly effective stimulator of cell death because in the absence of *p35*, virtually all infected cells die (Birnbaum et al., 1994). The first members of the IAP family, *Op-iap* and *Cp-iap*, were isolated from the genomes of baculoviruses CpGV and OpNPV because of their ability to rescue *p35* deficiency (Birnbaum et al., 1994; Crook, Clem, & Miller, 1993). Similar screens allowed the identification of additional viral and insect IAPs but, unfortunately, because mammalian IAPs cannot functionally substitute for *p35* absence, it has not been possible to screen for mammalian IAPs in this way (Seshagiri, Vucic, Lee, & Dixit, 1999). Following their identification in baculoviruses, IAPs were discovered in both invertebrates and vertebrates. The *Drosophila melanogaster* homolog of baculoviral IAPs, DIAP1, was identified in a screen for mutations that enhanced the effect of the endogenous IAP-antagonist Reaper in *Drosophila* developing eyes (Hay, Wassarman, & Rubin, 1995). The screen involved a collection of chromosomal deletions that covered a significant portion of *Drosophila* genome and one of the enhancers corresponded to lethal mutations in *thread* (*th*), a locus that encodes DIAP1. DIAP2, a related *Drosophila* IAP, was identified by a search of databases for sequences homologous to known IAPs (Hay et al., 1995).

Human IAP family members include X-chromosome-linked IAP (XIAP, also known as hILP, MIHA, and BIRC4; Duckett et al., 1996; Liston et al., 1996; Uren, Pakusch, Hawkins, Puls, & Vaux, 1996), cellular IAP1 (c-IAP1, also known as HIAP2, MIHB, and BIRC2; Liston et al., 1996; Rothe, Wong, Henzel, & Goeddel, 1994; Uren et al., 1996), c-IAP2 (also known as HIAP1, MIHC, and BIRC3; Liston et al., 1996; Rothe et al., 1994; Uren et al., 1996), neuronal apoptosis inhibitory protein (NAIP, also known as BIRC1; Roy et al., 1995), survivin (also known as TIAP and BIRC5; Ambrosini, 1997), Apollon (also known as Bruce and BIRC6; Chen et al., 1999; Hauser,

Bardroff, Pyrowolakis, & Jentsch, 1998), melanoma IAP (ML-IAP, also known as KIAP, livin, and BIRC7; Kasof & Gomes, 2001; Lin, Deng, Huang, & Morser, 2000; Vucic, Stennicke, Pisabarro, Salvesen, & Dixit, 2000), and IAP-like protein 2 (ILP2 and BIRC8; Lagace et al., 2001; Richter et al., 2001) (Fig. 2.1). Survivin and BRUCE appear to be most closely related to the yeast and *Caenorhabditis elegans* BIRC containing proteins and play a role in cytokinesis and, therefore, do not feature prominently in the rest of this review (Fraser, James, Evan, & Hengartner, 1999; Pohl & Jentsch, 2008; Speliotes, Uren, Vaux, & Horvitz, 2000; Uren et al., 2000). NAIP, the first characterized mammalian IAP family member, was identified during a search for the causative gene of spinal muscular atrophy (SMA; Roy et al., 1995). The investigators used positional cloning to identify *Naip*, although later it was discovered that the deletion of a neighboring gene (survival motor neuron, SMN) was in fact linked to SMA. Cellular IAP1 and IAP2 were originally identified as components of TNFR2-associated complex through their constitutive binding to TRAF1 and TRAF2 (Rothe, Pan, Henzel, Ayres, & Goeddel, 1995). This discovery required biochemical purification from the cellular lysate prepared from 120 l of cell culture and

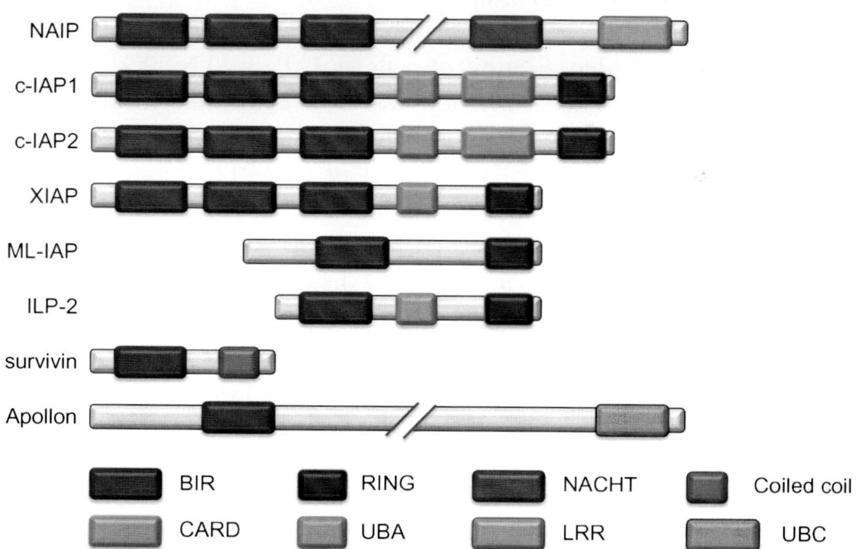

Figure 2.1 Schematic representation of human IAP proteins. BIR: baculovirus IAP repeat; CARD: caspase recruitment domain; LRR: leucine-rich repeat; NACHT: NAIP, CIITA, HET-E, and TP1; RING: really interesting gene; UBA: ubiquitin-associated domain; UBC: ubiquitin-conjugating domain. (See the color plate.)

passed through 25 ml of GST-TNFR2 column followed by a mass spectrometry analysis. Other IAPs were identified through homology searches for baculovirus IAP repeat (BIR) domains—a signature protein interaction region present in all IAP proteins (Salvesen & Duckett, 2002; Verhagen, Coulson, & Vaux, 2001).

1.2. Domain structure—BIRs

IAP proteins contain one to three 70–80 amino acid BIR domains that coordinate a zinc ion and are important for the IAP-antiapoptotic activity. Structurally, BIR domains contain a three-stranded antiparallel β-sheet and five α-helices with three cysteines and one histidine chelating a zinc atom (Franklin et al., 2003; Hinds, Norton, Vaux, & Day, 1999; Sun et al., 1999, 2000). The BIR domains are critical conduits for IAP-mediated interactions with multiple binding partners: c-IAP1/2 bind TRAF2 via BIR1, XIAP binds caspases-3 and -7 through the BIR2 domain and linker between BIR1 and BIR12 domains, and the third BIR domain of XIAP and c-IAPs as well as the single BIR domain of ML-IAP strongly bind Smac (second mitochondrial activator of caspases)/DIABLO (direct IAP-binding protein with low pI) and other IAP-binding motif (IBM)-containing proteins such as serine protease HtrA2. The interaction between c-IAP1/2 and TRAF2 allowed the original identification of cellular IAPs in TNFR2 pulldown (Rothe et al., 1995). This association was later confirmed in yeast two-hybrid system and subsequent structure/function experiments determined that amino-terminus of the BIR1 domain of c-IAP1/2 binds TRAF-N region of TRAF2 (Mace, Smits, Vaux, Silke, & Day, 2010; Uren et al., 1996; Varfolomeev, Wayson, Dixit, Fairbrother, & Vucic, 2006; Vince et al., 2009; Zheng, Kabaleeswaran, Wang, Cheng, & Wu, 2010). Similarly, structural and functional studies have definitively established that XIAP uses different BIR domains for inhibition of distinct classes of caspases; the second BIR domain together with the immediately preceding linker region binds and inhibits caspases-3 and -7, while the third BIR domain specifically inhibits caspase-9 (Salvesen & Duckett, 2002). The XIAP-mediated inhibition of these caspases is antagonized by the mitochondrial protein Smac, which is released into the cytoplasm in response to proapoptotic stimuli (Du, Fang, Li, Li, & Wang, 2000; Verhagen et al., 2000). However, Smac has higher affinity for the BIR3 domain than to the BIR2 domain, and hence, the inhibition of caspase-9 blocking activity is more pronounced (Huang, Rich, Myszka, & Wu, 2003).

1.3. Domain structure—RING and UBA

Several IAP proteins also possess a carboxy-terminal RING domain (Vaux & Silke, 2005). The RING domains of c-IAP1, c-IAP2, ML-IAP, and XIAP have ubiquitin E3 ligase activity and can promote autoubiquitylation as well as transubiquitylation of their binding partners (Vaux & Silke, 2005; Vucic, Dixit, & Wertz, 2011). A different IAP protein, Apollon/BRUCE, lacks a RING domain, but it contains a ubiquitin-conjugating domain (UBC), found in E2 enzymes, that can also mediate ubiquitylation (Chen et al., 1999; Hao et al., 2004). In addition to promoting ubiquitylation, several IAP proteins (c-IAP1, c-IAP2, XIAP, hILP2) can bind polyubiquitin chains composed of K11, K48, or K63 linkages via their UBA domain (Blankenship et al., 2009; Dynek et al., 2010; Gyrd-Hansen et al., 2008).

1.4. Domain structure—NACHT and enigmatic CARD

c-IAP1 and c-IAP2 also possess a caspase recruitment domain (CARD; Hofmann, Bucher, & Tschopp, 1997). The CARD domain in c-IAP1 and c-IAP2 was found based on the sequence homology 17 years ago, but the binding partner for this domain has still not been identified. The smallest IAP protein, survivin, plays a crucial role in cell division and contains a coiled coil domain responsible for its interaction with chromosomal passenger protein complex (Jeyaprakash et al., 2007; Uren et al., 2000). NAIP contains three BIR domains but also domains more typically found in members of the nuclear binding and oligomerization domain (NOD)-like intracellular pathogen receptor family: a leucine-rich repeat (LRR) domain and a NTPase NACHT domain after proteins NAIP, CIITA, HET-E, and TP1 (Inohara & Nunez, 2003; Koonin & Aravind, 2000) (Fig. 2.1).

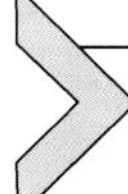

2. IAP PROTEINS AND CELL DEATH PATHWAYS

2.1. XIAP—Inhibitor of the intrinsic Bcl-2 blockable pathway

Their name and the route to their discovery strongly suggest that inhibiting apoptosis is a major function of the IAPs. While it is not the only function, the following section is focused on this aspect, particularly in mammalian cells, as well as their ability to inhibit the other major programmed cell death pathway, necroptosis. Because baculoviral IAPs prevented viral-induced apoptosis, mammalian (and baculoviral) IAPs were quickly tested for their ability to inhibit apoptosis in transient transfection assays in mammalian cells.

A common assay at this time was to transfect a fusion construct of *Caspase-1* and *Escherichia coli LacZ* (Jung, Miura, & Yuan, 1996). β-Galactosidase (the protein product of the *LacZ* gene) tetramerizes and promotes autoactivation of caspase-1. The other advantage of this fusion construct is that transfected cells could be readily identified by staining fixed cells with X-gal and scored by apoptotic morphology (Duckett et al., 1996; Ekert, Silke, & Vaux, 1999; Uren et al., 1996). Now it is usually easier (and more objective) to cotransfect cells with a fluorescent protein construct and use flow cytometry combined with Annexin V and propidium iodide (PI) staining. Annexin V binds to phosphatidylserine exposed on the surface of apoptotic cells, while PI only stains cells with a disrupted membrane. In retrospect, autoactivated caspase-1 may not induce a *bona fide* apoptotic death; however, this death could be blocked with known apoptotic inhibitors including the baculoviral protein p35, the cowpox viral protein CrmA, and even the baculoviral IAPs and seemed appropriate at the time (Duckett et al., 1996; Hawkins, Uren, Häcker, Medcalf, & Vaux, 1996). With the benefit of hindsight, it is also interesting that XIAP/MIHA/hILP also blocked this cell death (Duckett et al., 1996; Uren et al., 1996) because unlike CrmA and p35, XIAP is not a direct inhibitor of caspase-1 (Deveraux, Takahashi, Salvesen, & Reed, 1997; Garcia-Calvo et al., 1998). XIAP was recently shown to inhibit activation of the inflammasome and thereby activation of caspase-1, so there may be more to these experiments than initially meets the eye (Vince et al., 2012).

2.2. XIAP—Caspase inhibitor

Soon after recombinant XIAP was shown to be a potent low n*M* inhibitor of caspases-3, -7, and -9 but did not inhibit caspases-1, -6, and -8 (Deveraux et al., 1998, 1997). Recombinant GST-fused c-IAP1 and c-IAP2 were also shown to directly inhibit cleavage of fluorogenic peptide substrates, such as DEVD-AMC, by recombinant caspases (Roy, Deveraux, Takahashi, Salvesen, & Reed, 1997). Unfortunately, tagging with GST may lead to GST-dependent oligomerization that may create artifacts in such assays, although the relevance of this to the caspase inhibitory potential of c-IAP1 has been questioned (Burke, Smith, & Smith, 2010). When the untagged BIR domains of c-IAP1 and c-IAP2 were tested in these assays, they were unable to inhibit caspase cleavage of a fluorogenic peptide substrate at a physiologically meaningful concentration (Eckelman & Salvesen, 2006). Therefore, unlike the isolated BIRs of XIAP, the BIRs of c-IAP proteins are unable to inhibit caspase activity (Scott et al., 2005; Silke et al., 2001; Takahashi et al., 1998). Interestingly, the IBMs are conserved between XIAP and c-IAPs and, therefore, these BIR domains retain the ability to bind processed caspases-3, -7, and -9. Clearly, these experiments are not definitive; full-length c-IAPs are very difficult to generate as recombinant proteins and were not exhaustively tested in these assays (Eckelman & Salvesen, 2006); furthermore, a fluorogenic peptide is small and completely unlike a natural substrate, and finally, while the GST domain may promote oligomerization of c-IAPs, in their native signaling complexes c-IAPs may naturally form oligomers. It also remains an interesting speculation that c-IAPs might bind caspases and alter substrate specificity rather than blocking caspase activity altogether. Clearly, better tools are needed before these experiments can be attempted.

XIAP is therefore able to inhibit both initiator (caspase-9) and effector (caspase-3 and caspase-7) caspases activated during the intrinsic or Bcl-2 blockable apoptotic pathway with low nM K_i (Fig. 2.2). Furthermore, XIAP has been shown to bind to these caspases in the context of the apoptosome: the Apaf-1, caspase-9, caspase-3 complex activated by release of cytochrome *c* from the intermembrane space of mitochondria (Bratton, Lewis, Butterworth, Duckett, & Cohen, 2002; Bratton et al., 2001; Srinivasula et al., 2001). This targeting of XIAP to the activated apoptosome potentially lowers XIAPs real *in vivo* K_i (Bratton et al., 2002). Certainly, XIAP, when overexpressed, provides good short-term protection in transient transfection assays (Bratton et al., 2002; Deveraux et al., 1998, 1997; Silke et al., 2001, 2002). However, in clonogenic assays where Bcl-2 overexpression provides

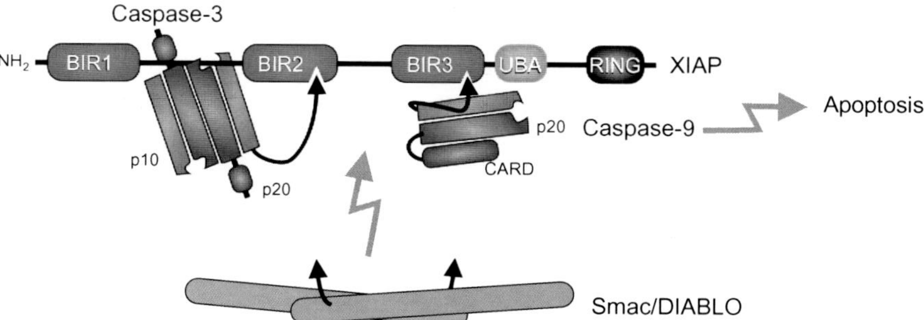

Figure 2.2 XIAP binds and inhibits processed caspases-3 and -9 and possibly functions as a safety catch to block limited activation of the apoptosome. If mitochondria are permeabilized, they release IAP-antagonist molecules including Smac/DIABLO and HtrA2, which liberate caspases from inhibition allowing apoptosis to proceed. (See the color plate.)

substantial protection, XIAP is a poor inhibitor of intrinsic apoptosis (J.S., personal observation). To perform a clonogenic assay, cells are usually treated with a death stimulus for a defined period of time, then washed and replated at low dilution and left to grow in tissue culture and usually subsequently fixed and stained with a dye-like crystal violet (see, e.g., Ekert et al., 1999; Geserick et al., 2009). It is a simple assay to perform yet uniquely powerful in its ability to assess long-term "functional" protection from a death stimulus and should be more widely used.

It is possible that normal levels of XIAP are usually insufficient to provide long-term protection from an intrinsic apoptotic stimuli for two reasons. First, if Bax/Bak disruption of mitochondria is sufficient and irreversible, and therefore the cells ability to generate energy irretrievably lost, it is hard to envisage how preventing caspase activity is able to provide long-term protection to cells. Second, disruption of the mitochondrial membrane can unleash a large number of potential IAP antagonists that all contain an IBM (Hegde et al., 2001; Martins et al., 2001; Suzuki et al., 2001; Verhagen, Silke, et al., 2001; Verhagen et al., 2007). Indeed, the first mammalian IAP antagonist, Smac/DIABLO, was identified by screening for a factor in the mitochondrial fraction that facilitated caspase activation (Du et al., 2000) or for an XIAP binding protein that was released using a detergent that solubilized the mitochondrial membrane (Verhagen et al., 2000). Therefore, it is possible that even overexpression levels of XIAP are overwhelmed by the wholesale release of a large amount of IAP antagonists.

XIAP can provide meaningful protection *in vivo* in the right conditions. Thus, wild-type but not *Bid* knockout mice die when injected with FasL as a consequence of massive hepatocyte apoptosis (Jost et al., 2009; Kaufmann et al., 2009). Bid is a BH3-containing protein that can antagonize Bcl-2 prosurvival proteins and lead to activation of Bax/Bak and mitochondrial permeabilization, cytochrome *c* release and cell death. Bid must, however, be processed by caspase-8 to reveal the BH3 and therefore provides an interface to the extrinsic/death receptor-induced apoptosis pathway (Czabotar, Lessene, Strasser, & Adams, 2013). The resistance of the *Bid*-deficient mice to FasL-induced death is, however, lost in the doubly deficient $Bid^{-/-}Xiap^{-/-}$ mice, demonstrating that in hepatocytes, in the absence of the mitochondrial arm of the FasL response, XIAP is able to provide resistance to FasL-induced apoptosis (Jost et al., 2009; Varfolomeev et al., 2009). XIAP ΔRING knock-in mice are also more susceptible to Eμ myc-driven lymphoma (Schile, García-Fernández, & Steller, 2008).

2.3. Inhibition of cell death by c-IAP1 and c-IAP2

Although c-IAPs are unlikely to inhibit the intrinsic/Bcl-2 blockable apoptosis pathway, they are potent inhibitors of the extrinsic pathway. Based on the foregoing discussion of *in vitro* caspase inhibition assays, it seems unlikely that they inhibit caspase-8 but that they act indirectly, and they appear to do this at several points (Fig. 2.3). A broad range of experimental approaches give great confidence in the notion that c-IAPs are essential for activation of nuclear factor kappa B (NF-κB) by TNF and that without them cells become sensitive to TNF-induced killing. First and foremost, the *c-IAP1* and *c-IAP2* double knockout MEFs fail to activate NF-κB normally in response to TNF (Feltham et al., 2010; Moulin et al., 2012) and are sensitive to TNF-induced cell death (Moulin et al., 2012). *c-Iap1*$^{-/-}$ MEFs are also very sensitive to TNF-induced death, suggesting that the key IAP in MEFs is c-IAP1 (Feltham et al., 2010; Moulin et al., 2012; Vince et al., 2008, 2007), although it has not been possible to look at c-IAP2 expression in these cells because of the lack of a good c-IAP2 antibody to detect mouse c-IAP2. Single c-IAP1 or c-IAP2 knockout cells with knockdown of the remaining c-IAP with siRNA are also defective in the NF-κB and TNF cell survival response (Mahoney et al., 2008; Varfolomeev et al., 2008) as are *Traf2* knockout cells reconstituted with a TRAF2 mutant that cannot recruit c-IAPs into the TNF signaling complex (Vince et al., 2009). Loss of detectable RIPK1 ubiquitylation is one of the most striking defects in these doubly deficient cells (Gerlach et al., 2011; Haas et al., 2009; Mahoney et al., 2008; Varfolomeev et al., 2008), as it is in *Traf2* knockout cells reconstituted with a c-IAP1 binding defective TRAF2 mutant (Vince et al., 2009). However, because RIPK1 is not absolutely required for TNF-induced activation of the canonical NF-κB pathway in all cells (Blackwell et al., 2013; Haas et al., 2009; Wong et al., 2010), there are likely to be other proteins within the signaling complex that fail to become ubiquitylated in c-IAP-deficient cells. Finally, many different Smac-mimetics (that promote degradation of c-IAP1 and c-IAP2) sensitize many different cell types and cell lines to TNF-induced killing and prevent TNF-induced NF-κB (Li et al., 2004; Varfolomeev et al., 2008; Vince et al., 2008, 2007; Wang, Du, & Wang, 2008).

A particularly important early observation in the field was that cancer cell lines were sensitive to treatment with TRAIL when combined with overexpression of a cytosolic form of Smac or Smac-mimetic peptides (Fulda, Wick, Weller, & Debatin, 2002). This finding has been confirmed

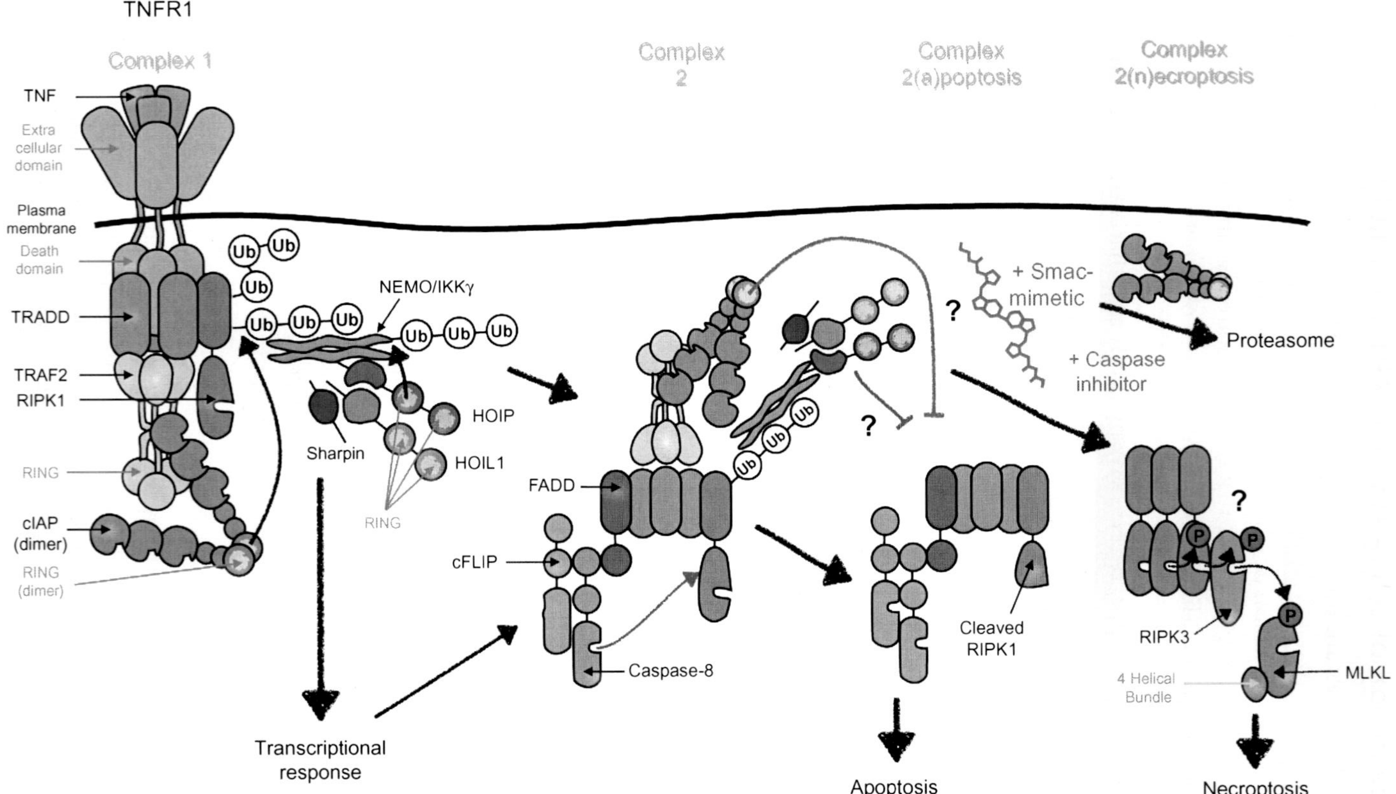

Figure 2.3 c-IAPs play critical roles in preventing the formation of cell death inducing complexes 2(a)poptosis and 2(n)ecroptosis. (See the color plate.)

independently many times with a host of different Smac-mimetic compounds (Gatti et al., 2013; Geserick et al., 2009; Li et al., 2004; Petrucci et al., 2012). An interesting methodological note on this experiment is that the IBM of Smac/DIABLO is only generated upon import of Smac/DIABLO into the intermembrane space of mitochondria by the inner membrane protease complex (Burri et al., 2005). If recombinant Smac/DIABLO is expressed as the processed form in the cytosol, it should nevertheless have an N-terminal methionine that should block the IBM and prevent it interacting with IAPs. However, the methionine of eukaryotic proteins can be removed by a methionine aminopeptidase (Polevoda & Sherman, 2003) probably accounting for the activity of Smac/DIABLO in this case. However, some researchers have used alternative systems, such as the ubiquitin fusion technique, to be assured of generating the IBM efficiently (Hegde et al., 2001; Hunter et al., 2003).

c-Iap1$^{-/-}$*c-Iap2*$^{-/-}$ MEFs, or cell lines treated with Smac-mimetics, are also sensitive to TRAIL- and FasL-induced cell death (Geserick et al., 2009). In both cases, it is unlikely the sensitivity is due to the lack of a protective NF-κB response (Peter et al., 2007). Rather, it appears to involve the formation of a secondary signaling complex emanating from the receptor that is called complex 2(a)poptosis (Geserick et al., 2009). This complex can be formed in response to a number of stimuli, including TNF, particularly when cells are treated with Smac-mimetics and has also been named the Ripoptosome (Feoktistova et al., 2011; Geserick et al., 2009; Tenev et al., 2011). The complex can be identified by immunoprecipitation of caspase-8 from cell lysates after immunoprecipitating the receptor complex. A particularly striking feature of this secondary complex when IAPs are absent is the accumulation of RIPK1. If caspase-8 activity in the complex is sufficient, RIPK1 is cleaved (Oberst et al., 2011; Vandenabeele, Galluzzi, Vanden Berghe, & Kroemer, 2010), preventing formation of complex 2(n)ecroptosis and cells die by apoptosis. However, if caspase-8 activity is impaired by addition of a synthetic caspase inhibitor, such as Q-VD-Oph or Z-VAD-FMK, or by expression of cFLIP$_S$, caspase-8 is no longer able to cleave and inactivate RIPK1 (Geserick et al., 2009; Vandenabeele et al., 2010). Intriguingly, expression of cFLIP$_L$ not only prevents caspase-8 cleavage of RIPK1 but also impairs recruitment of RIPK1 into complex 2a (Feoktistova et al., 2011; Geserick et al., 2009). Therefore, cFLIP$_L$ can inhibit caspase-8 and still prevent formation of complex 2n. Although the sequence of events at this stage is not clear, it is possible that accumulation of RIPK1 results in its autoactivation, that it then phosphorylates RIPK3, and that RIPK3 phosphorylates MLKL causing a conformational change that liberates the four helical bundle of MLKL resulting in membrane permeabilization by an as yet unidentified mechanism (Cai et al., 2014; Murphy et al., 2013; Sun et al., 2012; Vandenabeele et al., 2010) (Fig. 2.3). However, critical experimental evidence for all these steps is lacking.

The buildup of RIPK1 in complex 2a/n when Smac-mimetics are used suggests that the inhibitory action of c-IAPs in this complex is to reduce levels of RIPK1 presumably by ubiquitylation and proteasomal degradation of RIPK1 and it is the E3 ubiquitin ligase function of IAPs that we now discuss in more detail.

2.4. IAP proteins and ubiquitin

The controlled modification and degradation of cellular proteins by the ubiquitin–proteasome system influences a range of crucial cellular processes in normal and diseased cells (Hershko & Ciechanover, 1998). Ubiquitylation is executed through a multistep reaction involving an E1 ubiquitin-activating enzyme, an E2 ubiquitin-conjugating enzyme, and an E3 ubiquitin ligase (Schulman & Harper, 2009). RING domain-containing ubiquitin ligases bind the E2 and the substrate proteins as well as mediate transfer of the ubiquitin molecule from the E2 onto a lysine residue of the substrate protein (Deshaies & Joazeiro, 2009). Covalent attachment of a single ubiquitin molecule to the substrate results in monoubiquitylation, but a substrate may be modified by polyubiquitin chains, involving additional ubiquitin–ubiquitin linkages (Pickart & Fushman, 2004). A growing amount of evidence suggests that E3 ubiquitin ligase activity is an instrumental function of IAP proteins (Newton & Vucic, 2007; Vaux & Silke, 2005). IAP proteins can promote ubiquitylation and subsequent proteasomal degradation of themselves and several of their binding partners, including RIP1, TRAF2, and NF-κB-inducing kinase (NIK) (Bertrand et al., 2008; Conze et al., 2005; Li, Yang, & Ashwell, 2002; Park, Yoon, & Lee, 2004; Silke et al., 2005; Varfolomeev et al., 2007, 2008; Vince et al., 2008; Zarnegar et al., 2008). The *Drosophila* DIAP1 blocks cell death, at least in part, by mediating ubiquitylation and proteasomal degradation of the *Drosophila* caspases and IAP-antagonistic proteins: Reaper, Hid, and Grim (Holley et al., 2002; Wilson et al., 2002). All natural IAP antagonists tested, and most Smac-mimetics promote autoubiquitylation and proteasomal degradation of c-IAP1 and c-IAP2 in a process that is extremely rapid and very efficient

(Bertrand et al., 2008; Darding et al., 2011; Gaither et al., 2007; Varfolomeev et al., 2007; Vince et al., 2007; Yang & Du, 2004). Smac-mimetics have, therefore, proven to be enormously useful tools to deplete c-IAP1 and c-IAP2 in a large number of cell lines and cell types and thereby enable investigation of their function. While most Smac-mimetics induce rapid autoubiquitylation and degradation of c-IAPs, they are less good at promoting degradation of XIAP. Only AEG40730 appears to promote degradation that is not a consequence of apoptosis (Bertrand et al., 2008). On the other hand, the *Drosophila* IAP-antagonist Grim can promote efficient degradation of XIAP (Silke, Kratina, Ekert, Pakusch, & Vaux, 2004).

The RING domain from c-Cbl was shown to function as an E3 ligase in 1999 (Joazeiro et al., 1999) and soon after the Ashwell group established that RING domain-containing IAP proteins also possess E3 ligase activity that relies on the intact structure of their RING domains (Yang, Fang, Jensen, Weissman, & Ashwell, 2000). If provided with E1 and E2 enzymes as well as a source of ATP and zinc ions, the RING domains of IAP proteins can efficiently mediate autoubiquitylation or polyubiquitylation of their substrates—most commonly their binding partners (Vaux & Silke, 2005; Vucic et al., 2011). The RING domains of IAP proteins do not interact with all known E2 enzymes but show a clear preference for the UbcH5 and UbcH6 group of ubiquitin-conjugating enzymes (Dynek et al., 2010; Yang & Du, 2004). This was established *in vitro* and with yeast two-hybrid screens of the RING domains of c-IAP1, c-IAP2, XIAP, and ML-IAP in combination with 30 human E2 constructs and verified in functional cellular assays (Dynek et al., 2010; Yang & Du, 2004) (Fig. 2.4).

The E3 ligase activity of IAP proteins can be examined *in vitro* using bacterially produced and purified recombinant components (E1, E2, IAP full-length, or RING domain proteins) and quantifying autoubiquitylation (Blankenship et al., 2009; Feltham et al., 2011; Mace et al., 2008; Nakatani et al., 2013). In cells, a number of different approaches are possible. Overexpression of tagged-ubiquitin constructs has been popular and maybe combined with ubiquitin mutants that either have one of the essential lysines required for a particular ubiquitin-linked chain mutated to arginine, thereby preventing formation of a particular ubiquitin chain, or retain only one of the seven lysines, thereby only allowing formation of a particular ubiquitin-linked chain (Tang, Wang, Xiong, & Guan, 2003). One shortcoming of this approach is that endogenous ubiquitin is present at high levels and essential for cellular viability, so it is possible to get hybrid chains formed. To overcome this limitation, a tetracycline-inducible shRNA strategy to

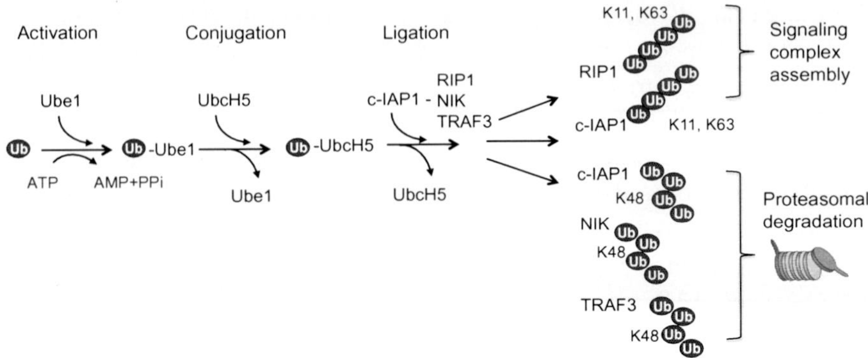

Figure 2.4 The enzymes and reactions of the IAP-mediated ubiquitylation. Activation reaction involves transfer of ubiquitin to an E1 enzyme (Ube1) in an ATP-dependent fashion and leads to transfer of activated ubiquitin to an E2 enzyme (UbcH5) in the conjugation reaction. The E2 with ubiquitin binds E3 ubiquitin ligase (c-IAP1), which can also bind a substrate—commonly via a different protein interaction domain—and thus allows the ubiquitin ligation to occur. When polyubiquitin chains are assembled, this process will be repeated with a lysine (K) residue of the ubiquitin molecule itself serving as a substrate. The assembly of K63- or K11-link polyubiquitin chains on RIP1 or c-IAP1 itself promotes the formation of signaling complexes, while K48-linked ubiquitylation of NIK, TRAF3, or c-IAP1 targets them for proteasomal degradation. (See the color plate.)

knockdown expression of endogenous ubiquitin from the four genes that encode it was combined with a K63R mutant ubiquitin that cannot form K63-linked chains. Somewhat remarkably cells tolerated up to 90% loss of endogenous ubiquitin in this system (Xu, Skaug, Zeng, & Chen, 2009). More recently, the development of ubiquitin chain-specific antibodies and mass spectrometry techniques has enabled the detection of endogenous ubiquitin chains on substrates and revealed the importance of IAPs in generating them (Blankenship et al., 2009; Dynek et al., 2010; Gerlach et al., 2011; Goncharov et al., 2013; Phu et al., 2011; Shin et al., 2003).

Ubiquitin chain-specific antibodies have also advanced studies of the IAP E3 ligase activity because they allow determination of modification of substrates by endogenous ubiquitin. The experiments with these antibodies revealed that during TNF signaling, c-IAP1 predominantly mediates RIPK1 polyubiquitylation with K11 and K63 linkages, while in TWEAK signaling, c-IAP1 mostly autoubiquitylates itself with K11, K48, and K63 ubiquitin chain linkages (Goncharov et al., 2013). Additionally, synthetic isotopically labeled internal standard peptides (AQUA peptides) have been used to quantify GG signature peptides in c-IAP1-mediated ubiquitylation, thus making the ubiquitin–AQUA approach a valuable tool to study IAP E3 ligase activity (Blankenship et al., 2009; Phu et al., 2011) (Fig. 2.4).

Another approach leading to an increased understanding of the function of the RING finger has been a structural approach. The structure of c-IAP2 RING domain alone or binding to the E2 UbcH5 confirmed earlier work, demonstrating the importance of the extreme C-terminal end of the RING finger for E3 activity (Mace et al., 2008; Silke et al., 2005). This study narrowed down amino acid residues within the C-terminal end of the RING, which are critical for RING dimerization and residues required for E2 interaction, thus paving the way for future examinations of IAP RING activity, including XIAP and ML-IAP (Feltham et al., 2011, 2010; Haas et al., 2009; Mace et al., 2008; Nakatani et al., 2013). It is now recognized that RINGs function as dimers in order to position the ubiquitin chain for transfer to a substrate (Dou, Buetow, Sibbet, Cameron, & Huang, 2012; Nakatani et al., 2013; Plechanovová, Jaffray, Tatham, Naismith, & Hay, 2012) and that the dimer interface and E2 interacting residues are highly conserved in many other RING fingers. Furthermore, when interrogating the requirement for RING finger E3 ligase activity in signaling pathways, it is preferable to mutate these residues, rather than the key structural Zn-coordinating residues, because mutation of these residues does not disrupt the fundamental structure of the RING, only function (Feltham et al., 2010).

IAP substrates have been mostly identified via a candidate (Bertrand et al., 2011, 2008; Damgaard et al., 2012; Jin et al., 2009; Li et al., 2002; Park et al., 2004; Tang et al., 2003) or immunoprecipitation mass spectrometry approach (Goncharov et al., 2013; Vaux & Silke, 2003). More recently, a screen designed to capture proteins modified by IAP E3 ligase activity has been developed. In this screen, a fusion between a NEDD8 E2-conjugating enzyme, Ubc12, and the ubiquitin ligases XIAP or c-IAP1 (Zhuang, Guan, Wang, Burlingame, & Wells, 2013) is generated and expressed in cells. This so-called NEDDylator enzyme covalently transfers the ubiquitin homolog NEDD8 to E3 substrates, enabling the purification for LC-MS/MS identification. This screen allowed the identification of multiple known and novel IAP substrates that contain the signature N-terminal IBMs. As an interesting aside, it appears that IAPs are also able to NEDDylate substrates without the necessity of generating a fusion with Ubc12 (Broemer et al., 2010).

The importance of IAP E3 ligase activity is further underlined by genetic evidence. First, while single IAP knockout mice are viable, combined knockout of c-IAP1 and c-IAP2 or c-IAP1 and XIAP is embryonic lethal (Moulin et al., 2012). Furthermore, mice with combined deficiency of c-IAP1 and c-IAP2 in the B-cell compartment or knock-in of an E3-inactive mutant c-IAP2, which appears to work as a dominant negative mutant also interfering with the activity of c-IAP1, accumulate abnormal B cells with increased noncanonical NF-κB signaling (Conze, Zhao, & Ashwell, 2010; Gardam et al., 2011). Furthermore, multiple myeloma patients with inactivating biallelic mutations in c-IAP1/2 have upregulated NIK levels and constitutive noncanonical NF-κB signaling, while the patients suffering from the X-linked lymphoproliferative syndrome type-2 (XLP2) harbor mutations in XIAP that abrogate XIAP expression or disable its E3 ligase activity (Annunziata et al., 2007; Damgaard et al., 2013; Keats et al., 2007; Rigaud et al., 2006).

2.5. Regulation of signaling pathways by IAP proteins

TNFR superfamily members, TLR and NOD receptor families, are essential for proper functioning of the immune system (Bodmer, Schneider, & Tschopp, 2002). These receptors mediate signaling by assembling protein complexes that activate NF-κB and mitogen-activated protein kinase (MAPK) pathways upon ubiquitin chain scaffolds (Locksley, Killeen, & Lenardo, 2001; Vucic et al., 2011). It is not surprising, therefore, that IAP proteins are their integral components. In TNF family, stimulated signaling E3 ligases c-IAP1 and c-IAP2 are recruited through the adaptor protein TRAF2 to receptor-associated complexes where they promote K63- and K11-linked polyubiquitylation of RIP1, NEMO, TRAF2, and themselves (Silke & Brink, 2010) (Fig. 2.4). These ubiquitylation events provide a platform for the recruitment of TAB2/3-TAK1 and IKK (IKKα/β/γ also known as IKK1/IKK2 and NEMO) kinase complexes as well as linear ubiquitylation assembly complex (LUBAC), thus allowing the formation of fully functional signaling complex and the activation of IKKβ kinase activity (Dejardin, 2006; Schmukle & Walczak, 2012; Silke, 2011; Vucic et al., 2011). IKKβ phosphorylates inhibitor of kappa B (IκBα) prompting the ubiquitylation and proteasomal degradation of IκBα to liberate p50/RelA dimer and activation of canonical NF-κB pathway (Scheidereit, 2006). While cellular IAP proteins are required for the activation of TNF family-induced canonical NF-κB signaling, they are negative regulators of the noncanonical NF-κB pathway (Bertrand et al., 2008; Blackwell et al., 2013; Mahoney et al., 2008; Varfolomeev et al., 2007, 2008; Vince et al., 2008, 2007). In unstimulated cells, c-IAP1/2 form a complex with adaptor proteins TRAF2 and TRAF3 and mediate degradative ubiquitylation of NIK (Vallabhapurapu et al., 2008; Varfolomeev et al., 2007; Vince et al., 2007). Following stimulation of any one of several TRAF3-binding TNF superfamily receptors, this cytoplasmic complex is disrupted by the membrane recruitment and degradation of c-IAP1/2 and TRAF2/3, thereby permitting accumulation of NIK (Matsuzawa et al., 2008; Varfolomeev et al., 2007, 2012; Vince et al., 2008, 2007). NIK phosphorylates IKKα/IKK1, leading to NF-κB2/p100 phosphorylation, ubiquitylation, and partial proteasomal degradation, which results in the translocation of p52/Relb dimers to the nucleus and activation of the noncanonical NF-κB signaling (Dejardin, 2006).

The role of c-IAP1/2 in the activation of NF-κB and MAPK signaling is evident from the depletion of c-IAPs, either by si/shRNA-mediated knockdown, genetic knockout, or by using Smac-mimetic IAP antagonists that cause their proteasomal degradation. In the absence of c-IAPs, RIPK1, TRAF2, and (self-evidently) c-IAPs cannot undergo ubiquitylation, which prevents the formation of distal kinase and LUBACs (Blackwell et al., 2013; Dynek et al., 2010; Gerlach et al., 2011; Haas et al., 2009; Mahoney et al., 2008; Varfolomeev et al., 2008; Vince et al., 2009). Examining TNF receptor-associated signaling complexes following stimulation with their respective ligands from the cells expressing or lacking c-IAP proteins can assess this role of c-IAPs. In addition, to complement receptor recruitment assay, investigators can also perform secondary immunoprecipitation for ubiquitinated proteins within the receptor complex using lysine chain-specific antibodies or tandem ubiquitin-binding entities (TUBES; Damgaard et al., 2012; Dynek et al., 2010; Goncharov et al., 2013). Such assays can reveal if the absence of c-IAPs affected the recruitment and/or ubiquitylation of various signaling proteins within the receptor-associated signaling complexes. In cellular studies, lack of c-IAPs or reconstitution of c-IAP-deficient cells with mutant versions that cannot be recruited to signaling complex or lack E3 ligase activity greatly abrogated NF-κB and MAPK signaling. This was demonstrated by Western blotting through the lack of phosphorylation of MAP kinases JNK and p38 and the absence of phosphorylation and degradation of IκBα in canonical NF-κB signaling (Moulin et al., 2012; Varfolomeev et al., 2012; Vince et al., 2009). As these signaling pathways lead to the expression of a number of proinflammatory cytokines and antiapoptotic proteins, the role of c-IAPs can be functionally evaluated by cytokine ELISA or by the real-time quantitative RT-PCR (Blackwell et al., 2013; Moulin et al., 2012; Varfolomeev et al., 2012).

In addition to their well-established roles in regulating TNF, TLR, and NOD receptor-mediated signaling, IAP proteins may also play a role in DNA damage, Wnt signaling, cell motility and migration, and autophagy pathways. Thus, during DNA damage-induced NF-κB activation, c-IAP1 can form a signaling complex with ATM-TRAF6 to promote NEMO ubiquitylation (Hinz et al., 2010), while XIAP-mediated ubiquitylation of ELKS can stimulate TAK1 activation (Wu et al., 2010). Upon activation of the Wnt pathway, XIAP binds TCF/Lef and promotes monoubiquitylation of Groucho (Gro/TLE; Hanson et al., 2012). IAP proteins may also regulate cell migration through their E3 ligase activity. XIAP and c-IAP1 bind Rac1 in a nucleotide-independent manner to promote its polyubiquitylation and subsequent proteasomal degradation leading to enhanced cell migration (Oberoi et al., 2012). On the other hand, it has been reported that loss of c-IAP1 suppresses cell migration (Lopez et al., 2011). XIAP also associates with the Rho GDP-dissociation inhibitor (RhoGDI) through its RING domain to control cell motility (Liu et al., 2012). Lastly, there is the recent observation that XIAP inhibits autophagy via an Mdm2–p53 axis in tumors (Huang, Wu, Mei, & Wu, 2013). One point to bear in mind with the XIAP studies is that the XIAP knockout mice are, as far as we know, phenotypically normal, without increased risk of tumorigenesis or developmental defects due to problems in cell migration or Wnt signaling; therefore, the role of XIAP in these pathways maybe limited in their operation or redundant.

2.6. Targeting IAP proteins

IAP proteins are implicated in a number of human pathologies, especially in cancer where elevated levels of IAP mRNA and protein levels have been documented in various tumor types (Fulda & Vucic, 2012; Hunter, LaCasse, & Korneluk, 2007; LaCasse, Baird, Korneluk, & MacKenzie, 1998). Survivin and ML-IAP, for example, are scarcely expressed in normal tissues but can be detected in a number of tumor types (Fulda & Vucic, 2012; Ryan, O'Donovan, & Duffy, 2009). In several tumor types, the expression of XIAP, c-IAP1, and c-IAP2 is associated with poor prognosis. Furthermore, IAP proteins have also been implicated in tumor cell mobility, invasion, and metastasis, and antagonism of IAP proteins can block tumor cell migration and invasion, see earlier (Lopez et al., 2011; Mehrotra et al., 2010; Oberoi et al., 2012; Tchoghandjian, Jennewein, Eckhardt,

Rajalingam, & Fulda, 2013). In addition to promoting tumor progression and contributing to resistance to anticancer therapies (Fulda & Vucic, 2012), IAP proteins are also important regulators of survival signaling pathways (Silke & Brink, 2010). For all these reasons, IAP proteins are attractive targets for therapeutic intervention.

Among several strategies employed in targeting IAP proteins, the most advanced and attractive involve antisense oligonucleotides and Smac-mimicking small-molecule IAP antagonists. The main focus of the antisense approach is XIAP and survivin that have been targeted with oligonucleotides AEG35156 and LY2181308, respectively. AEG35156 efficiently decreased XIAP levels at low nanomolar concentrations and showed antitumor activity *in vivo* in xenograft models in combination with chemotherapeutic agents, death receptor agonists, or radiation therapy, which prompted the advancement of AEG35156 to clinical studies (LaCasse et al., 2006; Shaw, Lacasse, Durkin, & Vanderhyden, 2008). Although Phase I/II nonrandomized study of AEG35156 confirmed its tolerability in patients, dose-dependent knockdown of XIAP and some induction of apoptosis (Carter et al., 2011) in randomized Phase II trial in patients with refractory AML AEG35156 did not provide antitumor benefit in the reinduction chemotherapy protocol (Schimmer et al., 2011). As pharmacodynamic studies were not performed in this Phase II trial, it is not clear if efficient knockdown of XIAP was achieved and whether the lack of efficacy could be explained by insufficient XIAP downregulation.

Smac-mimicking small-molecule IAP antagonists were initially derived from amino-terminal peptides of mature active Smac. Smac peptides were extensively used to validate targeting of IAP proteins as they demonstrated the ability of Smac-derived peptides to promote cell death in cancer cells and in *in vivo* models in combination with chemotherapeutics or the death receptor ligand TRAIL/Apo2L (Arnt, Chiorean, Heldebrant, Gores, & Kaufmann, 2002; Fulda et al., 2002; Vucic & Fairbrother, 2007). At the same time, they paved the way for the generation of high-affinity binders for select BIR domains of IAP proteins (Ndubaku et al., 2009). As a result of these efforts, a few different classes of IAP antagonists have emerged: monovalent and bivalent antagonists as well as compounds with selectivity for particular IAP or a group of IAP proteins and/or their BIR domains (Ndubaku et al., 2009) (Fig. 2.5). Monovalent IAP antagonists contain a single Smac AVPI-like motif, while bivalent antagonists consist of two such motifs connected by a chemical linker (Fig. 2.5). Importantly, bivalent antagonists can

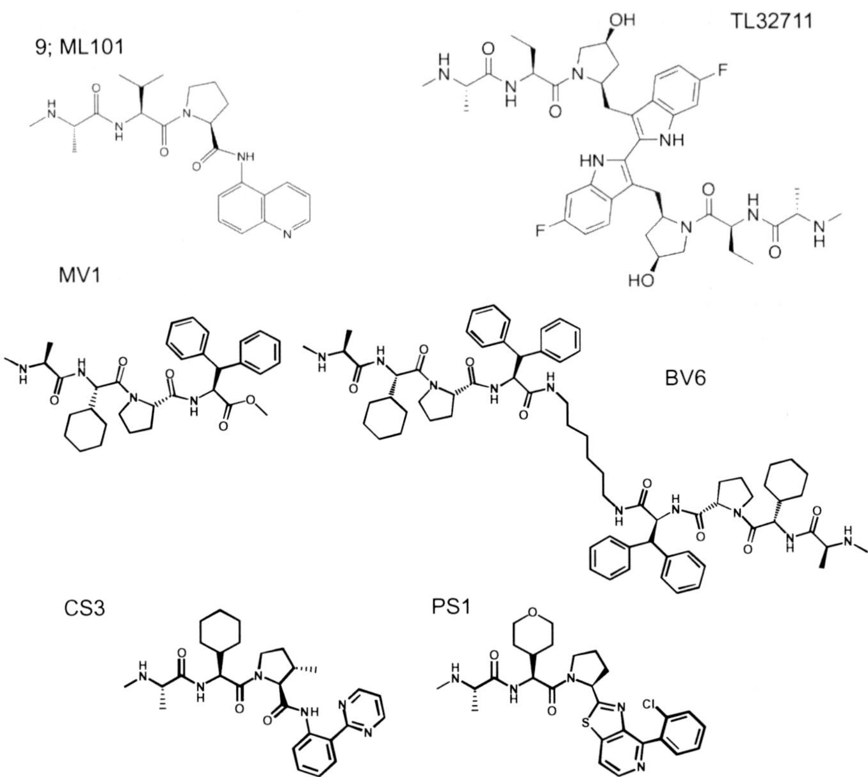

Figure 2.5 Structure of IAP antagonists. Examples of monovalent (MV1 (Varfolomeev et al., 2007) and PS1 (Ndubaku et al., 2009)), bivalent (BV6 (Varfolomeev et al., 2007) and TL32711 (Fulda & Vucic, 2012)), and c-IAP selective (CS3 (Ndubaku et al., 2009) and XIAP selective (9′ ML101 (González-López et al., 2011))) antagonists.

simultaneously bind the BIR2 and BIR3 domains of XIAP leading to enhanced activation of caspases and enhanced cell killing (Gao et al., 2007; Li et al., 2004; Varfolomeev et al., 2009, 2007).

Understanding of the structural properties of particular BIR domains and the functional roles of IAP proteins has enabled development of selective antagonists. CS3 (c-IAP selective 3), an antagonist that is over 2000-fold more selective for c-IAP1 over XIAP, activates cell death in sensitive cell lines and promotes c-IAP1 and c-IAP2 degradation and stimulation of canonical and noncanonical NF-κB signaling (Ndubaku et al., 2009). However, cell death induction by CS3 was much weaker in comparison to chemically related pan-IAP antagonist PS1, indicating that antagonism of both XIAP and c-IAP proteins is needed for the efficient activation of cell death

(Ndubaku et al., 2009) (Fig. 2.5). Development of XIAP selective compounds has been focused on preventing XIAP from binding to and inhibiting caspase-3. Several molecules have emerged from those efforts with some of them capable of sensitize otherwise resistant cancer cell to TRAIL treatment (Gao et al., 2007; González-López et al., 2011; Wu, Wagner, Bursulaya, Schultz, & Deveraux, 2003).

The exploration of the mechanism of IAP antagonism has lead researches to several unexpected discoveries. Treatment of cells with IAP antagonists leads to rapid autoubiquitylation proteasomal degradation of c-IAP1 and c-IAP2 proteins (Gaither et al., 2007; Varfolomeev et al., 2007; Vince et al., 2007). Autoubiquitylation is activated by a simple but elegant mechanism whereby the RING finger, which is normally held in check by binding to the IBM binding groove of BIR3, is released by IAP-antagonist binding, leading to opening of the c-IAP1 structure, RING domain dimerization and activation of its E3 ligase activity (Dueber et al., 2011; Feltham et al., 2011). Rapid degradation of c-IAP proteins results in stabilization of NIK because NIK levels are normally kept to undetectably low levels by the action of an enzyme complex consisting of TRAF2, TRAF3, c-IAPs, and NIK (Vallabhapurapu et al., 2008; Varfolomeev et al., 2007; Vince et al., 2007; Zarnegar et al., 2008). Indeed, this is how natural signaling complexes such as TWEAK/Fn14 and LIGHT/LT-βR activate the noncanonical signaling pathway (Gardam et al., 2011; Varfolomeev et al., 2007, 2012; Vince et al., 2008). However, as discussed, loss of c-IAPs also prevents canonical signaling via TNF (Mahoney et al., 2008; Varfolomeev et al., 2008), thus providing an unanticipated link between regulation of canonical and noncanonical NF-κB. Finally, IAP antagonists sensitize cells to death receptor-induced necroptosis and have proven an invaluable tool to investigate this novel signaling pathway.

Nevertheless, our understanding of IAP proteins and cellular processes they regulate is far from complete and future studies, both in basic research and in clinical settings, should focus on fully unraveling the biological roles of IAP proteins, and thereby open up additional therapeutic strategies.

REFERENCES

Ambrosini, G., Adida, C., & Altieri, D. C. (1997). A novel anti-apoptosis gene, survivin, expressed in cancer and lymphoma. *Nature Medicine, 3*, 917–921.

Annunziata, C. M., Davis, R. E., Demchenko, Y., Bellamy, W., Gabrea, A., Zhan, F., et al. (2007). Frequent engagement of the classical and alternative NF-kappaB pathways by diverse genetic abnormalities in multiple myeloma. *Cancer Cell, 12*, 115–130.

Arnt, C. R., Chiorean, M. V., Heldebrant, M. P., Gores, G. J., & Kaufmann, S. H. (2002). Synthetic Smac/DIABLO peptides enhance the effects of chemotherapeutic agents by binding XIAP and cIAP1 in situ. *The Journal of Biological Chemistry*, *277*, 44236–44243.

Bertrand, M. J., Lippens, S., Staes, A., Gilbert, B., Roelandt, R., De Medts, J., et al. (2011). cIAP1/2 are direct E3 ligases conjugating diverse types of ubiquitin chains to receptor interacting proteins kinases 1 to 4 (RIP1-4). *PLoS One*, *6*, e22356.

Bertrand, M. J., Milutinovic, S., Dickson, K. M., Ho, W. C., Boudreault, A., Durkin, J., et al. (2008). cIAP1 and cIAP2 facilitate cancer cell survival by functioning as E3 ligases that promote RIP1 ubiquitination. *Molecular Cell*, *30*, 689–700.

Birnbaum, M. J., Clem, R. J., & Miller, L. K. (1994). An apoptosis-inhibiting gene from a nuclear polyhedrosis virus encoding a polypeptide with Cys/His sequence motifs. *Journal of Virology*, *68*, 2521–2528.

Blackwell, K., Zhang, L., Workman, L. M., Ting, A. T., Iwai, K., & Habelhah, H. (2013). Two coordinated mechanisms underlie tumor necrosis factor alpha-induced immediate and delayed IκB kinase activation. *Molecular and Cellular Biology*, *33*, 1901–1915.

Blankenship, J. W., Varfolomeev, E., Goncharov, T., Fedorova, A. V., Kirkpatrick, D. S., Izrael-Tomasevic, A., et al. (2009). Ubiquitin binding modulates IAP antagonist-stimulated proteasomal degradation of c-IAP1 and c-IAP2. *The Biochemical Journal*, *417*, 149–160.

Bodmer, J. L., Schneider, P., & Tschopp, J. (2002). The molecular architecture of the TNF superfamily. *Trends in Biochemical Sciences*, *27*, 19–26.

Bratton, S. B., Lewis, J., Butterworth, M., Duckett, C. S., & Cohen, G. M. (2002). XIAP inhibition of caspase-3 preserves its association with the Apaf-1 apoptosome and prevents CD95- and Bax-induced apoptosis. *Cell Death and Differentiation*, *9*, 881–892.

Bratton, S. B., Walker, G., Srinivasula, S. M., Sun, X., Butterworth, M., Alnemri, E. S., et al. (2001). Recruitment, activation and retention of caspases-9 and -3 by Apaf-1 apoptosome and associated XIAP complexes. *The EMBO Journal*, *20*, 998–1009.

Broemer, M., Tenev, T., Rigbolt, K. T., Hempel, S., Blagoev, B., Silke, J., et al. (2010). Systematic in vivo RNAi analysis identifies IAPs as NEDD8-E3 ligases. *Molecular Cell*, *40*, 810–822.

Burke, S. P., Smith, L., & Smith, J. B. (2010). cIAP1 cooperatively inhibits procaspase-3 activation by the caspase-9 apoptosome. *The Journal of Biological Chemistry*, *285*, 30061–30068.

Burri, L., Strahm, Y., Hawkins, C. J., Gentle, I. E., Puryer, M. A., Verhagen, A., et al. (2005). Mature DIABLO/Smac is produced by the IMP protease complex on the mitochondrial inner membrane. *Molecular Biology of the Cell*, *16*, 2926–2933.

Cai, Z., Jitkaew, S., Zhao, J., Chiang, H. C., Choksi, S., Liu, J., et al. (2014). Plasma membrane translocation of trimerized MLKL protein is required for TNF-induced necroptosis. *Nature Cell Biology*, *16*, 55–65.

Carter, B. Z., Mak, D. H., Morris, S. J., Borthakur, G., Estey, E., Byrd, A. L., et al. (2011). XIAP antisense oligonucleotide (AEG35156) achieves target knockdown and induces apoptosis preferentially in CD34+38- cells in a phase 1/2 study of patients with relapsed/refractory AML. *Apoptosis*, *16*, 67–74.

Chen, Z., Naito, M., Hori, S., Mashima, T., Yamori, T., & Tsuruo, T. (1999). A human IAP-family gene, apollon, expressed in human brain cancer cells. *Biochemical and Biophysical Research Communications*, *264*, 847–854.

Conze, D. B., Albert, L., Ferrick, D. A., Goeddel, D. V., Yeh, W. C., Mak, T., et al. (2005). Posttranscriptional downregulation of c-IAP2 by the ubiquitin protein ligase c-IAP1 in vivo. *Molecular and Cellular Biology*, *25*, 3348–3356.

Conze, D. B., Zhao, Y., & Ashwell, J. D. (2010). Non-canonical NF-κB activation and abnormal B cell accumulation in mice expressing ubiquitin protein ligase-inactive c-IAP2. *PLoS Biology*, *8*, e1000518.

Crook, N. E., Clem, R. J., & Miller, L. K. (1993). An apoptosis inhibiting baculovirus gene with a zinc finger like motif. *Journal of Virology*, *67*, 2168–2174.

Czabotar, P. E., Lessene, G., Strasser, A., & Adams, J. M. (2013). Control of apoptosis by the BCL-2 protein family: Implications for physiology and therapy. *Nature Reviews Molecular Cell Biology*, *15*, 49–63.

Damgaard, R. B., Fiil, B. K., Speckmann, C., Yabal, M., zur Stadt, U., Bekker-Jensen, S., et al. (2013). Disease-causing mutations in the XIAP BIR2 domain impair NOD2-dependent immune signalling. *EMBO Molecular Medicine*, *5*, 1278–1295.

Damgaard, R. B., Nachbur, U., Yabal, M., Wong, W. W., Fiil, B. K., Kastirr, M., et al. (2012). The ubiquitin ligase XIAP recruits LUBAC for NOD2 signaling in inflammation and innate immunity. *Molecular Cell*, *46*, 746–758.

Darding, M., Feltham, R., Tenev, T., Bianchi, K., Benetatos, C., Silke, J., et al. (2011). Molecular determinants of Smac mimetic induced degradation of cIAP1 and cIAP2. *Cell Death and Differentiation*, *18*, 1376–1386.

Dejardin, E. (2006). The alternative NF-kappaB pathway from biochemistry to biology: Pitfalls and promises for future drug development. *Biochemical Pharmacology*, *72*, 1161–1179.

Deshaies, R. J., & Joazeiro, C. A. (2009). RING domain E3 ubiquitin ligases. *Annual Review of Biochemistry*, *78*, 399–434.

Deveraux, Q. L., Roy, N., Stennicke, H. R., Van Arsdale, T., Zhou, Q., Srinivasula, S. M., et al. (1998). IAPs block apoptotic events induced by caspase-8 and cytochrome c by direct inhibition of distinct caspases. *EMBO Journal*, *17*, 2215–2223.

Deveraux, Q. L., Takahashi, R., Salvesen, G. S., & Reed, J. C. (1997). X-linked IAP is a direct inhibitor of cell-death proteases. *Nature*, *388*, 300–304.

Dou, H., Buetow, L., Sibbet, G. J., Cameron, K., & Huang, D. T. (2012). BIRC7-E2 ubiquitin conjugate structure reveals the mechanism of ubiquitin transfer by a RING dimer. *Nature Structural and Molecular Biology*, *19*, 876–883.

Du, C., Fang, M., Li, Y., Li, L., & Wang, X. (2000). Smac, a mitochondrial protein that promotes cytochrome c-dependent caspase activation by eliminating IAP inhibition. *Cell*, *102*, 33–42.

Duckett, C. S., Nava, V. E., Gedrich, R. W., Clem, R. J., Vandongen, J. L., Gilfillan, M. C., et al. (1996). A conserved family of cellular genes related to the baculovirus iap gene and encoding apoptosis inhibitors. *The EMBO Journal*, *15*, 2685–2694.

Dueber, E. C., Schoeffler, A. J., Lingel, A., Elliott, J. M., Fedorova, A. V., Giannetti, A. M., et al. (2011). Antagonists induce a conformational change in cIAP1 that promotes autoubiquitination. *Science*, *334*, 376–380.

Dynek, J. N., Goncharov, T., Dueber, E. C., Fedorova, A. V., Izrael-Tomasevic, A., Phu, L., et al. (2010). c-IAP1 and UbcH5 promote K11-linked polyubiquitination of RIP1 in TNF signalling. *The EMBO Journal*, *29*, 4198–4209.

Eckelman, B. P., & Salvesen, G. S. (2006). The human anti-apoptotic proteins cIAP1 and cIAP2 bind but do not inhibit caspases. *The Journal of Biological Chemistry*, *281*, 3254–3260.

Ekert, P. G., Silke, J., & Vaux, D. L. (1999). Inhibition of apoptosis and clonogenic survival of cells expressing crmA variants: Optimal caspase substrates are not necessarily optimal inhibitors. *The EMBO Journal*, *18*, 330–338.

Feltham, R., Bettjeman, B., Budhidarmo, R., Mace, P. D., Shirley, S., Condon, S. M., et al. (2011). Smac mimetics activate the E3 ligase activity of cIAP1 protein by promoting RING domain dimerization. *The Journal of Biological Chemistry*, *286*, 17015–17028.

Feltham, R., Moulin, M., Vince, J. E., Mace, P. D., Wong, W. W., Anderton, H., et al. (2010). Tumor necrosis factor (TNF) signaling, but not TWEAK (TNF-like weak inducer of apoptosis)-triggered cIAP1 (cellular inhibitor of apoptosis protein 1) degradation, requires cIAP1 RING dimerization and E2 binding. *The Journal of Biological Chemistry*, *285*, 17525–17536.

Feoktistova, M., Geserick, P., Kellert, B., Dimitrova, D. P., Langlais, C., Hupe, M., et al. (2011). cIAPs block ripoptosome formation, a RIP1/caspase-8 containing intracellular cell death complex differentially regulated by cFLIP isoforms. *Molecular Cell*, *43*, 449–463.

Franklin, M. C., Kadkhodayan, S., Ackerly, H., Alexandru, D., Distefano, M. D., Elliott, L. O., et al. (2003). Structure and function analysis of peptide antagonists of melanoma inhibitor of apoptosis (ML-IAP). *Biochemistry, 42*, 8223–8231.

Fraser, A. G., James, C., Evan, G. I., & Hengartner, M. O. (1999). Caenorhabditis elegans inhibitor of apoptosis protein (IAP) homologue BIR-1 plays a conserved role in cytokinesis. *Current Biology, 9*, 292–301.

Fulda, S., & Vucic, D. (2012). Targeting IAP proteins for therapeutic intervention in cancer. *Nature Reviews. Drug Discovery, 11*, 109–124.

Fulda, S., Wick, W., Weller, M., & Debatin, K. M. (2002). Smac agonists sensitize for Apo2L/TRAIL- or anticancer drug-induced apoptosis and induce regression of malignant glioma in vivo. *Nature Medicine, 8*, 808–815.

Gaither, A., Porter, D., Yao, Y., Borawski, J., Yang, G., Donovan, J., et al. (2007). A Smac mimetic rescue screen reveals roles for inhibitor of apoptosis proteins in tumor necrosis factor-alpha signaling. *Cancer Research, 67*, 11493–11498.

Gao, Z., Tian, Y., Wang, J., Yin, Q., Wu, H., Li, Y. M., et al. (2007). A dimeric Smac/diablo peptide directly relieves caspase-3 inhibition by XIAP. Dynamic and cooperative regulation of XIAP by Smac/Diablo. *The Journal of Biological Chemistry, 282*, 30718–30727.

Garcia-Calvo, M., Peterson, E. P., Leiting, B., Ruel, R., Nicholson, D. W., & Thornberry, N. A. (1998). Inhibition of human caspases by peptide-based and macromolecular inhibitors. *Journal of Biological Chemistry, 273*, 32608–32613.

Gardam, S., Turner, V. M., Anderton, H., Limaye, S., Basten, A., Koentgen, F., et al. (2011). Deletion of cIAP1 and cIAP2 in murine B lymphocytes constitutively activates cell survival pathways and inactivates the germinal center response. *Blood, 117*, 4041–4051.

Gatti, L., De Cesare, M., Ciusani, E., Corna, E., Arrighetti, N., Cominetti, D., et al. (2014). Antitumor activity of a novel homodimeric SMAC mimetic in ovarian carcinoma. *Molecular Pharmaceutics, 11*, 283–293.

Gerlach, B., Cordier, S. M., Schmukle, A. C., Emmerich, C. H., Rieser, E., Haas, T. L., et al. (2011). Linear ubiquitination prevents inflammation and regulates immune signalling. *Nature, 471*, 591–596.

Geserick, P., Hupe, M., Moulin, M., Wong, W. W., Feoktistova, M., Kellert, B., et al. (2009). Cellular IAPs inhibit a cryptic CD95-induced cell death by limiting RIP1 kinase recruitment. *The Journal of Cell Biology, 187*, 1037–1054.

Goncharov, T., Niessen, K., de Almagro, M. C., Izrael-Tomasevic, A., Fedorova, A. V., Varfolomeev, E., et al. (2013). OTUB1 modulates c-IAP1 stability to regulate signalling pathways. *The EMBO Journal, 32*, 1103–1114.

González-López, M., Welsh, K., Finlay, D., Ardecky, R. J., Ganji, S. R., Su, Y., et al. (2011). Design, synthesis and evaluation of monovalent Smac mimetics that bind to the BIR2 domain of the anti-apoptotic protein XIAP. *Bioorganic and Medicinal Chemistry Letters, 21*, 4332–4336.

Gyrd-Hansen, M., Darding, M., Miasari, M., Santoro, M. M., Zender, L., Xue, W., et al. (2008). IAPs contain an evolutionarily conserved ubiquitin-binding domain that regulates NF-kappaB as well as cell survival and oncogenesis. *Nature Cell Biology, 10*, 1309–1317.

Haas, T. L., Emmerich, C. H., Gerlach, B., Schmukle, A. C., Cordier, S. M., Rieser, E., et al. (2009). Recruitment of the linear ubiquitin chain assembly complex stabilizes the TNF-R1 signaling complex and is required for TNF-mediated gene induction. *Molecular Cell, 36*, 831–844.

Hanson, A. J., Wallace, H. A., Freeman, T. J., Beauchamp, R. D., Lee, L. A., & Lee, E. (2012). XIAP monoubiquitylates Groucho/TLE to promote canonical Wnt signaling. *Molecular Cell, 45*, 619–628.

Hao, Y., Sekine, K., Kawabata, A., Nakamura, H., Ishioka, T., Ohata, H., et al. (2004). Apollon ubiquitinates SMAC and caspase-9, and has an essential cytoprotection function. *Nature Cell Biology, 6*, 849–860.

Hauser, H. P., Bardroff, M., Pyrowolakis, G., & Jentsch, S. (1998). A giant ubiquitin-conjugating enzyme related to IAP apoptosis inhibitors. *The Journal of Cell Biology, 141*, 1415–1422.

Hawkins, C. J., Uren, A. G., Häcker, G., Medcalf, R. L., & Vaux, D. L. (1996). Inhibition of interleukin 1 beta-converting enzyme-mediated apoptosis of mammalian cells by baculovirus IAP. *Proceedings of the National Academy of Sciences of the United States of America, 93*, 13786–13790.

Hay, B. A., Wassarman, D. A., & Rubin, G. M. (1995). Drosophila homologs of baculovirus inhibitor of apoptosis proteins function to block cell death. *Cell, 83*, 1253–1262.

Hegde, R., Srinivasula, S. M., Zhang, Z., Wassell, R., Mukattash, R., Cilenti, L., et al. (2001). Identification of Omi/HtrA2 as a mitochondrial apoptotic serine protease that disrupts inhibitor of apoptosis protein-caspase interaction. *The Journal of Biological Chemistry, 277*, 432–438.

Hershko, A., & Ciechanover, A. (1998). The ubiquitin system. *Annual Review of Biochemistry, 67*, 425–479.

Hinds, M. G., Norton, R. S., Vaux, D. L., & Day, C. L. (1999). Solution structure of a baculoviral inhibitor of apoptosis (IAP) repeat. *Nature Structural Biology, 6*, 648–651.

Hinz, M., Stilmann, M., Arslan, S. Ç., Khanna, K. K., Dittmar, G., & Scheidereit, C. (2010). A cytoplasmic ATM-TRAF6-cIAP1 module links nuclear DNA damage signaling to ubiquitin-mediated NF-κB activation. *Molecular Cell, 40*, 63–74.

Hofmann, K., Bucher, P., & Tschopp, J. (1997). The CARD domain: A new apoptotic signalling motif. *Trends in Biochemical Sciences, 22*, 155–156.

Holley, C. L., Olson, M. R., Colon-Ramos, D. A., & Kornbluth, S. (2002). Reaper eliminates IAP proteins through stimulated IAP degradation and generalized translational inhibition. *Nature Cell Biology, 4*, 439–444.

Huang, Y., Rich, R. L., Myszka, D. G., & Wu, H. (2003). Requirement of both the second and third BIR domains for the relief of X-linked inhibitor of apoptosis protein (XIAP)-mediated caspase inhibition by Smac. *The Journal of Biological Chemistry, 278*, 49517–49522.

Huang, X., Wu, Z., Mei, Y., & Wu, M. (2013). XIAP inhibits autophagy via XIAP-Mdm2-p53 signalling. *The EMBO Journal, 32*, 2204–2216.

Hunter, A. M., Kottachchi, D., Lewis, J., Duckett, C. S., Korneluk, R. G., & Liston, P. (2003). A novel ubiquitin fusion system bypasses the mitochondria and generates biologically active Smac/DIABLO. *The Journal of Biological Chemistry, 278*, 7494–7499.

Hunter, A. M., LaCasse, E. C., & Korneluk, R. G. (2007). The inhibitors of apoptosis (IAPs) as cancer targets. *Apoptosis, 12*, 1543–1568.

Inohara, N., & Nunez, G. (2003). NODs: Intracellular proteins involved in inflammation and apoptosis. *Nature Reviews. Immunology, 3*, 371–382.

Jeyaprakash, A. A., Klein, U. R., Lindner, D., Ebert, J., Nigg, E. A., & Conti, E. (2007). Structure of a survivin-borealin-INCENP core complex reveals how chromosomal passengers travel together. *Cell, 131*, 271–285.

Jin, H. S., Lee, D. H., Kim, D. H., Chung, J. H., Lee, S. J., & Lee, T. H. (2009). cIAP1, cIAP2, and XIAP act cooperatively via nonredundant pathways to regulate genotoxic stress-induced nuclear factor-kappaB activation. *Cancer Research, 69*, 1782–1791.

Joazeiro, C. A., Wing, S. S., Huang, H., Leverson, J. D., Hunter, T., & Liu, Y. C. (1999). The tyrosine kinase negative regulator c-Cbl as a RING-type, E2-dependent ubiquitin-protein ligase. *Science, 286*, 309–312.

Jost, P. J., Grabow, S., Gray, D., McKenzie, M. D., Nachbur, U., Huang, D. C., et al. (2009). XIAP discriminates between type I and type II FAS-induced apoptosis. *Nature, 460*, 1035–1039.

Jung, Y., Miura, M., & Yuan, J. (1996). Suppression of interleukin-1 beta-converting enzyme-mediated cell death by insulin-like growth factor. *The Journal of Biological Chemistry, 271*, 5112–5117.

Kasof, G. M., & Gomes, B. C. (2001). Livin, a novel inhibitor of apoptosis protein family member. *The Journal of Biological Chemistry*, *276*, 3238–3246.

Kaufmann, T., Jost, P. J., Pellegrini, M., Puthalakath, H., Gugasyan, R., Gerondakis, S., et al. (2009). Fatal hepatitis mediated by tumor necrosis factor TNFalpha requires caspase-8 and involves the BH3-only proteins Bid and Bim. *Immunity*, *30*, 56–66.

Keats, J. J., Fonseca, R., Chesi, M., Schop, R., Baker, A., Chng, W. J., et al. (2007). Promiscuous mutations activate the noncanonical NF-kappaB pathway in multiple myeloma. *Cancer Cell*, *12*, 131–144.

Koonin, E. V., & Aravind, L. (2000). The NACHT family—A new group of predicted NTPases implicated in apoptosis and MHC transcription activation. *Trends in Biochemical Sciences*, *25*, 223–224.

LaCasse, E. C., Baird, S., Korneluk, R. G., & MacKenzie, A. E. (1998). The inhibitors of apoptosis (IAPs) and their emerging role in cancer. *Oncogene*, *17*, 3247–3259.

LaCasse, E. C., Cherton-Horvat, G. G., Hewitt, K. E., Jerome, L. J., Morris, S. J., Kandimalla, E. R., et al. (2006). Preclinical characterization of AEG35156/GEM 640, a second-generation antisense oligonucleotide targeting X-linked inhibitor of apoptosis. *Clinical Cancer Research*, *12*, 5231–5241.

Lagace, M., Xuan, J. Y., Young, S. S., McRoberts, C., Maier, J., Rajcan-Separovic, E., et al. (2001). Genomic organization of the X-linked inhibitor of apoptosis and identification of a novel testis-specific transcript. *Genomics*, *77*, 181–188.

Li, L., Thomas, R. M., Suzuki, H., De Brabander, J. K., Wang, X., & Harran, P. G. (2004). A small molecule Smac mimic potentiates TRAIL- and TNFα-mediated cell death. *Science*, *305*, 1471–1474.

Li, X., Yang, Y., & Ashwell, J. D. (2002). TNF-RII and c-IAP1 mediate ubiquitination and degradation of TRAF2. *Nature*, *416*, 345–347.

Lin, J. H., Deng, G., Huang, Q., & Morser, J. (2000). KIAP, a novel member of the inhibitor of apoptosis protein family. *Biochemical and Biophysical Research Communications*, *279*, 820–831.

Liston, P., Roy, N., Tamai, K., Lefebvre, C., Baird, S., Chertonhorvat, G., et al. (1996). Suppression of apoptosis in mammalian cells by NAIP and a related family of IAP genes. *Nature*, *379*, 349–353.

Liu, J., Zhang, D., Luo, W., Yu, J., Li, J., Yu, Y., et al. (2012). E3 ligase activity of XIAP RING domain is required for XIAP-mediated cancer cell migration, but not for its RhoGDI binding activity. *PLoS One*, *7*, e35682.

Locksley, R. M., Killeen, N., & Lenardo, M. J. (2001). The TNF and TNF receptor superfamilies: Integrating mammalian biology. *Cell*, *104*, 487–501.

Lopez, J., John, S. W., Tenev, T., Rautureau, G. J., Hinds, M. G., Francalanci, F., et al. (2011). CARD-mediated autoinhibition of cIAP1's E3 ligase activity suppresses cell proliferation and migration. *Molecular Cell*, *42*, 569–583.

Mace, P. D., Linke, K., Feltham, R., Schumacher, F. R., Smith, C. A., Vaux, D. L., et al. (2008). Structures of the cIAP2 RING domain reveal conformational changes associated with ubiquitin-conjugating enzyme (E2) recruitment. *The Journal of Biological Chemistry*, *283*, 31633–31640.

Mace, P. D., Smits, C., Vaux, D. L., Silke, J., & Day, C. L. (2010). Asymmetric recruitment of cIAPs by TRAF2. *Journal of Molecular Biology*, *400*, 8–15.

Mahoney, D. J., Cheung, H. H., Mrad, R. L., Plenchette, S., Simard, C., Enwere, E., et al. (2008). Both cIAP1 and cIAP2 regulate TNFalpha-mediated NF-kappaB activation. *Proceedings of the National Academy of Sciences of the United States of America*, *105*, 11778–11783.

Martins, L. M., Iaccarino, I., Tenev, T., Gschmeissner, S., Totty, N. F., Lemoine, N. R., et al. (2001). The serine protease Omi/HtrA2 regulates apoptosis by binding XIAP through a Reaper-like motif. *The Journal of Biological Chemistry*, *277*, 439–444.

Matsuzawa, A., Tseng, P. H., Vallabhapurapu, S., Luo, J. L., Zhang, W., Wang, H., et al. (2008). Essential cytoplasmic translocation of a cytokine receptor-assembled signaling complex. *Science, 321*, 663–668.

Mehrotra, S., Languino, L. R., Raskett, C. M., Mercurio, A. M., Dohi, T., & Altieri, D. C. (2010). IAP regulation of metastasis. *Cancer Cell, 17*, 53–64.

Moulin, M., Anderton, H., Voss, A. K., Thomas, T., Wong, W. W., Bankovacki, A., et al. (2012). IAPs limit activation of RIP kinases by TNF receptor 1 during development. *The EMBO Journal, 31*, 1679–1691.

Murphy, J. M., Czabotar, P. E., Hildebrand, J. M., Lucet, I. S., Zhang, J., Avarez-Diaz, S., et al. (2013). The pseudokinase MLKL mediates necroptosis via a molecular switch mechanism. *Immunity, 39*, 443–453.

Nakatani, Y., Kleffmann, T., Linke, K., Condon, S. M., Hinds, M. G., & Day, C. L. (2013). Regulation of ubiquitin transfer by XIAP, a dimeric RING E3 ligase. *The Biochemical Journal, 450*, 629–638.

Ndubaku, C., Varfolomeev, E., Wang, L., Zobel, K., Lau, K., Elliott, L. O., et al. (2009). Antagonism of c-IAP and XIAP proteins is required for efficient induction of cell death by small-molecule IAP antagonists. *ACS Chemical Biology, 4*, 557–566.

Newton, K., & Vucic, D. (2007). Ubiquitin ligases in cancer: Ushers for degradation. *Cancer Investigation, 25*, 502–513.

Oberoi, T. K., Dogan, T., Hocking, J. C., Scholz, R. P., Mooz, J., Anderson, C. L., et al. (2012). IAPs regulate the plasticity of cell migration by directly targeting Rac1 for degradation. *The EMBO Journal, 31*, 14–28.

Oberst, A., Dillon, C. P., Weinlich, R., McCormick, L. L., Fitzgerald, P., Pop, C., et al. (2011). Catalytic activity of the caspase-8-FLIP(L) complex inhibits RIPK3-dependent necrosis. *Nature, 471*, 363–367.

Park, S. M., Yoon, J. B., & Lee, T. H. (2004). Receptor interacting protein is ubiquitinated by cellular inhibitor of apoptosis proteins (c-IAP1 and c-IAP2) in vitro. *FEBS Letters, 566*, 151–156.

Peter, M. E., Budd, R. C., Desbarats, J., Hedrick, S. M., Hueber, A. O., Newell, M. K., et al. (2007). The CD95 receptor: Apoptosis revisited. *Cell, 129*, 447–450.

Petrucci, E., Pasquini, L., Bernabei, M., Saulle, E., Biffoni, M., Accarpio, F., et al. (2012). A small molecule SMAC mimic LBW242 potentiates TRAIL- and anticancer drug-mediated cell death of ovarian cancer cells. *PLoS One, 7*, e35073.

Phu, L., Izrael-Tomasevic, A., Matsumoto, M. L., Bustos, D., Dynek, J. N., Fedorova, A. V., et al. (2011). Improved quantitative mass spectrometry methods for characterizing complex ubiquitin signals. *Molecular and Cellular Proteomics, 10* M110.003756.

Plechanovová, A., Jaffray, E. G., Tatham, M. H., Naismith, J. H., & Hay, R. T. (2012). Structure of a RING E3 ligase and ubiquitin-loaded E2 primed for catalysis. *Nature, 489*, 115–120.

Pickart, C. M., & Fushman, D. (2004). Polyubiquitin chains: polymeric protein signals. *Current Opinion Chemical Biology, 8*, 610–616.

Pohl, C., & Jentsch, S. (2008). Final stages of cytokinesis and midbody ring formation are controlled by BRUCE. *Cell, 132*, 832–845.

Polevoda, B., & Sherman, F. (2003). N-terminal acetyltransferases and sequence requirements for N-terminal acetylation of eukaryotic proteins. *Journal of Molecular Biology, 325*, 595–622.

Richter, B. W., Mir, S. S., Eiben, L. J., Lewis, J., Reffey, S. B., Frattini, A., et al. (2001). Molecular cloning of ILP-2, a novel member of the inhibitor of apoptosis protein family. *Molecular and Cellular Biology, 21*, 4292–4301.

Rigaud, S., Fondanèche, M. C., Lambert, N., Pasquier, B., Mateo, V., Soulas, P., et al. (2006). XIAP deficiency in humans causes an X-linked lymphoproliferative syndrome. *Nature, 444*, 110–114.

Rothe, M., Pan, M. G., Henzel, W. J., Ayres, T. M., & Goeddel, D. V. (1995). The TNF-R2-TRAF signaling complex contains two novel proteins related to baculoviral-inhibitor of apoptosis proteins. *Cell*, *83*, 1243–1252.

Rothe, M., Wong, S. C., Henzel, W. J., & Goeddel, D. V. (1994). A novel family of putative signal transducers associated with the cytoplasmic domain of the 75 kda tumor necrosis factor receptor. *Cell*, *78*, 681–692.

Roy, N., Deveraux, Q. L., Takahashi, R., Salvesen, G. S., & Reed, J. C. (1997). The c-IAP-1 and c-IAP-2 proteins are direct inhibitors of specific caspases. *The EMBO Journal*, *16*, 6914–6925.

Roy, N., Mahadevan, M. S., Mclean, M., Shutler, G., Yaraghi, Z., Farahani, R., et al. (1995). The gene for neuronal apoptosis inhibitory protein is partially deleted in individuals with spinal muscular atrophy. *Cell*, *80*, 167–178.

Ryan, B. M., O'Donovan, N., & Duffy, M. J. (2009). Survivin: A new target for anti-cancer therapy. *Cancer Treatment Reviews*, *35*, 553–562.

Salvesen, G. S., & Duckett, C. S. (2002). IAP proteins: Blocking the road to death's door. *Nature Reviews Molecular Cell Biology*, *3*, 401–410.

Scheidereit, C. (2006). IkappaB kinase complexes: Gateways to NF-kappaB activation and transcription. *Oncogene*, *25*, 6685–6705.

Schile, A. J., García-Fernández, M., & Steller, H. (2008). Regulation of apoptosis by XIAP ubiquitin-ligase activity. *Genes and Development*, *22*, 2256–2266.

Schimmer, A. D., Herr, W., Hänel, M., Borthakur, G., Frankel, A., Horst, H. A., et al. (2011). Addition of AEG35156 XIAP antisense oligonucleotide in reinduction chemotherapy does not improve remission rates in patients with primary refractory acute myeloid leukemia in a randomized phase II study. *Clinical Lymphoma, Myeloma & Leukemia*, *11*, 433–438.

Schmukle, A. C., & Walczak, H. (2012). No one can whistle a symphony alone—How different ubiquitin linkages cooperate to orchestrate NF-κB activity. *Journal of Cell Science*, *125*, 549–559.

Schulman, B. A., & Harper, J. W. (2009). Ubiquitin-like protein activation by E1 enzymes: The apex for downstream signalling pathways. *Nature Reviews Molecular Cell Biology*, *10*, 319–331.

Scott, F. L., Denault, J. B., Riedl, S. J., Shin, H., Renatus, M., & Salvesen, G. S. (2005). XIAP inhibits caspase-3 and -7 using two binding sites: Evolutionarily conserved mechanism of IAPs. *The EMBO Journal*, *24*, 645–655.

Seshagiri, S., Vucic, D., Lee, J., & Dixit, V. M. (1999). Baculovirus-based genetic screen for antiapoptotic genes identifies a novel IAP. *The Journal of Biological Chemistry*, *274*, 36769–36773.

Shaw, T. J., Lacasse, E. C., Durkin, J. P., & Vanderhyden, B. C. (2008). Downregulation of XIAP expression in ovarian cancer cells induces cell death in vitro and in vivo. *International Journal of Cancer*, *122*, 1430–1434.

Shin, H., Okada, K., Wilkinson, J. C., Solomon, K. M., Duckett, C. S., Reed, J. C., et al. (2003). Identification of ubiquitination sites on the X-linked inhibitor of apoptosis protein. *The Biochemical Journal*, *373*, 965–971.

Silke, J. (2011). The regulation of TNF signalling: What a tangled web we weave. *Current Opinion in Immunology*, *23*, 620–626.

Silke, J., & Brink, R. (2010). Regulation of TNFRSF and innate immune signalling complexes by TRAFs and cIAPs. *Cell Death and Differentiation*, *17*, 35–45.

Silke, J., Ekert, P. G., Day, C. L., Hawkins, C. J., Baca, M., Chew, J., et al. (2001). Direct inhibition of caspase 3 is dispensable for the anti-apoptotic activity of XIAP. *The EMBO Journal*, *20*, 3114–3123.

Silke, J., Hawkins, C. J., Ekert, P. G., Chew, J., Day, C. L., Pakusch, M., et al. (2002). The anti-apoptotic activity of XIAP is retained upon mutation of both the caspase 3- and caspase 9-interacting sites. *The Journal of Cell Biology*, *157*, 115–124.

Silke, J., Kratina, T., Chu, D., Ekert, P. G., Day, C. L., Pakusch, M., et al. (2005). Determination of cell survival by RING-mediated regulation of inhibitor of apoptosis (IAP) protein abundance. *Proceedings of the National Academy of Sciences of the United States of America, 102*, 16182–16187.

Silke, J., Kratina, T., Ekert, P. G., Pakusch, M., & Vaux, D. L. (2004). Unlike Diablo/Smac, Grim promotes global ubiquitination and specific degradation of XIAP and neither cause apoptosis. *The Journal of Biological Chemistry, 279*, 4313–4321.

Speliotes, E. K., Uren, A., Vaux, D., & Horvitz, H. R. (2000). The survivin-like *C. elegans* BIR-1 protein acts with the Aurora-like kinase AIR-2 to affect chromosomes and the spindle midzone. *Molecular Cell, 6*, 211–223.

Srinivasula, S. M., Hegde, R., Saleh, A., Datta, P., Shiozaki, E., Chai, J., et al. (2001). A conserved XIAP-interaction motif in caspase-9 and Smac/DIABLO regulates caspase activity and apoptosis. *Nature, 410*, 112–116.

Sun, C., Cai, M., Gunasekera, A. H., Meadows, R. P., Wang, H., Chen, J., et al. (1999). NMR structure and mutagenesis of the inhibitor-of-apoptosis protein XIAP. *Nature, 401*, 818–821.

Sun, C., Cai, M., Meadows, R. P., Xu, N., Gunasekera, A. H., Herrmann, J., et al. (2000). NMR structure and mutagenesis of the third BIR domain of the inhibitor of apoptosis protein XIAP. *The Journal of Biological Chemistry, 275*, 33777–33781.

Sun, L., Wang, H., Wang, Z., He, S., Chen, S., Liao, D., et al. (2012). Mixed lineage kinase domain-like protein mediates necrosis signaling downstream of RIP3 kinase. *Cell, 148*, 213–227.

Suzuki, Y., Imai, Y., Nakayama, H., Takahashi, K., Takio, K., & Takahashi, R. (2001). A serine protease, HtrA2, is released from the mitochondria and interacts with XIAP, inducing cell death. *Molecular Cell, 8*, 613–621.

Takahashi, R., Deveraux, Q., Tamm, I., Welsh, K., Assamunt, N., Salvesen, G. S., et al. (1998). A single BIR domain of XIAP sufficient for inhibiting caspases. *The Journal of Biological Chemistry, 273*, 7787–7790.

Tang, E. D., Wang, C. Y., Xiong, Y., & Guan, K. L. (2003). A role for NF-kappaB essential modifier/IkappaB kinase-gamma (NEMO/IKKgamma) ubiquitination in the activation of the IkappaB kinase complex by tumor necrosis factor-alpha. *The Journal of Biological Chemistry, 278*, 37297–37305.

Tchoghandjian, A., Jennewein, C., Eckhardt, I., Rajalingam, K., & Fulda, S. (2013). Identification of non-canonical NF-κB signaling as a critical mediator of Smac mimetic-stimulated migration and invasion of glioblastoma cells. *Cell Death and Disease, 4*, e564.

Tenev, T., Bianchi, K., Darding, M., Broemer, M., Langlais, C., Wallberg, F., et al. (2011). The Ripoptosome, a signaling platform that assembles in response to genotoxic stress and loss of IAPs. *Molecular Cell, 43*, 432–448.

Uren, A. G., Pakusch, M., Hawkins, C. J., Puls, K. L., & Vaux, D. L. (1996). Cloning and expression of apoptosis inhibitory protein homologs that function to inhibit apoptosis and/or bind TRAFs. *Proceedings of the National Academy of Sciences of the United States of America, 93*, 4974–4978.

Uren, A. G., Wong, L., Pakusch, M., Fowler, K. J., Burrows, F. J., Vaux, D. L., et al. (2000). Survivin and the inner centromere protein INCENP show similar cell-cycle localization and gene knockout phenotype. *Current Biology, 10*, 1319–1328.

Vallabhapurapu, S., Matsuzawa, A., Zhang, W., Tseng, P. H., Keats, J. J., Wang, H., et al. (2008). Nonredundant and complementary functions of TRAF2 and TRAF3 in a ubiquitination cascade that activates NIK-dependent alternative NF-kappaB signaling. *Nature Immunology, 9*, 1364–1370.

Vandenabeele, P., Galluzzi, L., Vanden Berghe, T., & Kroemer, G. (2010). Molecular mechanisms of necroptosis: An ordered cellular explosion. *Nature Reviews Molecular Cell Biology, 11*, 700–714.

Varfolomeev, E., Alicke, B., Elliott, J. M., Zobel, K., West, K., Wong, H., et al. (2009). X chromosome-linked inhibitor of apoptosis regulates cell death induction by proapoptotic receptor agonists. *The Journal of Biological Chemistry*, *284*, 34553–34560.

Varfolomeev, E., Blankenship, J. W., Wayson, S. M., Fedorova, A. V., Kayagaki, N., Garg, P., et al. (2007). IAP antagonists induce autoubiquitination of c-IAPs, NF-kappaB activation, and TNFalpha-dependent apoptosis. *Cell*, *131*, 669–681.

Varfolomeev, E., Goncharov, T., Fedorova, A. V., Dynek, J. N., Zobel, K., Deshayes, K., et al. (2008). c-IAP1 and c-IAP2 are critical mediators of tumor necrosis factor alpha (TNFalpha)-induced NF-kappaB activation. *The Journal of Biological Chemistry*, *283*, 24295–24299.

Varfolomeev, E., Goncharov, T., Maecker, H., Zobel, K., Kömüves, L. G., Deshayes, K., et al. (2012). Cellular inhibitors of apoptosis are global regulators of NF-κB and MAPK activation by members of the TNF family of receptors. *Science Signaling*, *5*, ra22.

Varfolomeev, E., Wayson, S. M., Dixit, V. M., Fairbrother, W. J., & Vucic, D. (2006). The inhibitor of apoptosis protein fusion c-IAP2.MALT1 stimulates NF-κB activation independently of TRAF1 and TRAF2. *The Journal of Biological Chemistry*, *281*, 29022–29029.

Vaux, D. L., & Silke, J. (2003). Mammalian mitochondrial IAP binding proteins. *Biochemical and Biophysical Research Communications*, *304*, 499–504.

Vaux, D. L., & Silke, J. (2005). IAPs, RINGs and ubiquitylation. *Nature Reviews Molecular Cell Biology*, *6*, 287–297.

Verhagen, A. M., Coulson, E. J., & Vaux, D. L. (2001). Inhibitor of apoptosis proteins and their relatives: IAPs and other BIRPs. *Genome Biology*, *2*, REVIEWS3009.

Verhagen, A. M., Ekert, P. G., Pakusch, M., Silke, J., Connolly, L. M., Reid, G. E., et al. (2000). Identification of DIABLO, a mammalian protein that promotes apoptosis by binding to and antagonizing IAP proteins. *Cell*, *102*, 42–53.

Verhagen, A. M., Kratina, T. K., Hawkins, C. J., Silke, J., Ekert, P. G., & Vaux, D. L. (2007). Identification of mammalian mitochondrial proteins that interact with IAPs via N-terminal IAP binding motifs. *Cell Death and Differentiation*, *14*, 348–357.

Verhagen, A. M., Silke, J., Ekert, P. G., Pakusch, M., Kaufmann, H., Connolly, L. M., et al. (2001). HtrA2 promotes cell death through its serine protease activity and its ability to antagonize inhibitor of apoptosis proteins. *The Journal of Biological Chemistry*, *277*, 445–454.

Vince, J. E., Chau, D., Callus, B., Wong, W. W., Hawkins, C. J., Schneider, P., et al. (2008). TWEAK-FN14 signaling induces lysosomal degradation of a cIAP1-TRAF2 complex to sensitize tumor cells to TNFalpha. *The Journal of Cell Biology*, *182*, 171–184.

Vince, J. E., Pantaki, D., Feltham, R., Mace, P. D., Cordier, S. M., Schmukle, A. C., et al. (2009). TRAF2 must bind to cellular inhibitors of apoptosis for tumor necrosis factor (tnf) to efficiently activate nf-{kappa}b and to prevent tnf-induced apoptosis. *The Journal of Biological Chemistry*, *284*, 35906–35915.

Vince, J. E., Wong, W. W., Gentle, I., Lawlor, K. E., Allam, R., O'Reilly, L., et al. (2012). Inhibitor of apoptosis proteins limit RIP3 kinase-dependent interleukin-1 activation. *Immunity*, *36*, 215–227.

Vince, J. E., Wong, W. W., Khan, N., Feltham, R., Chau, D., Ahmed, A. U., et al. (2007). IAP antagonists target cIAP1 to induce TNFalpha-dependent apoptosis. *Cell*, *131*, 682–693.

Vucic, D., Dixit, V. M., & Wertz, I. E. (2011). Ubiquitylation in apoptosis: A post-translational modification at the edge of life and death. *Nature Reviews Molecular Cell Biology*, *12*, 439–452.

Vucic, D., & Fairbrother, W. J. (2007). The inhibitor of apoptosis proteins as therapeutic targets in cancer. *Clinical Cancer Research*, *13*, 5995–6000.

Vucic, D., Stennicke, H. R., Pisabarro, M. T., Salvesen, G. S., & Dixit, V. M. (2000). ML-IAP, a novel inhibitor of apoptosis that is preferentially expressed in human melanomas. *Current Biology*, *10*, 1359–1366.

Wang, L., Du, F., & Wang, X. (2008). TNF-alpha induces two distinct caspase-8 activation pathways. *Cell*, *133*, 693–703.

Wilson, R., Goyal, L., Ditzel, M., Zachariou, A., Baker, D. A., Agapite, J., et al. (2002). The DIAP1 RING finger mediates ubiquitination of Dronc and is indispensable for regulating apoptosis. *Nature Cell Biology*, *4*, 445–450.

Wong, W. W., Gentle, I. E., Nachbur, U., Anderton, H., Vaux, D. L., & Silke, J. (2010). RIPK1 is not essential for TNFR1-induced activation of NF-kappaB. *Cell Death and Differentiation*, *17*, 482–487.

Wu, T. Y., Wagner, K. W., Bursulaya, B., Schultz, P. G., & Deveraux, Q. L. (2003). Development and characterization of nonpeptidic small molecule inhibitors of the XIAP/caspase-3 interaction. *Chemistry and Biology*, *10*, 759–767.

Wu, Z. H., Wong, E. T., Shi, Y., Niu, J., Chen, Z., Miyamoto, S., et al. (2010). ATM- and NEMO-dependent ELKS ubiquitination coordinates TAK1-mediated IKK activation in response to genotoxic stress. *Molecular Cell*, *40*, 75–86.

Xu, M., Skaug, B., Zeng, W., & Chen, Z. J. (2009). A ubiquitin replacement strategy in human cells reveals distinct mechanisms of IKK activation by TNFalpha and IL-1beta. *Molecular Cell*, *36*, 302–314.

Yang, Q. H., & Du, C. (2004). Smac/DIABLO selectively reduces the levels of c-IAP1 and c-IAP2 but not that of XIAP and livin in HeLa cells. *The Journal of Biological Chemistry*, *279*, 16963–16970.

Yang, Y., Fang, S., Jensen, J. P., Weissman, A. M., & Ashwell, J. D. (2000). Ubiquitin protein ligase activity of IAPs and their degradation in proteasomes in response to apoptotic stimuli. *Science*, *288*, 874–877.

Zarnegar, B. J., Wang, Y., Mahoney, D. J., Dempsey, P. W., Cheung, H. H., He, J., et al. (2008). Noncanonical NF-kappaB activation requires coordinated assembly of a regulatory complex of the adaptors cIAP1, cIAP2, TRAF2 and TRAF3 and the kinase NIK. *Nature Immunology*, *9*, 1371–1378.

Zheng, C., Kabaleeswaran, V., Wang, Y., Cheng, G., & Wu, H. (2010). Crystal structures of the TRAF2: cIAP2 and the TRAF1: TRAF2: cIAP2 complexes: Affinity, specificity, and regulation. *Molecular Cell*, *38*, 101–113.

Zhuang, M., Guan, S., Wang, H., Burlingame, A. L., & Wells, J. A. (2013). Substrates of IAP ubiquitin ligases identified with a designed orthogonal E3 ligase, the NEDDylator. *Molecular Cell*, *49*, 273–282.

CHAPTER THREE

Activation of the NLRP3 Inflammasome by Proteins That Signal for Necroptosis

Tae-Bong Kang[*,†], **Seung-Hoon Yang**[*], **Beata Toth**[*], **Andrew Kovalenko**[*], **David Wallach**[*,1]

[*]Department of Biological Chemistry, The Weizmann Institute of Science, Rehovot, Israel
[†]Department of Biotechnology, College of Biomedical and Health Science, Konkuk University, Chung-Ju, Republic of Korea
[1]Corresponding author: e-mail address: d.wallach@weizmann.ac.il

Contents

Methods in Enzymology, Volume 545
ISSN 0076-6879
http://dx.doi.org/10.1016/B978-0-12-801430-1.00003-2

Abstract

Necroptosis—a form of programmed necrotic cell death—and its resulting release of damage-associated molecular patterns (DAMPs) are believed to participate in the triggering of inflammatory processes. To assess the relative contribution of this cell death mode to inflammation, we need to know what other cellular effects can be exerted by molecules shown to trigger necrotic death, and the extent to which those effects might themselves contribute to inflammation. Here, we describe the technical approaches that have been applied to assess the impact of the main signaling molecules known to mediate activation of necroptosis upon generation of inflammatory cytokines in LPS-treated mouse bone marrow-derived dendritic cells. The findings obtained by this assessment indicated that signaling molecules known to initiate necroptosis can also initiate activation of the NLRP3 inflammasome, thereby inducing inflammation independently of cell death by triggering the generation of proinflammatory cytokines such as IL-1β.

1. INTRODUCTION

Although the molecular mechanisms that execute necroptosis—a form of programmed necrotic cell death—and the effector molecules that initiate them are still unknown, several signaling molecules that trigger this process have been identified. Such triggering requires protein phosphorylation by RIPK1 (Degterev et al., 2008; Holler et al., 2000), as well as activities of the protein kinase RIPK3 (Cho et al., 2009; He et al., 2009; Zhang et al., 2009) and the pseudokinase MLKL (Sun et al., 2012; Zhao et al., 2012). In contrast, the proteolytic activity of caspase-8 blocks the induction of necroptosis (Holler et al., 2000; Oberst et al., 2011). A number of other signaling proteins also contribute to the initiation of necroptosis or to its regulation (Zhou, Han, & Han, 2012).

Because necrotic death results in rupture of the cell membrane and various intracellular components released from dying cells are known to trigger inflammation (Rock & Kono, 2008), the induction of necroptosis *in vivo* may well inflict inflammation. Therefore, the findings from several animal studies that inflammation is influenced by the abovementioned signaling proteins have been suggested to demonstrate a causal role for necroptosis in inflammation (Cho et al., 2009; Degterev et al., 2005; Duprez et al., 2011; He et al., 2009; Mocarski, Upton, & Kaiser, 2012; Smith et al., 2007; Zhang et al., 2009). However, because signaling proteins often serve more than one function, it is also possible that the inflammation induced by proteins known to participate in the initiation of necrotic death is actually mediated through other activities of these proteins. Since necrotic cell death can be not only a cause but also a result of inflammation, the mere detection of cells that have died necrotically in inflamed tissues does not suffice to define necrosis as the cause of this inflammation (Wallach, Kovalenko, & Kang, 2011).

Our interest in the causal role of necroptosis in inflammation and the signaling proteins controlling it was aroused by findings of our group and others on the *in vivo* function of caspase-8. This protein serves as the proximal enzyme in the induction of apoptotic cell death via the extrinsic cell death pathway (Boldin, Goncharov, Goltsev, & Wallach, 1996; Muzio et al., 1996). However, it was found to serve other functions as well. In several transgenic mouse models, deficiency of this enzyme was found to trigger chronic inflammation (e.g., see Kang, Yang, Toth, Kovalenko, & Wallach, 2013; Wallach et al., 2010 and references therein). Since caspase-8 had been found to block necroptotic death induction (Holler et al., 2000; Oberst et al., 2011), it seemed possible that inflammation incurred by its deficiency is a result of spontaneous necroptotic death.

The experiments described in this chapter were initiated following our serendipitous finding that generation of the inflammatory mediators IL-1β and IL 18 in response to treatment with bacterial endotoxin is greatly enhanced in mice whose dendritic cells (DCs) are deficient in caspase-8 (Kang et al., 2013). These experiments revealed that: (a) this enhancement reflects activation of the NLRP3 inflammasome; (b) the activation is dependent on caspase-8 deficiency; and (c) it also depends (as does the induction of necroptosis) on the functions of the protein kinases, RIPK1 and RIPK3, and of two other proteins, the pseudokinase MLKL and the phosphatase PGAM5 (Sun et al., 2012; Wang, Jiang, Chen, Du, & Wang, 2012; Zhao et al., 2012). Yet, activation of the inflammasome in this setting did not correlate to induction of necrotic death, and it occurred independently of the release of damage-associated molecular patterns (DAMPs). These findings implied that the signaling proteins so far known to participate in initiation of necroptotic cell death also serve other functions, through which they can also contribute to the induction of inflammation (Kang et al., 2013).

Here, we provide a more detailed description of the experimental setup in which these findings were reached.

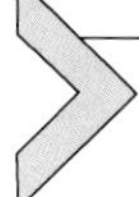

2. ALTERED EXPRESSION OR FUNCTION OF ENZYMES THAT CONTROL INDUCTION OF NECROPTOSIS RESULTS IN ALTERED GENERATION OF IL-1β AND IL-18 BY MOUSE DCs

This research was prompted by the serendipitous finding that the serum levels of IL-1β after injection of bacterial endotoxin (LPS) are substantially higher in mice whose DC are deficient in caspase-8 (Kang et al., 2013). In exploring the mechanism underlying this elevation in IL-1β, we applied cultured DC derived from mouse bone marrow precursor cells (BMDCs) to assess the impact of caspase-8 deficiency on the generation of inflammatory cytokines in these cells in response to LPS treatment. Because caspase-8 deficiency has also been shown to facilitate necroptotic death, we examined the effect of modulation of the function or expression of several other enzymes that participate in necroptosis induction on IL-1β expression in the cultured BMDCs.

Induction of IL-1β by LPS in mouse normally requires further stimulation of these cells by an "activating agent" such as ATP (Schroder & Tschopp, 2010). Deficiency of caspase-8 in the BMDC dramatically altered this pattern of induction. While not affecting the amounts of IL-1β induced by combined stimulation with LPS and ATP, it largely alleviated the need for application of ATP or any other activating agent to the cells. Mere stimulation by LPS (or by some other Toll-like receptor 4 (TLR4) ligands) sufficed to induce generation of IL-1β by the caspase-8-deficient cells. Assessment of the effect of caspase-8 deficiency on the induction of other cytokines in the BMDC showed that it similarly alleviated the need for further treatment by an activation agent to induce IL-18. However, the caspase-8 deficiency had no effect on the induction of TNF or IL-6 by LPS.

Since caspase-8 deficiency is known to facilitate necroptotic death in some cells, we assessed the effect of modulation of the expression and function of other genes known to contribute to the induction of necroptosis on the induction of IL-1β generation by LPS. As reported for the regulation of necroptosis, the generation of IL-1β observed in caspase-8-deficient BMDC when these cells were treated by LPS alone could be blocked by necrostatin-1 (Nec-1; Sigma), a chemical inhibitor of the kinase function of RIPK1. It was also greatly decreased by knockdown of MLKL or PGAM5, two proteins reported to participate in the induction of necroptosis. Moreover, in mice with a RIPK3-null background caspase-8 deficiency had no effect on the generation of IL-1β or IL-18 by the BMDC in response to LPS, implying that RIPK3 function is crucial for the enhanced generation of these cytokines just as it is for the induction of necroptosis.

2.1. Generation of mouse bone marrow-derived DCs

We generated BMDC of C57BL/6 mice as previously described (Inaba et al., 1992), with some modifications. Bone marrow progenitor cells from mouse femurs and tibias were cultured (5×10^6 cells in 10 ml/100-mm petri dish) in culture medium (RPMI-1640 medium supplemented with 10% (vol/vol) fetal bovine serum, 100 units/ml penicillin, 50 μg/ml streptomycin, 1 m*M* sodium pyruvate, 2 m*M* L-glutamine, 1% nonessential amino acid solution, 50 μ*M* 2-mercaptoethanol, and 20 ng/ml granulocyte–macrophage colony-stimulating factor (GM-CSF; PeproTech, catalog no. 315-03)). On day 3 of culture, the same volume of culture medium was added. On day 6, 50% of the nonadherent cell population was removed and centrifuged at 300 × *g* for 5 min at room temperature. Cells were then resuspended in culture medium and replated in the original tissue-culture dish. On the 8th day of culturing, the nonadherent cells were harvested, washed once with culture medium, and seeded for the experiments at a density of 1×10^6 cells/ml in fresh culture medium.

2.2. Use of transgenic mice to obtain DCs deficient in caspase-8 or RIPK3

Mouse strains harboring a knockout *Casp8* allele (*Casp8*$^{-/+}$) (Varfolomeev et al., 1998) and a conditional *Casp8* allele (*Casp8*$^{F/+}$) (Kang et al., 2004) were established in our laboratory. Mice deficient in RIPK3 were obtained from Dr. Vishva Dixit (Newton, Sun, & Dixit, 2004) and mice expressing Cre under control of the Itgax (CD11c) promoter (Itgax-Cre) were from Dr. Boris Reizis (Caton, Smith-Raska, & Reizis, 2007). In all experiments, the genotype of mice with caspase-8-deficient BMDC was *Casp8*$^{fl/-}$: *Itgax-Cre,* and in most of them, the control mice were *Casp8*$^{fl/fl}$. In preliminary experiments, we observed no differences between *Casp8*$^{fl/fl}$ and *Casp8*$^{fl/+}$*:Itgax-Cre* mice in the response to LPS injection, nor in the amount of IL-1β induced in their BMDC *in vitro* by LPS alone or by LPS + ATP.

The *Itgax* (integrin alpha X chain protein) gene is expressed in differentiated DC but not in their bone marrow precursors. Accordingly, the expression of Cre that was dictated in the *Casp8*$^{fl/-}$*:Itgax-Cre* and *Casp8*$^{fl/+}$ *Itgax-Cre* cells by the *Itgas* promoter occurred only when the cultured BMDC began to differentiate, toward the 8th day of their incubation. Western analysis of these cells using anti-caspase-8 antibody (3B10; Alexis) confirmed that the resulting deletion of the floxed *caspase-8* gene was almost complete. In all immunoblot analyses, the primary antibodies were detected by horseradish peroxidase-conjugated anti-mouse or anti-rabbit IgG antibodies. Blots were developed with the enhanced chemiluminescence SuperSignal West Pico or SuperSignal West Femto Chemiluminescent

Substrate system (Thermo Scientific) according to the manufacturer's instructions.

2.3. Knockdown of proteins signaling for necroptosis in DCs

BMDCs were plated in 24-well plates at a concentration of 1×10^6 cells/well and transfected using an Amaxa Nucleofector kit (Amaxa) with 500 nmol of siRNA duplex per well (Dharmacon, Thermo Scientific). The following siRNAs from Dharmacon were used: caspase-8 (E-043044-00-0005), MLKL (L-061420-00-0005), and PGAM5 (L-052506-01-0005). After 48 h, the cells were processed for analysis.

2.4. Induction of cytokines in DCs by agents inducing and activating the NLRP3 inflammasome

Ultra-pure *Escherichia coli* LPS, *Porphyromonas gingivalis* LPS (PG-LPS), and the synthetic triacylated lipoprotein Pam3CSK4 were purchased from InvivoGen. These agents were normally applied to the cells for 3 h, each at 1 μg/ml. In cells further activated by ATP (5 m*M*), the latter was applied for the last 15 min of this incubation period. To inhibit the kinase function of RIPK1, Nec-1 (50 μ*M*) was added to the medium prior to application of LPS.

2.5. Quantification of the induced cytokines

IL-1β, TNF-α, IL-18, and IL-6 ELISA kits were from eBioscience or R&D Systems.

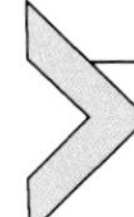

3. SIGNALING PROTEINS CONTROLLING NECROPTOSIS AFFECT ASSEMBLY OF THE NLRP3 INFLAMMASOME

Generation of IL-1β in BMDC is controlled at two main mechanistic levels: (a) synthesis of the IL-1β precursor protein and some of the proteins that mediate its processing, and (b) assembly of inflammasomes, the macromolecular complexes that mediate activation of caspase-1, the enzyme that processes the IL-1β precursor protein (Schroder & Tschopp, 2010). In wild-type BMDC, inducing agents like LPS usually affect only the first mechanistic level, whereas activating agents like ATP affect the second. Thus, the fact that caspase-8 deficiency enhanced the generation of IL-1β in BMDC treated by LPS alone, but not the generation induced by their combined treatment with LPS and ATP, suggested that the impact of caspase-8 deficiency is restricted to the second mechanistic level, assembly of the inflammasome. When we quantified the mRNA for IL-1β, as well as for the IL-1β precursor protein and the NLRP3 inflammasome components that mediate its processing, we indeed found that caspase-8 deficiency did not enhance the induction of these proteins by LPS. Listed below are the different approaches we applied in seeking more direct evidence that modulation of the expression of IL-1β by caspase-8 and by other enzymes known to control necroptosis reflects induced changes in assembly of the inflammasome.

3.1. Assessment of the proteolytic processing of caspase-1 and of the IL-1β precursor protein

Assembly of the inflammasomes results in activation of pro-caspase-1 molecules associated with these complexes. As a result, the pro-caspase molecules initially self-process, yielding two fragments of the caspase molecule that assemble in a proteolytically active complex containing two copies of each fragment. This caspase-1 complex then cleaves substrate proteins, such as the IL-1β and IL-18 precursor proteins. One of the ways we monitored generation of the NLRP3 inflammasome was by quantifying the proteolytic products of the activation of these caspase-1 complexes. Generation of the processed form (17 kDa) of IL-1β was monitored by Western blotting of the culture medium to which the processed cytokine was released (using an anti-mouse IL-1β antibody from R&D Systems, catalog no. AF401-NA). To avoid interference with the serum proteins by this Western analysis, incubation of the BMDC with inflammasome-activating agents was carried out in this case in serum-reduced Opti-MEM medium (Invitrogen). Generation of the processed form of caspase-1 was done by Western analysis of cellular lysates using the anti-caspase-1 antibody sc-514 from Santa Cruz Biotechnology.

3.2. Confirmation of the requirement for NLRP3 and ASC for IL-1β generation

By applying siRNAs to knock down NLRP3 and ASC, we were able to confirm that these two protein constituents of the NLRP3 inflammasome are required for the secretion of IL-1β by the LPS-treated caspase-8-deficient BMDC. The siRNAs (from Dharmacon) that we used were of NLRP3 (L-053455-00-0005) and ASC (L-051439-01-0005).

3.3. Assessment of the assembly of the inflammasome by measuring the detergent solubility of the inflammasome components

The inflammasome is a large complex that does not dissociate in response to nonionic detergents. One of the ways we studied its formation was, therefore, by estimating the extent of the transition of two of the NLRP3 inflammasome components, pro-caspase-1 and ASC, to detergent-insoluble forms. BMDCs were lysed in Nonidet P-40 buffer (1% NP-40, 50 m*M* Tris–HCl pH 7.5, 150 m*M* NaCl, 1 m*M* EDTA, and complete protease inhibitor cocktail) on ice for 15 min, and the lysates were then spun for 15 min at 13,000 rpm. The insoluble residue was lysed by boiling for 3 min in SDS-PAGE sample buffer containing β-mercaptoethanol (5%) and estimated quantitatively by immunoblotting using anti-caspase-1 (M20; Santa Cruz Biotechnology) and anti-ASC (AL177; Enzo Life Sciences) antibodies.

3.4. Assessment of the assembly of the inflammasome by the use of cross-linking reagents

Using a reversible chemical cross-linking reagent, we further verified the interactions among the inflammasome components. After trying several different cross-linking agents, we opted for 3,3′-dithiobis (sulfosuccinimidylpropionate) (DTSSP; Pierce). We applied it at 2 m*M* to intact BMDC or to the insoluble residue remaining after their extraction with NP40. In the first approach, aliquots of 1.5×10^7 cells were suspended in PBS containing 2 m*M* DTSSP. In the second, aliquots of 2.5×10^7 cells were first extracted with a buffer containing 1% NP40, 20 m*M* HEPES pH 7.5, and 150 m*M* NaCl, and the lysates were then spun at 4 °C for 10 min at 6000 rpm. The pellets were washed once with PBS and then resuspended in PBS containing 2 m*M* DTSSP. In both cases, the cross-linking was allowed to proceed for 2 h at 4 °C and was terminated by the addition of Tris pH 7.4 to a final concentration of 20 m*M*. This was followed by further incubation for 10 min at room temperature. The protein preparations were then resuspended with the aid of a 21-gauge needle in buffer containing 1% SDS, 50 m*M* Tris pH 7.4, 150 m*M* NaCl, and 1 m*M* EDTA and were then spun at 14,000 rpm at 4 °C for 15 min.

One portion of the clear supernatant was used to determine the total amount of NLRP3 inflammasome components in this preparation. We reversed the cross-linking of the proteins in it by boiling the preparation in SDS-PAGE sample buffer containing β-mercaptoethanol. Another portion was used to assess the association of the proteins in it by immunoprecipitation. To protect the antibodies applied at that step from the denaturing effect of the SDS, these samples were first diluted twofold in a buffer containing 4% Triton X-100 (which forms mixed micelles with SDS and thus protects proteins from SDS-induced denaturation), 50 m*M* Tris pH 7.4, 150 m*M* NaCl, and 1 m*M* EDTA. The samples were then precleared with protein G agarose beads (GE Healthcare Life Sciences) at 4 °C for 2 h. The precleared samples were incubated overnight with anti-ASC antibody (N15; Santa Cruz Biotechnology) at 4 °C, and protein G beads were added for 4 h. The immunoprecipitated proteins were then denatured, and their cross-linking was reversed by boiling in SDS-PAGE sample buffer containing β-mercaptoethanol. All samples were then analyzed quantitatively by Western blotting using antibodies against NLRP3 (Cryo-2; Enzo Life Sciences), caspase-1 (M20; Santa Cruz Biotechnology), and ASC (AL177; Enzo).

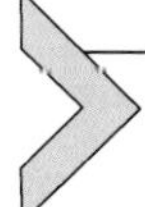

4. DOES THE SIMILARITY BETWEEN THE REGULATION OF NECROPTOSIS AND OF ASSEMBLY OF THE NLRP3 INFLAMMASOME REFLECT ACTIVATION OF THE INFLAMMASOME BY PRODUCTS OF NECROTIC CELLS?

One possible mechanistic explanation for our finding that signaling proteins known to control necroptotic death also regulate activation of the inflammasome could be that activation of the inflammasome in our experiments occurs as a consequence of necroptosis induction. This explanation seems particularly plausible in view of the fact that some of the DAMPs released from cells that die necrotically are known to trigger inflammasome assembly. ATP is one example of such a DAMP. To examine this possibility, we quantified, using different techniques, the extent of death of the BMDC in our cultures. We also measured the generation of reactive oxygen species (ROS) in the BMDC in response to LPS, and examined how this was affected by caspase-8 deficiency. Such generation, according to some studies, would be expected to occur at an early stage of necroptotic death and also to trigger inflammasome generation (reviewed in Franchi, Munoz-Planillo, & Nunez, 2012; Green, Oberst, Dillon, Weinlich, & Salvesen, 2011; Mankan, Kubarenko, & Hornung, 2012; Vandenabeele, Galluzzi, Vanden Berghe, & Kroemer, 2010). We also examined the release of ATP by the BMDC: first by direct quantification of ATP in the culture media and then by measuring the extent of generation of large pores in the tested cells, an effect that released ATP would be expected to have on them (Nihei, de Carvalho, Savino, & Alves, 2000; Virginio, MacKenzie, Rassendren, North, & Surprenant, 1999). Finally, we examined whether coculturing of caspase-8-deficient BMDC with normal BMDC facilitates IL-1β induction by LPS in the latter cells. Our reasoning for this was that if IL-1β generation in cultures of these cells is triggered by released DAMPs, these DAMPs should also facilitate generation of IL-1β in cocultured caspase-8-expressing cells.

4.1. Viability tests applied to DCs

Viability of the cultured BMDC was assessed by staining with the Live/Dead Viability/Cytotoxicity Assay Kit (Invitrogen) as specified by the manufacturer and visualized by standard fluorescence microscopy with a band-pass filter. We quantified cell death by using a cytotoxicity detection kit (Roche Applied Science) to determine the concentration of lactic dehydrogenase in the cell media and measuring the staining obtained with the Topro-3 fluorescent dye (50 n*M*). To assess mitochondrial transmembrane potential (which is expected to collapse when cells die), we harvested the cells using a cell lifter, washed them in PBS, stained them with TMRE (Abcam) according to the manufacturer's instructions, and then analyzed them by FACS. The results indicated that mouse BMDC is quite resistant to the induction of necroptosis. We observed no signs of death of the caspase-8-deficient BMDC within the period of LPS treatment required for their maximal IL-1β induction (less than 3 h). Signs of minor induction of cell death were first discerned in these cells only about 24 h after LPS was applied. In contrast, treatment with Z-Val-Ala-Asp-FMK (Z-VAD; EMD Biosciences)—a pan-caspase inhibitor often used to sensitize cells to inducers of necroptosis—did result in rapid death of the BMDC. Interestingly, this death was greatly enhanced by caspase-8 deficiency, suggesting that sensitization of cells to necroptotic death by Z-VAD reflects some additional activity of this inhibitor that is distinct from caspase inhibition.

4.2. Assessment of ROS generation in the DCs

Cellular ROS were quantitatively determined by staining with H_2DCFDA (Invitrogen) according to the manufacturer's protocol. Briefly, cells cultured with various stimulants were harvested using a cell lifter, washed once with warm PBS, incubated with 20 μ*M* H_2DCFDA for 30 min, and then analyzed by FACS. While rotenone (Sigma) applied at 10 μ*M*, which served as a positive control, triggered massive generation of ROS in the BMDC, treatment with LPS resulted in only slight enhancement of ROS generation. This increase was not affected by caspase-8 deficiency.

4.3. Assessment of the release of inflammasome-activating agents by the DCs

4.3.1 ATP release by the DCs

Concentrations of released ATP in the growth media of the cells in the presence of the ecto-ATPase inhibitor ARL 67156 (200 m*M*, as in Piccini et al., 2008) were determined using a commercial kit (A22066; Invitrogen). Neither the amounts of ATP that accumulated in the caspase-8-deficient BMDC cultures nor the extent of their increase by LPS was any greater than those observed in the wild-type cultures.

4.3.2 Assessment of permeability changes in the DCs as a marker for exposure to ATP

BMDCs were incubated with ethidium bromide (10 μ*M*) for 15 min at 37 °C. Uptake of the dye was assayed by FACS as described (Nihei et al., 2000). Whereas application of ATP to BMDC induced the generation of cell membrane pores (Nihei et al., 2000; Virginio et al., 1999), as reflected in the dramatic enhancement of ethidium bromide uptake, only very little dye uptake both in the caspase-8-deficient and in the wild-type cells resulted from treatment with LPS.

4.3.3 Assessment of the ability of DC lysates to activate IL-1β generation

To examine the possibility that caspase-8 deficiency in the BMDC facilitates the generation of some inflammasome-activating DAMPs other than ATP, we generated lysates of wild-type and caspase-8-deficient BMDC by subjecting suspensions of these cells in culture medium (1×10^6 cells/ml) to five freeze–thaw cycles at −80 and 37 °C. This procedure is often applied to test the ability of intracellular components to act as DAMPs and trigger inflammation. Samples of lysates derived from 3×10^5 wild-type cells and from the same number of caspase-8-deficient cells were each incubated with 1×10^6 BMDC in a final volume of 1 ml. None of the lysate samples had any effect on the generation of IL-1β by either the wild-type or the caspase-8-deficient BMDCs.

4.3.4 Coculturing of wild-type and caspase-8-deficient cells

The next experiment was designed to further examine the possibility that caspase-8 deficiency in BMDC promotes the release of some kind of activating agent from these cells that might underlie their ability to generate IL-1β in response to LPS in the absence of added activating agent. We compared the yield of IL-1β with and without stimulation for 3 h by LPS from cultures of 1×10^6 wild-type BMDC to the yield obtained from a similar number of caspase-8-deficient BMDC, and further compared it to the yield from a mixture of the two cell types (5×10^5 cells each). The IL-1β yield induced in the mixture of wild-type and caspase-8-deficient cells was identical to that produced upon incubation of the caspase-8-deficient cells alone, excluding a role for any secreted activating agent in the generation of IL-1β by the caspase-8-deficient cells.

5. CONCLUDING REMARKS

Our assessment of the function of caspase-8-deficient BMDC revealed activation of the NLRP3 inflammasome in these cells by a signaling pathway whose proximal components are indistinguishable from those known to initiate necrotic cell death. This pathway can be activated by LPS and some other TLR4 ligands. We found that it is also activated by TNF (in cells treated with SMAC (second mitochondria-derived activator of caspases)-mimetic agents to block cIAP (cellular inhibitor of apoptosis) function; Kang et al., 2013)). The activation is blocked by caspase-8, as indicated by the fact that we could observe it only in caspase-8-deficient cells or in cells treated with a caspase-8 inhibitor. It depends on the kinase function of RIPK1 (as indicated by its arrest in the presence of Nec-1) as well as on the expression of RIPK3, MLKL, and PGAM5.

The possible interrelationships between IL-1β induction and necrotic cell death in these cells should be considered with caution. As mentioned earlier, using several different techniques to assess the viability of the BMDC, we could not discern significant induction of cell death in the caspase-8-deficient BMDC under conditions where LPS effectively induced IL-1β generation. Since death did occur when the cells were treated with Z-VAD, an agent widely used to assist experimental induction of necroptosis, it seems likely that induction of necroptotic cell death and induction of inflammasome activation in BMDC by the signaling proteins listed earlier involve distinct effector mechanisms and occur independently of each other.

It should be noted, however, that whereas at least some of the techniques we applied to assess cell viability monitored changes in individual cells of the populations that we studied, both IL-1β generation and inflammasome activation were quantified by assessing overall changes in entire cell populations. Therefore, we cannot at this point exclude the possibility that IL-1β generation occurred in just a few of the cells we studied, and that in those few cells, death did occur in association with the generation of IL-1β. This possibility must be seriously considered, particularly in view of the evidence that activation of the inflammasome is indeed often associated with pyroptosis, a particular form of necrotic death (von Moltke, Ayres, Kofoed, Chavarria-Smith, & Vance, 2013). To find out whether such association occurs in the caspase-8-deficient BMDC, it will be necessary to develop techniques for monitoring IL-1β generation at the level of the individual cell, simultaneously with examining the viability of that cell.

It should be further noted, however, that irrespective of whether activation of the NLRP3 in caspase-8-deficient cells will eventually be found to be associated with death of these cells, our findings clearly exclude a causal relationship between these two processes. Several different tests that we applied seem to exclude any cellular component released on death from participating in activation of the NLRP3 inflammasome by LPS in caspase-8-deficient cells. Our data clearly indicated that activation of the NLRP3 inflammasome in the caspase-8-deficient cells occurred in a cell-autonomous manner, and thus, irrespective of whether it will eventually be found to be associated with death or, as now appears more likely, to occur without it, activation of the inflammasome by LPS in caspase-8-deficient cells is clearly not a consequence of DAMP release as a result of their necrotic death, but rather constitutes a distinct mechanism by which the proteins known to mediate necroptosis can trigger inflammation.

ACKNOWLEDGMENTS

This study was supported in part by grants from Ares Trading SA, Switzerland, a Center of Excellence Grant from the Flight Attendant Medical Research Institute (FAMRI), the Kekst Family Center for Medical Genetics, and the Shapell Family Center for Genetic Disorders Research at The Weizmann Institute of Science. D. W. is the incumbent of the Joseph and Bessie Feinberg Professorial Chair at The Weizmann Institute of Science.

REFERENCES

Boldin, M. P., Goncharov, T. M., Goltsev, Y. V., & Wallach, D. (1996). Involvement of MACH, a novel MORT1/FADD-interacting protease, in Fas/APO1- and TNF receptor-induced cell death. *Cell*, *85*, 803–815.

Caton, M. L., Smith-Raska, M. R., & Reizis, B. (2007). Notch-RBP-J signaling controls the homeostasis of CD8- dendritic cells in the spleen. *Journal of Experimental Medicine*, *204*(7), 1653–1664.

Cho, Y. S., Challa, S., Moquin, D., Genga, R., Ray, T. D., Guildford, M., et al. (2009). Phosphorylation-driven assembly of the RIP1-RIP3 complex regulates programmed necrosis and virus-induced inflammation. *Cell*, *137*(6), 1112–1123.

Degterev, A., Hitomi, J., Germscheid, M., Ch'en, I. L., Korkina, O., Teng, X., et al. (2008). Identification of RIP1 kinase as a specific cellular target of necrostatins. *Nature Chemical Biology*, *4*(5), 313–321.

Degterev, A., Huang, Z., Boyce, M., Li, Y., Jagtap, P., Mizushima, N., et al. (2005). Chemical inhibitor of nonapoptotic cell death with therapeutic potential for ischemic brain injury. *Nature Chemical Biology*, *1*(2), 112–119.

Duprez, L., Takahashi, N., Van Hauwermeiren, F., Vandendriessche, B., Goossens, V., Vanden Berghe, T., et al. (2011). RIP kinase-dependent necrosis drives lethal systemic inflammatory response syndrome. *Immunity*, *35*(6), 908–918.

Franchi, L., Munoz-Planillo, R., & Nunez, G. (2012). Sensing and reacting to microbes through the inflammasomes. *Nature Immunology*, *13*(4), 325–332.

Green, D. R., Oberst, A., Dillon, C. P., Weinlich, R., & Salvesen, G. S. (2011). RIPK-dependent necrosis and its regulation by caspases: A mystery in five acts. *Molecular Cell*, *44*(1), 9–16.

He, S., Wang, L., Miao, L., Wang, T., Du, F., Zhao, L., et al. (2009). Receptor interacting protein kinase-3 determines cellular necrotic response to TNF-alpha. *Cell*, *137*(6), 1100–1111.

Holler, N., Zaru, R., Micheau, O., Thome, M., Attinger, A., Valitutti, S., et al. (2000). Fas triggers an alternative, caspase-8-independent cell death pathway using the kinase RIP as effector molecule. *Nature Immunology*, *1*(6), 489–495.

Inaba, K., Inaba, M., Romani, N., Aya, H., Deguchi, M., Ikehara, S., et al. (1992). Generation of large numbers of dendritic cells from mouse bone marrow cultures supplemented with granulocyte/macrophage colony-stimulating factor. *Journal of Experimental Medicine*, *176*(6), 1693–1702.

Kang, T. B., Ben-Moshe, T., Varfolomeev, E. E., Pewzner-Jung, Y., Yogev, N., Jurewicz, A., et al. (2004). Caspase-8 serves both apoptotic and nonapoptotic roles. *Journal of Immunology*, *173*(5), 2976–2984.

Kang, T. B., Yang, S. H., Toth, B., Kovalenko, A., & Wallach, D. (2013). Caspase-8 blocks kinase RIPK3-mediated activation of the NLRP3 inflammasome. *Immunity*, *38*(1), 27–40.

Mankan, A. K., Kubarenko, A., & Hornung, V. (2012). Immunology in clinic review series; focus on autoinflammatory diseases: Inflammasomes: Mechanisms of activation. *Clinical and Experimental Immunology*, *167*(3), 369–381.

Mocarski, E. S., Upton, J. W., & Kaiser, W. J. (2012). Viral infection and the evolution of caspase 8-regulated apoptotic and necrotic death pathways. *Nature Reviews Immunology*, *12*(2), 79–88.

Muzio, M., Chinnaiyan, A. M., Kischkel, F. C., O'Rourke, K., Shevchenko, A., Ni, J., et al. (1996). FLICE, a novel FADD-homologous ICE/CED-3-like protease, is recruited to the CD95 (Fas/APO-1) death-inducing signaling complex. *Cell*, *85*, 817–827.

Newton, K., Sun, X., & Dixit, V. M. (2004). Kinase RIP3 is dispensable for normal NF-kappa Bs, signaling by the B-cell and T-cell receptors, tumor necrosis factor receptor 1, and Toll-like receptors 2 and 4. *Molecular and Cellular Biology*, *24*(4), 1464–1469.

Nihei, O. K., de Carvalho, A. C., Savino, W., & Alves, L. A. (2000). Pharmacologic properties of P(2Z)/P2X(7)receptor characterized in murine dendritic cells: Role on the induction of apoptosis. *Blood*, *96*(3), 996–1005.

Oberst, A., Dillon, C. P., Weinlich, R., McCormick, L. L., Fitzgerald, P., Pop, C., et al. (2011). Catalytic activity of the caspase-8-FLIP(L) complex inhibits RIPK3-dependent necrosis. *Nature*, *471*(7338), 363–367.

Piccini, A., Carta, S., Tassi, S., Lasiglie, D., Fossati, G., & Rubartelli, A. (2008). ATP is released by monocytes stimulated with pathogen-sensing receptor ligands and induces IL-1beta and IL-18 secretion in an autocrine way. *Proceedings of the National Academy of Sciences of the United States of America*, *105*(23), 8067–8072.

Rock, K. L., & Kono, H. (2008). The inflammatory response to cell death. *Annual Review of Pathology*, *3*, 99–126.

Schroder, K., & Tschopp, J. (2010). The inflammasomes. *Cell*, *140*(6), 821–832.

Smith, C. C., Davidson, S. M., Lim, S. Y., Simpkin, J. C., Hothersall, J. S., & Yellon, D. M. (2007). Necrostatin: A potentially novel cardioprotective agent? *Cardiovascular Drugs and Therapy*, *21*(4), 227–233.

Sun, L., Wang, H., Wang, Z., He, S., Chen, S., Liao, D., et al. (2012). Mixed lineage kinase domain-like protein mediates necrosis signaling downstream of RIP3 kinase. *Cell*, *148*(1–2), 213–227.

Vandenabeele, P., Galluzzi, L., Vanden Berghe, T., & Kroemer, G. (2010). Molecular mechanisms of necroptosis: An ordered cellular explosion. *Nature Reviews Molecular Cell Biology*, *11*(10), 700–714.

Varfolomeev, E. E., Schuchmann, M., Luria, V., Chiannilkulchai, N., Beckmann, J. S., Mett, I. L., et al. (1998). Targeted disruption of the mouse Caspase 8 gene ablates cell death induction by the TNF receptors, Fas/Apo1, and DR3 and is lethal prenatally. *Immunity*, *9*(2), 267–276.

Virginio, C., MacKenzie, A., Rassendren, F. A., North, R. A., & Surprenant, A. (1999). Pore dilation of neuronal P2X receptor channels. *Nature Neuroscience*, *2*(4), 315–321.

von Moltke, J., Ayres, J. S., Kofoed, E. M., Chavarria-Smith, J., & Vance, R. E. (2013). Recognition of bacteria by inflammasomes. *Annual Review of Immunology*, *31*, 73–106.

Wallach, D., Kang, T. B., Rajput, A., Kim, J. C., Bogdanov, K., Yang, S. H., et al. (2010). Anti-inflammatory functions of the "apoptotic" caspases. *Annals of the New York Academy of Sciences*, *1209*, 17–22.

Wallach, D., Kovalenko, A., & Kang, T. B. (2011). 'Necrosome'-induced inflammation: Must cells die for it? *Trends in Immunology*, *32*(11), 505–509.

Wang, Z., Jiang, H., Chen, S., Du, F., & Wang, X. (2012). The mitochondrial phosphatase PGAM5 functions at the convergence point of multiple necrotic death pathways. *Cell*, *148*(1–2), 228–243.

Zhang, D. W., Shao, J., Lin, J., Zhang, N., Lu, B. J., Lin, S. C., et al. (2009). RIP3, an energy metabolism regulator that switches TNF-induced cell death from apoptosis to necrosis. *Science*, *325*(5938), 332–336.

Zhao, J., Jitkaew, S., Cai, Z., Choksi, S., Li, Q., Luo, J., et al. (2012). Mixed lineage kinase domain-like is a key receptor interacting protein 3 downstream component of TNF-induced necrosis. *Proceedings of the National Academy of Sciences of the United States of America*, *109*(14), 5322–5327.

Zhou, Z., Han, V., & Han, J. (2012). New components of the necroptotic pathway. *Protein & Cell*, *3*(11), 811–817.

CHAPTER FOUR

Characterization of the Ripoptosome and Its Components: Implications for Anti-inflammatory and Cancer Therapy

Ramon Schilling, Peter Geserick, Martin Leverkus[1]

Section of Molecular Dermatology, Department of Dermatology, Venereology, and Allergology, Medical Faculty Mannheim, University Heidelberg, Heidelberg, Germany

[1]Corresponding author: e-mail address: martin.leverkus@medma.uni-heidelberg.de

Contents

Abstract

Most intracellular signaling cascades rely on the formation of multiprotein signaling complexes assembled in large protein signaling platforms. Especially in cell death signaling, there is a large variety of these complexes, including the apoptosome, the necrosome, or the death-inducing signaling complex (DISC), to name only a few. During the last years, a number of cellular conditions were identified that lead to the formation of another signaling platform, the so-called ripoptosome. Diverse stimuli such as genotoxic stress, death receptor or Toll-like-receptor (TLR) ligation, or degradation of cellular

Methods in Enzymology, Volume 545
ISSN 0076-6879
http://dx.doi.org/10.1016/B978-0-12-801430-1.00004-4

inhibitor of apoptosis proteins (cIAPs) are able to induce ripoptosome formation. The ripoptosome is tightly regulated by cIAPs that control intracellular RIP1 assembly and the association with other cell death-regulating proteins, most likely by ubiquitin linkage. The suppression of cIAP activity results in accumulation of RIP1 platforms that ultimately triggers necroptosis by activation of RIP3-MLKL-dependent necrosis signaling pathways. The ripoptosome is a 2-MDa protein complex, which consists of the core components caspase-8, FADD, different cFLIP isoforms, and RIP1. It represents one of the rheostats in cell death signaling, as it can activate apoptotic and necroptotic cell death responses. The specific formation and activation of the ripoptosome in cancer but not in primary cells suggests that this complex is a potential novel target for cancer or anti-inflammatory therapy, as suggested by the potential proinflammatory effects of necroptosis. Therefore, the better understanding and characterization of this signaling platform is of enormous importance for the development of novel cancer therapeutics. In this chapter, we describe several methods for purification and investigation of the ripoptosome in human cells. We also describe methods for monitoring apoptotic as well as necroptotic cell death.

1. INTRODUCTION

Multicellular organisms are able to remove excessive or potentially dangerous cells in a directed and coordinated manner. Therefore, they use a highly regulated cell death program, named apoptosis. This form of cell death has been well characterized in the past and relies mainly on the activation of cysteinyl-aspartate specific proteases (caspases). During the last years, solid evidence arose that there is also a regulated form of programmed necrosis, namely necroptosis (Geserick et al., 2009). This form of cell death is caspase independent and depends mainly on the kinase activity of the proteins RIP1 and RIP3 (Zhou, Han, & Han, 2012). Through the necroptotic pathway, RIP1 is able to phosphorylate RIP3, which then further phosphorylate the recently identified protein MLKL (mixed lineage kinase domain-like protein) (Sun et al., 2012). The role of MLKL in necroptosis is undeniable, as the knockdown or the inhibition of the kinase activity protects from necroptotic cell death (Zhao et al., 2012), but further downstream targets remain elusive to date. Although the necroptotic signaling pathway is still unclear, it was shown that this form of cell death is highly relevant in physiology, pathology, and embryonic development (Dunai, Bauer, & Mihalik, 2011). Tumorigenesis, immune surveillance of cancer cells, and pathogen-induced disease, to name only a few, are affected by the mode of cell death, meaning that the better understanding of necroptosis opens up new possibilities to target apoptosis-resistant cancer cells.

As many signaling cascades in living cells, cell death is driven by the formation of intracellular signaling complexes. During the activation of the intrinsic apoptotic pathway, cytochrome *c* is released from mitochondria and forms the apoptosome together with Apaf1 and procaspase-9 (Cain et al., 2000). When cytochrome *c* is released from the mitochondria, procaspase-9 is activated within the apoptosome and initiates further downstream propagation of the apoptotic death signal (Reubold & Eschenburg, 2012). In contrast, the extrinsic cell death activation through stimulation of death receptors, for example, CD95, TRAIL-R1, and TRAIL-R2, by their ligands, leads to the formation of the DISC. Thereby, the adapter molecule FADD is recruited along with the inactive form of caspase-8 (procaspase-8) and other modulator proteins like cFLIP (Dickens, Powley, Hughes, & MacFarlane, 2012). This platform initiates the further downstream propagation of the cell death signal, for example, through the activation of caspase-8. With the identification of the regulated necroptotic pathway, the formation of a RIP1/RIP3 signaling complex, the so called necrosome, was identified. Within this complex, RIP1 can phosphorylate RIP3 and thereby activate the protein for further downstream signaling. One central complex at the crossroad between necroptotic and apoptotic death signaling is the so-called ripoptosome (Bertrand & Vandenabeele, 2011). This complex consists of the core components FADD, caspase-8, cFLIP, and the name giving RIP1. The formation of the ripoptosome is driven by DNA damage, the accumulation of the kinase RIP1, or the depletion of inhibitors of apoptosis proteins (cIAPs) through genotoxic stress, the addition of SMAC mimetics, or treatment with etoposide/tenoposide (Tenev et al., 2011). The formation of the ripoptosome is highly dependent on the balance between the intracellular levels of cIAPs and RIP1 (Feoktistova et al., 2011). Thereby, cIAPs function as negative regulators of the complex as they ubiquitylate RIP1 and mark it for proteasomal degradation. Interestingly, the ripoptosome can lead to apoptotic as well as necroptotic cell death depending on the stoichiometric composition (Feoktistova, Geserick, Panayotova-Dimitrova, & Leverkus, 2012) and thereby the activation of the respective enzymatic or kinase activities within the complex (Fig. 4.1).

High levels of procaspase-8 promote apoptosis because they allow the self-cleavage into its active form. In parallel, RIP1 is cleaved and thereby inactivated, resulting in the dissolution of the ripoptosome. The active caspase-8 is now able to cleave Bid (a member of the Bcl-2 family) to truncated Bid, which activates the release of proapoptotic proteins

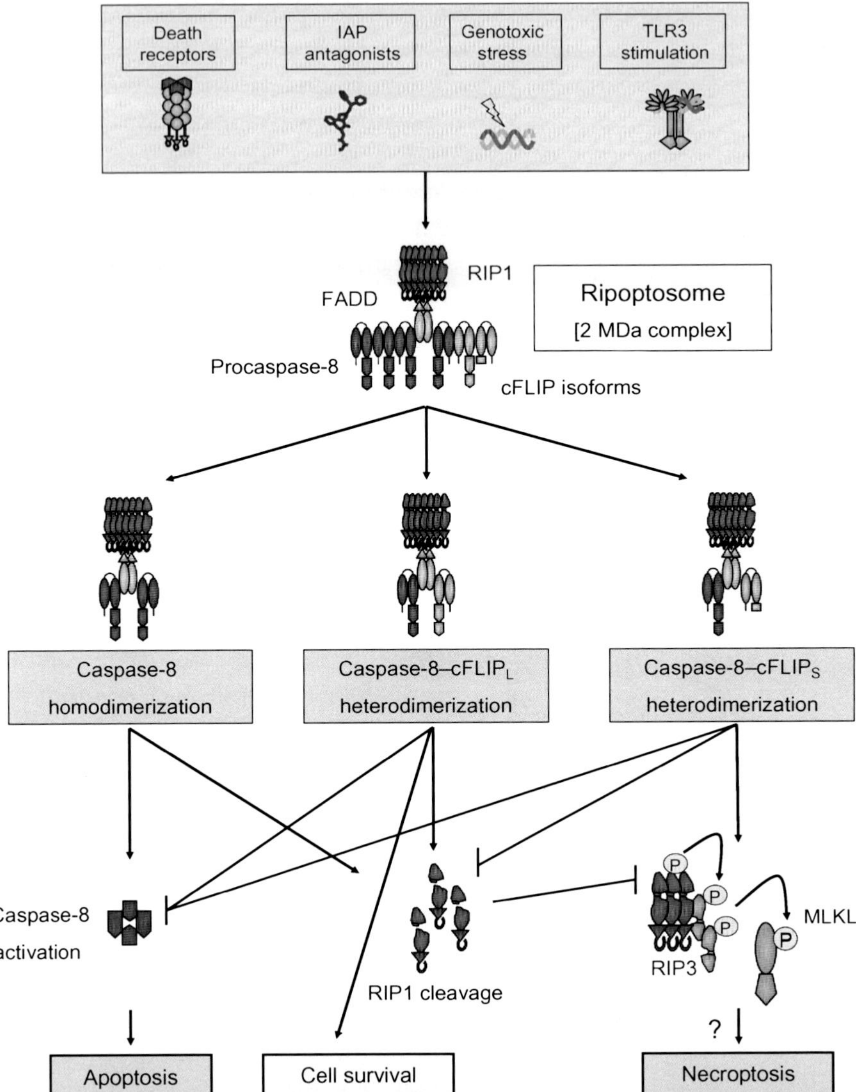

Figure 4.1 The stoichiometric composition of the ripoptosome determines the mode of cell death. Depletion of cIAPs, genotoxic stress, death receptor activation, or TLR3 stimulation can lead to the formation of the ripoptosome, consisting of the proteins FADD, caspase-8, RIP1, and cFLIP isoforms. Within the ripoptosome RIP1 and caspase-8 assemble in chains. Thereby, the stoichiometric composition determines the downstream signaling. Caspase-8 homodimerization results in the propagation of apoptosis. cFLIP$_L$ caspase-8 heterodimerization leads to RIP1 cleavage and thereby promotes cell survival. The heterodimerization of caspase-8 and cFLIP$_S$ blocks apoptotic cell death and leads to increased level of active RIP1, which can then phosphorylate RIP3, resulting in MLKL-dependent necroptotic cell death. (See the color plate.)

like SMAC/DIABLO or cytochrome *c* from the mitochondria. Caspase-8 can also directly cleave caspase-3, which leads to apoptotic execution (Hengartner, 2000). On the other hand, the formation of a $cFLIP_L$/caspase-8 heterodimer within the ripoptosome leads to a full inhibition of cell death. The heterodimer shows enough protease activity to cleave and inactivate RIP1 but not to activate procaspase-8, resulting in a blockage of the apoptotic and necroptotic pathways. Under conditions where high level of $cFLIP_S$ enables a heterodimerization with procaspase-8, apoptotic cell death is blocked and necroptosis is promoted in the absence of active cIAPs. The heterodimer is not able to either self-cleave caspase-8 or inactivate RIP1, because it lacks protease activity. Thereby, RIP1 is stabilized and is able to interact with RIP3 to promote necroptosis (Declercq, Vanden Berghe, & Vandenabeele, 2009).

The pathological consequences of necroptosis have become evident in a number of *in vivo* models. It was possible to show that the embryonic lethality of caspase-8 deficiency could be rescued with the additional deletion of RIP3, which is the key player of necroptotic cell death (Kaiser et al., 2011; Oberst et al., 2011). One common hypothesis is that necroptosis has developed as a backup mechanism to kill the cells whenever apoptosis is blocked. Therefore, a better understanding of the necroptotic cell death pathway and the role of the ripoptosome within this cell death process, as well as the composition and function of this signaling platform, is of great interest for the development of novel cancer therapeutics. Here, we demonstrate different methods for analyzing ripoptosome formation in human cells, the composition of this protein complex, as well as the functional relevance of necroptotic cell death *in vitro*.

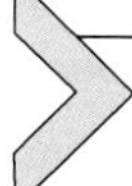

2. THE RIPOPTOSOME: CELLULAR MODEL SYSTEMS TO STUDY ITS FORMATION

The relevance of ripoptosome formation in cancer therapy is undeniable. As an example, a recent report showed that CLL cells lack sensitivity to IAP antagonists because of the inability to form a ripoptosome (Maas et al., 2013). One intriguing possibility is that high levels of cFLIP in CLL cells could be responsible for the failure to form a ripoptosome. Therefore, cellular model systems that enable studies of ripoptosome formation, composition, and disassembly are important tools for further examination of this death signaling platform. The ripoptosome can be formed through a variety of cellular conditions like DNA damage and genotoxic stress as well as extracellular stimuli like TLR3 activation through viral infections, the binding of death ligands to their receptors, or IAP depletion by drugs like tenoposide/etoposide or SMAC mimetics. In this chapter, we describe possibilities of ripoptosome induction in skin tumor cells as well as functional methods for the quantitative and qualitative analysis of ripoptosome-dependent cell death.

2.1. Induction of ripoptosome formation by IAP antagonists (SMAC mimetics)

IAP antagonists, which are also called SMAC mimetics, are small compounds which enable the proteasomal degradation of cIAP1/2 by autoubiquitination (Varfolomeev et al., 2007; Vince et al., 2007). These proteins are intracellular regulators of RIP1 kinase, as they most likely constitutively mark this protein for proteasomal degradation via ubiquitination, although other forms of ubiquitination have been described (Bertrand et al., 2011; Moulin et al., 2012). As this lack of ubiquitination was most convincingly shown within the TNF complex I (Emmerich, Schmukle, & Walczak, 2011; Feoktistova et al., 2011; Tenev et al., 2011), the depletion of cIAP1/2 by IAP antagonists results in an increase in nonubiquitinated cellular RIP1, leading to spontaneous formation of the ripoptosome in cells expressing $cFLIP_S$. Thereby, the cells are sensitized to TLR agonist (Poly (I:C)) or death ligand (TNF or CD95L)-induced cell death.

2.1.1 Experimental procedure

The cells of interest are prestimulated for 1 h with 100 n*M* of an IAP antagonist. In our studies, we have used compound A (provided by Tetralogics Corp., Malvern, PA, USA; Vince et al., 2007) in combination with 10 μ*M* of the caspase inhibitor zVAD-fmk. With this treatment, a dramatic loss of cellular IAP1/2 levels is reached, and the ripoptosome is stabilized, most likely by protection from caspase-dependent apoptosis. Afterward, a death ligand of interest, in our studies, for example, TNF or CD95L or the TLR3 agonist Poly I:C is added, and the cells are incubated for 4–6 h at 37 °C. The appropriate concentration and incubation time of the death ligand has to be determined for each death-inducing compound and cell line. After a 4–6 h incubation period, in most transformed cell lines examined, the intracellular ripoptosome formation is increased, and cells can be used for further analysis (Fig. 4.2).

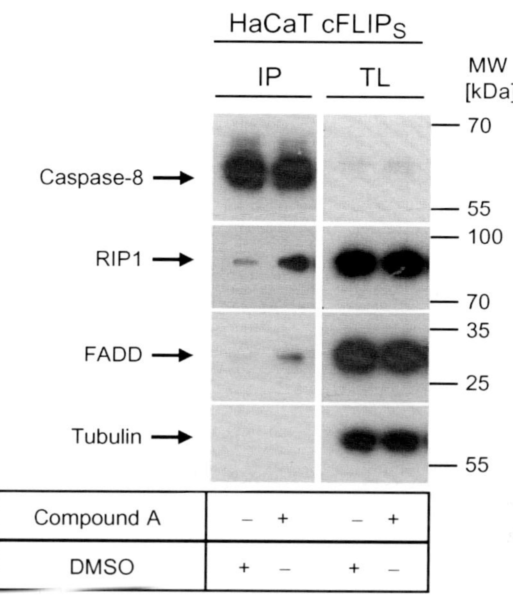

Figure 4.2 Ripoptosome induction after SMAC-mimetic treatment in cFLIP$_S$-expressing cells. HaCaT cells expressing cFLIP$_S$ were treated with the SMAC-mimetic compound A (Tetralogics) or DMSO as control for 4 h at 37 °C. Afterward the cells were lysed and a caspase-8 IP was performed. The total lysate and IP samples were analyzed by SDS-PAGE and Western blot.

2.2. Induction of ripoptosome formation via RIP1 over expression

An alternative method for ripoptosome induction is a stoichiometric change in the balance of RIP1 and cIAPs. To achieve this, we have established a model system in which inducible overexpression of wild-type RIP1 kinase can be achieved. As the constitutive over expression of this kinase is lethal, we took advantage of the previously published lentiviral inducible vector system (Diessenbacher et al., 2008; Vince et al., 2007). With these cells (Feoktistova et al., 2012), the stoichiometry of RIP1 and cIAPs is rapidly shifted toward increased RIP1 levels, whenever 4-hydroxytamoxifen (4-HT) is added. This results in the spontaneous formation of the ripoptosome. To inhibit apoptotic cell death induced by RIP1 expression and to stabilize the complex, it is necessary to interfere with caspase-dependent cell death by using the caspase inhibitor zVAD-fmk as described in the previous chapter.

2.2.1 Experimental procedure

The cells are prestimulated with 10 μ*M* zVAD-fmk for 1 h. Afterward RIP1 protein expression is induced by the addition of 100 n*M* 4-HT, and the cells are incubated for 4–6 h, depending on the experimental question, to allow complex formation (Fig. 4.3).

2.3. Quantitative and qualitative analysis of ripoptosome-mediated cell death

2.3.1 Screening test: Crystal violet cell death assay

The staining of adherent cells via crystal violet is commonly used in molecular biology. This method can be used as an indirect measurement of cell death in attached cells, as dying cells loose their adherence to the plate through cell death processes. This method is easy and reliable to monitor

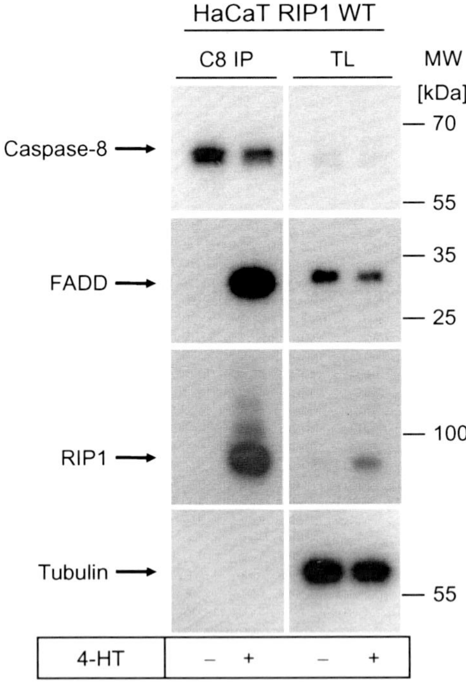

Figure 4.3 Ripoptosome formation in HaCaT cells upon inducible RIP1 expression. Inducible RIP1 HaCaT cells were prestimulated with 10 μ*M* of zVAD-fmk, followed by 6 h incubation with 100 n*M* 4-hydroxytamoxifen. The cells were lysed, total lysate (TL) samples were taken, and 4 mg of the remaining protein lysate was used for a caspase-8 IP. The samples were separated by SDS-PAGE followed by Western blot analysis.

the influence of numerous compounds such as chemotherapeutic drugs, IAP antagonists, or other compounds to investigate cell survival. The disadvantage of the crystal violet assay is that this assay cannot distinguish between cell survival and inhibition of proliferation. However, for the distinction between cell survival and lack of proliferation, additional methods (described in later chapters) can easily be used.

2.3.1.1 Experimental procedure

Depending on the cell type, $1–2 \times 10^4$ cells are seeded per well of a 96-well plate and incubated for 12–24 h at 37 °C. The medium is discarded, and the cells are stimulated with the compound of interest for 12–24 h at 37 °C. The medium is again aspired, and the cells are washed 2 × with 200 µl PBS per well. After washing, the plate is inversed on a filter paper. Flap the plate slightly to get rid of the remaining liquid.

Note: Some cells (e.g., keratinocytes) adhere to the surface of cell culture plates quite robustly, although others (like HeLa or 293T) are easily detached from the plastic surface. Thus, the washing step and flapping of the plate increases the specificity of the assay but can be skipped in case cells do not adhere well to the plastic plate.

Afterward 50 µl of crystal violet solution (0.5 g crystal violet powder in 80 ml water and 20 ml methanol) is added in each well, and the plate is incubated for 20 min at room temperature by tumbling. The plate is then washed 4 × with a stream of tap water. The plate is dried for at least 2 h at room temperature with open lid. Afterward 200 µl of methanol is added to each well to resolve the crystal violet dye, and the plate is again incubated for 20 min by tumbling. The OD of each well is measured with a plate reader (OD_{570nm}). The OD of the nonstimulated cells is set to 100% and compared with the stimulated samples.

2.3.2 Specific test to assay the quality of cell death: Annexin V/propidium iodide staining followed by FACS analysis

One early event of apoptotic cell death is the translocation and therefore the externalization of phosphatidylserine from the inner to the outer plasma membrane. This phospholipid switch can be visualized by the staining with the phospholipid-binding protein Annexin V-Cy5, which has a high affinity for phosphatidylserine. Additionally, the cells are stained with the standard flow cytometric viability probe propidium iodide (PI) to distinguish between dead and living cells. PI is not able to cross the membranes of viable cells, whereas the membranes of dying and damaged cells are permeable for PI that allows influx into the cell and staining of the DNA. The outcome can easily be measured by flow cytometry. Cells that stain positive for Annexin V and negative for PI are undergoing early apoptosis. Cells that are positive for PI are considered dead (secondary necrosis), undergoing necrosis, necroptosis, or late stages of apoptosis. It is important to study the kinetics of the cell death process to get insight what is the initial cell biological process, as secondary necrosis is, at late stages, not distinguishable from late apoptotic cell death.

2.3.2.1 Experimental procedure

Depending on the cell type, 2–4 × 10^5 cells are seeded in one well of a six-well plate. *Note*: In order to calibrate your FACS machine, the FACS settings of unstained, only Annexin V stained (stimulated and unstimulated), and only PI stained (stimulated and unstimulated) cells are needed. On the second day, the cells are stimulated with the apoptosis inducing compound of interest for 3–4 h at 37 °C. The cells are trypsinized and washed two times with PBS. Afterward they are washed with 1 × Annexin V-Cy5 binding buffer (10 × Annexin V-Cy5 binding buffer: 0.1 *M* HEPES pH 7.4, 1.4 *M* NaCl, 25 m*M* $CaCl_2$); 1 × 10^6 cells/ml are resuspended in Annexin V-Cy5 binding buffer. Then, 5 μl of Annexin V-Cy5 (BD Pharmingen, cat no. 559934) and 5 μl of PI solution (Sigma, cat no. P4864) are added to 100 μl of cell suspension. The cells are vortexed immediately and incubated at room temperature and light protected for 15 min. Finally, 400 μl of Annexin V-Cy5 binding buffer is added, and the cells are analyzed by flow cytometry.

2.3.3 Morphological analysis: Sytox Green/Hoechst staining followed by fluorescent microscopy

During late stages of cell death, dying cells lose their cell membrane integrity. This allows the penetration of different dyes like PI or other DNA-binding fluorochromes such as Hoechst and SYTOX Green to enter these cells and thereby stain them. As these dyes have different penetration abilities, it is possible to distinguish between different phases of apoptosis, necroptosis, and viable cells by visualizing the staining under a fluorescent microscope. Moreover, microscopic analysis at the single-cell level allows obtaining more extensive morphological information of the cell death. In this respect, kinetic or even time-lapse methodology can be applied to this easy method to stain dying cells, excellently described elsewhere (Wallberg, Tenev, & Meier, 2013).

2.3.3.1 Experimental procedure

The $3–5 \times 10^4$ cells (depending on the cell type, confluence should be 70–80% the following day) are seeded in one well of a 12-well plate and incubated at 37 °C for 24 h. On the second day, the cells are stimulated with the compounds of interest, according to the experimental aims. Then, Hoechst 33342 (5 μg/ml) and SYTOX® Green (5 p*M*) are added to the cells and incubated for 15 min at 37 °C. The cells should now immediately be analyzed under the fluorescent microscope.

2.4. Signaling pathway analysis: siRNA knockdown of target proteins involved in ripoptosome-mediated cell death

One common method to study the influence of certain proteins in ripoptosome-mediated cell death is the systematic knockdown of these target genes. Known critical proteins are the adapter molecule FADD that allows binding between RIP1 and caspase-8 as well as caspase-8 that influences RIP1 cleavage after activation. The different cFLIP isoforms are also critically involved in regulation of the ripoptosome activity and therefore for cell death outcome. After the depletion of the proteins, the cells can be further analyzed for changes in cell death responses compared with control cells. Here, we describe the knockdown via siRNAs specific for the target gene of interest via reverse transfection. It is recommended to test a number of siRNAs for a specific target as it might well be possible that some siRNAs are mediating off-target side effects (Horn & Boutros, 2013).

2.4.1 Experimental procedure

For reverse transfection, cells in optimal growth phase are prepared so that it is possible to seed 2×10^5 cells for each transfection in one well of a six-well plate (always transfect one well with scrambled siRNA as control). On the day of transfection, for each well 5 μl of lipofectamine 2000 (Invitrogen) is mixed with 300 μl DMEM without any supplements (*Note*: No FCS in the medium as this inhibits lipofectamine and siRNA complex formation) and incubated for 10 min at room temperature. Meanwhile, 5 μl of each 20 μ*M* siRNA stock you wish to transfect is mixed with 300 μl DMEM without any supplements. After the 10 min incubation, lipofectamine and siRNA solution are combined, and the total of 600 μl is pipetted in the corresponding well of the six-well plate and incubated for 30 min at room temperature. Meanwhile, a single-cell suspension of the cells you wish to transfect is prepared and 2×10^5 cells in 1.4 ml DMEM with FCS are added to the siRNA lipofectamine solution. The next day the medium is changed as lipofectamine is toxic to the cells during longer incubation times. After 48 h incubation, the knockdown generally achieved should be sufficient and the cells can be used in additional cell death assays as mentioned earlier.

3. BIOCHEMICAL ANALYSIS OF THE RIPOPTOSOME: ANALYSIS OF RIPOPTOSOME FORMATION AND IDENTIFICATION OF NOVEL COMPONENTS VIA IMMUNOPRECIPITATION AND MASS SPECTROMETRY

The current evidence suggests the ripoptosome as a FADD, caspase-8, cFLIP, and RIP1-containing signaling complex. Although these proteins are the core components, it is likely that, based on the high molecular weight of the complex in gel-filtration analysis (Feoktistova et al., 2012; Tenev et al., 2011), additional compounds which are so far not described are components of the complex and regulate the ripoptosome. For the better understanding of the structure and composition of the ripoptosome, coimmunoprecipitations are the commonly used methods. In these IPs, the stability of the protein–protein interaction within the complex is utilized for the copurification of the whole complex based upon the isolation of one compound by a specific primary antibody. In this chapter, we describe different ways of IPs combined with other purification methods, which can be used for the isolation of the ripoptosome and associated proteins.

3.1. Caspase-8 immunoprecipitation

As described earlier, a number of intracellular and extracellular stimuli induce the formation of the ripoptosome as a DISC. In this chapter, we describe how the formation of this complex can be tested via a caspase-8 antibody-mediated purification from cellular lysates, followed by the analysis of the complex and copurified proteins by SDS-PAGE and Western blotting (Fig. 4.2).

3.1.1 Experimental procedure

Cells are seeded in one T175 cell culture flask (use one flask per experimental condition) and grown to a confluence of 80–90%. The cells are stimulated according the experimental aim. A possible stimulation for ripoptosome formation is, for example, the prestimulation with 100 n*M* IAP antagonist for

1 h followed by 6 h incubation with Poly I:C (final conc. 20 μg/μl) (Feoktistova et al., 2011). After the stimulation, the cells are collected via trypsinization and washed once with PBS. The samples are then lysed with 1.5 ml of DISC lysis buffer (30 m*M* Tris–HCl, pH 7.5, 120 m*M* NaCl, 10% glycerol, 1% Triton X-100, complete protease inhibitor cocktail (Roche)) for 45 min on ice. To get rid of the cell debris, the samples are centrifuged for 30 min at 20,000 × *g*. Afterward the protein concentration of the cleared lysate is determined by a Bradford protein assay. For Western blot analysis, 100 μg of each sample is adjusted with DISC lysis buffer and gel-loading buffer (Laemmli buffer) to a final protein concentration of 1 μg/μl as total lysate sample. The samples have to be boiled for 10 min at 95 °C to denature the proteins before they can be loaded on a SDS Gel. For the caspase-8 immunoprecipitation (IP), 40 μl IgG Agarose beads slurry is prepared for each sample. The beads are washed 3 × with DISC lysis buffer to equilibrate the bead matrix. Afterward 2–4 mg of the cleared protein lysate (it is eligible to use as much protein as possible, as this will increase the complex signal for the Western blot analysis) is added to the beads. Finally, 5 μl (conc. 200 μg/μl) of the caspase-8 antibody is added to each sample and the binding is performed over night at 4 °C by rotating.

On the following day, the beads are collected via centrifugation for 2 min with 400 × *g*. The supernatant is discarded and the beads are washed 4 × with DISC lysis buffer. After the last washing step, the liquid is completely aspirated with a cannula and the beads are resuspended in 1 × Laemmli buffer. The samples are then boiled for 10 min to dissociate the bound proteins. The samples can then be loaded on a SDS Gel for Western blot analysis of the constituents of the induced complex.

3.2. Ripoptosome purification by tandem-affinity purification

As described in the previous chapter, IP is a common tool for the purification and analysis of signaling complexes like the ripoptosome. The isolation of the ripoptosome through a caspase-8 IP is robustly working in living cells and published in detail elsewhere (Geserick et al., 2009). Under most conditions, caspase-8 co-IP is sufficient for most experimental questions. However, it might be possible that detection of low-abundance proteins within the complex requests for a higher purity of the signaling complex. Therefore, it is possible to perform a two-step or so-called tandem-affinity purification of the complex. For this procedure, it is important that the protein of interest or another component of the complex of interest is linked to a cleavable tag. During the first step of the purification, the IP is directed against a cleavable tag, usually performed comparable to the procedure described in the previous paragraph. Following the successful purification of tag-bound precipitate, the gel matrix is incubated with a protease able to digest the tag, as previously published (Haas et al., 2009). The eluates of the beads can now be used in a second round of affinity purification in order to reach a higher purity. The resulting precipitate is then used for mass spectrometric analysis, Western blot analysis, or other downstream applications.

3.3. Complex gel filtration combined with mass spectrometry

In order to determine the composition of protein complexes in general and particularly the compounds of the ripoptosome, the analysis of the purified complex by mass spectrometry is indispensable. For this method, it is important that the isolated samples are of great purity, as contaminations will lead to false positive results. Gel filtration allows the separation of proteins and protein complexes by size, as published for the ripoptosome (Feoktistova et al., 2011; Tenev et al., 2011). The idea of this methodology is that small molecules can enter the pores of the gel matrix whereas the big protein complexes cannot and will therefore elute faster, allowing for protein complex separation under functionally distinct conditions (Hughes, Langlais, Cain, & MacFarlane, 2013). Thereby, the ripoptosome can be separated from smaller complexes as well as from the free pool of its components, like RIP1 or caspase-8, found under nonstimulated conditions (Hughes et al., 2013). To isolate the ripoptosome from other high molecular weight complexes, additional IP steps are required to generate a highly purified complex which can then be analyzed via mass spectrometry. Details for this methodology can be found elsewhere (Hughes et al., 2013).

3.4. Advantages and disadvantages of these procedures

All described methods are important and reliable tools in order to study the formation, composition, and functionality of the ripoptosome. The advantages and disadvantages of the different methods rely mainly on the scientific question, especially on the required purity for the experimental aim. The described caspase-8 IP enables the easiest and fastest way to test different cellular conditions or stimulations for their ability to form the ripoptosome. This method is sufficient for most experimental procedures like SDS-PAGE and Western blot analysis. However, one has to keep in mind that there is always a certain contamination within the purified sample. For higher purity, the tandem-affinity purification is used. But the purification process is more complicated and will lead to a stronger loss of sample with every additional purification step. Thus, these technologies may not be possible with spurious cell samples, for example, from primary cells.

The advantage of the gel filtration is that this method enables the highest complex pureness which makes it suitable for mass spectrometric analysis in order to identify each component of the ripoptosome (Hughes et al., 2013). The general principle of this methodology has been published elsewhere (Hughes et al., 2013). The disadvantage of this method is its complexity and that a certain expertise is needed to perform such purifications. Taken together, all different described methods are suitable for other tasks to further delineate assembly, composition, stability, enzymatic activity, and downstream cell biological events. Only the combination of all of them allows to further study the composition and function of the ripoptosome as a fascinating cell DISC.

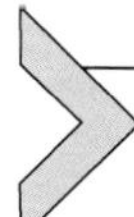

4. OUTLOOK: FUTURE IMPLICATIONS OF THE FUNCTION AND REGULATION OF THE RIPOPTOSOME

The ripoptosome as a FADD–caspase-8-cFLIP–RIP1-containing cytoplasmic complex was first identified during CD95-induced cell death after depletion of cIAPs by IAP antagonists (Geserick et al., 2009). When further experiments revealed that this signaling platform is also forming in the cytoplasm after TLR3 stimulation (Feoktistova et al., 2011) or genotoxic stress through treatment with chemotherapeutics (e.g., etoposides) (Tenev et al., 2011), it became apparent that it is a rather general stress-responsive signaling platform that regulates cell death pathways (Bertrand & Vandenabeele, 2011). As the proteins found in association with the ripoptosome, namely FADD, caspase-8, RIP1, and cFLIP, are also constituents of the TNF-induced complex II, it could be postulated that the ripoptosome is induced by numerous stimuli. However, ripoptosome formation is not influenced in cells that lack TNFR1 expression or by inhibition of TNF signaling with TNF-blocking receptor fusion proteins. These findings support the idea that the ripoptosome is an independent signaling platform. However, it remains an important question, what is the physiological function of the ripoptosome? All studies thus far connect the ripoptosome to either apoptotic or necroptotic cell death. In extrinsic apoptotic cell death pathways, we suggest that the ripoptosome might be preformed in the cytosol whenever cIAPs are lost, and may then be secondarily recruited to the death receptor. This facilitated recruitment may allow rapid caspase-8 cleavage and thereby a fast downstream propagation of the apoptotic death signal. On the other hand, it might also be possible that the ripoptosome is a sole cytosolic phenomenon which allows an accelerating transduction of the apoptotic death signal after receptor stimulation, and the detected association is secondary to a strong intracellular association of RIP1 with caspase-8 (Cullen et al., 2013). Whenever the apoptotic cell death pathway is blocked, the cell can die through necroptosis which we believe to function as a backup mechanism. Here, the ripoptosome is also one of the key regulators; whenever the cleavage of caspase-8 and RIP1 within the complex is inhibited, the active (phosphorylated) form of RIP1 can accumulate and phosphorylate its downstream target RIP3. This leads to the further propagation of the death signal via MLKL phosphorylation and necroptotic execution (Murphy et al., 2013; Sun et al., 2012).

Several *in vivo* studies have indicated the physiological relevance of ripoptosome-associated proteins in embryonic development. The knockout of caspase-8 (Varfolomeev et al., 1998), FADD (Zhang, Cado, Chen, Kabra, & Winoto, 1998), or RIP1 (Kelliher et al., 1998) is embryonic lethal. Intriguingly, FADD-deficient mice can be rescued through the additional loss of RIP1, arguing for direct interaction of these two molecules (Zhang et al., 2011). Studies with caspase-8 knockout mice could convincingly show that additional loss of the necroptosis-relevant RIP3 can rescue the lethality of caspase-8 deficiency (Kaiser et al., 2011; Oberst et al., 2011). Taken together, these findings support the important role of these ripoptosome core components for embryonic development as well as downstream cell death signaling, although future studies will have to elucidate under which conditions the cell biological response of necroptosis is of relevance *in vivo*. As we showed that the ripoptosome is highly regulated by the different cFLIP isoforms (Geserick et al., 2009), the stoichiometric balance of cFLIP isoforms dictates the extent of necroptosis (promoted by $cFLIP_S$) or apoptosis and its inhibition (by $cFLIP_L$). It has been long known that cFLIP knockout mice died through embryogenesis at E10.5 similar to caspase-8 knockout animals (Varfolomeev et al., 1998; Yeh et al., 2000).

Recent mouse studies have further highlighted the critical role of cFLIP for cell death regulation and have investigated the role of FADD, caspase-8, and cFLIP during embryogenesis. It was shown that these three proteins have a survival function during embryogenesis (Dillon et al., 2012).

We surprisingly found that constitutive knockout of cFLIP in the skin also results in embryonic lethality around E10–E11. Studies using an acute inducible knockout of cFLIP in the skin showed a rapid and severe inflammatory response that finally leads to apoptosis-dependent destruction of the skin through autocrine TNF induction (Panayotova-Dimitrova et al., 2013; Weinlich et al., 2013). These data highlight the critical role of cFLIP as a ripoptosome-regulating protein in cell death regulation and possibly inflammatory processes in the skin, other epithelial tissues (Wittkopf et al., 2013), and during embryogenesis.

One striking feature of cancer cells is resistance toward apoptotic and possibly necroptotic cell death through deregulation of the respective signaling pathways. Here, the ripoptosome is a potential target for new therapies as the complex forms predominantly in cancer cells but not in primary cells (Feoktistova et al., 2011). Furthermore, it was shown that many cancer cell lines, for example, glioblastoma cells (Wagner et al., 2013), HeLa cells (Feoktistova et al., 2011), or SCC cells (Geserick et al., 2009), could be sensitized to chemotherapeutics through IAP antagonist or genotoxic stress-induced ripoptosome formation. Of interest, it was shown that other types of cancer such as breast and prostate show resistance to cell death induction. In these types of cancer, treatment with the death ligand TRAIL requires cotreatment with IAP antagonist to allow for cell death induction (Lu et al., 2011). All these findings underline the important role of the ripoptosome in SMAC-mimetic-induced apoptosis and necroptosis. We speculate that induction of the ripoptosome can specifically sensitize treatment-resistant cancer cells to cell death. It might well be that the preclinical therapeutic success of SMAC mimetics is mediated at least in part by ripoptosome-induced cell death.

Although the core components of the ripoptosome seem to be identified, it is not clear if there are additional proteins within this complex. Therefore, future work might focus on the identification of novel components of the ripoptosome or proteins which are only transiently interacting with this signaling complex. These interaction partners might even play major roles in other signaling pathways which are so far not directly connected to the ripoptosome, for example, interferon signaling, NF-κB activation, or inflammatory responses. The exact determination of the composition and the interacting protein of the ripoptosome are of great interest to understand the regulatory function of this complex for cell death activation or inhibition and finally the role in pathophysiological disorders. The second most striking future implication is the usage of the ripoptosome as a model system to study necroptotic cell death. As described earlier, necroptosis is very important for developmental processes, inflammatory responses, and activation of the immune system to overcome cancer. It could be speculated and remains to be formally tested that necroptosis induction in tumor cells is a way to overcome immune evasion during tumorigenesis. However, the downstream events after MLKL phosphorylation through RIP3 are still unclear. With ripoptosome induction and interference with apoptosis signaling that may well occur *in vivo*, the downstream events of necroptotic cell death could be further studied. Taken together, numerous open questions remain to be answered concerning the ripoptosome, its downstream signaling, and most importantly its physiological and pathophysiological relevance in mice and man.

REFERENCES

Bertrand, M. J., Lippens, S., Staes, A., Gilbert, B., Roelandt, R., De Medts, J., et al. (2011). cIAP1/2 are direct E3 ligases conjugating diverse types of ubiquitin chains to receptor interacting proteins kinases 1 to 4 (RIP1-4). *PLoS One, 6*, e22356.

Bertrand, M. J., & Vandenabeele, P. (2011). The Ripoptosome: Death decision in the cytosol. *Molecular Cell, 43*, 323–325.

Cain, K., Bratton, S. B., Langlais, C., Walker, G., Brown, D. G., Sun, X. M., et al. (2000). Apaf-1 oligomerizes into biologically active approximately 700-kDa and inactive approximately 1.4-MDa apoptosome complexes. *Journal of Biological Chemistry, 275*, 6067–6070.

Cullen, S. P., Henry, C. M., Kearney, C. J., Logue, S. E., Feoktistova, M., Tynan, G. A., et al. (2013). Fas/CD95-induced chemokines can serve as "find-me" signals for apoptotic cells. *Molecular Cell, 49*, 1034–1048.

Declercq, W., Vanden Berghe, T., & Vandenabeele, P. (2009). RIP kinases at the crossroads of cell death and survival. *Cell, 138*, 229–232.

Dickens, L. S., Powley, I. R., Hughes, M. A., & MacFarlane, M. (2012). The 'complexities' of life and death: Death receptor signalling platforms. *Experimental Cell Research, 318*, 1269–1277.

Diessenbacher, P., Hupe, M., Sprick, M. R., Kerstan, A., Geserick, P., Haas, T. L., et al. (2008). NF-kappaB inhibition reveals differential mechanisms of TNF versus TRAIL-induced apoptosis upstream or at the level of caspase-8 activation independent of cIAP2. *Journal of Investigative Dermatology, 128*, 1134–1147.

Dillon, C. P., Oberst, A., Weinlich, R., Janke, L. J., Kang, T. B., Ben-Moshe, T., et al. (2012). Survival function of the FADD-CASPASE-8-cFLIP(L) complex. *Cell Reports, 1*, 401–407.

Dunai, Z., Bauer, P. I., & Mihalik, R. (2011). Necroptosis: Biochemical, physiological and pathological aspects. *Pathology and Oncology Research, 17*, 791–800.

Emmerich, C. H., Schmukle, A. C., & Walczak, H. (2011). The emerging role of linear ubiquitination in cell signaling. *Science Signaling, 4*, re5.

Feoktistova, M., Geserick, P., Kellert, B., Dimitrova, D. P., Langlais, C., Hupe, M., et al. (2011). cIAPs block Ripoptosome formation, a RIP1/caspase-8 containing intracellular cell death complex differentially regulated by cFLIP isoforms. *Molecular Cell, 43*, 449–463.

Feoktistova, M., Geserick, P., Panayotova-Dimitrova, D., & Leverkus, M. (2012). Pick your poison: The Ripoptosome, a cell death platform regulating apoptosis and necroptosis. *Cell Cycle*, *11*, 460–467.

Geserick, P., Hupe, M., Moulin, M., Wong, W. W., Feoktistova, M., Kellert, B., et al. (2009). Cellular IAPs inhibit a cryptic CD95-induced cell death by limiting RIP1 kinase recruitment. *Journal of Cell Biology*, *187*, 1037–1054.

Haas, T. L., Emmerich, C. H., Gerlach, B., Schmukle, A. C., Cordier, S. M., Rieser, E., et al. (2009). Recruitment of the linear ubiquitin chain assembly complex stabilizes the TNF-R1 signaling complex and is required for TNF-mediated gene induction. *Molecular Cell*, *36*, 831–844.

Hengartner, M. O. (2000). The biochemistry of apoptosis. *Nature*, *407*, 770–776.

Horn, T., & Boutros, M. (2013). Design of RNAi reagents for invertebrate model organisms and human disease vectors. *Methods in Molecular Biology*, *942*, 315–346.

Hughes, M. A., Langlais, C., Cain, K., & MacFarlane, M. (2013). Isolation, characterisation and reconstitution of cell death signalling complexes. *Methods*, *61*, 98–104.

Kaiser, W. J., Upton, J. W., Long, A. B., Livingston-Rosanoff, D., Daley-Bauer, L. P., Hakem, R., et al. (2011). RIP3 mediates the embryonic lethality of caspase-8-deficient mice. *Nature*, *471*, 368–372.

Kelliher, M. A., Grimm, S., Ishida, Y., Kuo, F., Stanger, B. Z., & Leder, P. (1998). The death domain kinase RIP mediates the TNF-induced NF-kappaB signal. *Immunity*, *8*, 297–303.

Lu, J., McEachern, D., Sun, H., Bai, L., Peng, Y., Qiu, S., et al. (2011). Therapeutic potential and molecular mechanism of a novel, potent, nonpeptide, Smac mimetic SM-164 in combination with TRAIL for cancer treatment. *Molecular Cancer Therapeutics*, *10*, 902–914.

Maas, C., Tromp, J. M., Van Laar, J., Thijssen, R., Elias, J. A., Malara, A., et al. (2013). CLL cells are resistant to smac mimetics because of an inability to form a ripoptosome complex. *Cell Death and Disease*, *4*, e782.

Moulin, M., Anderton, H., Voss, A. K., Thomas, T., Wong, W. W., Bankovacki, A., et al. (2012). IAPs limit activation of RIP kinases by TNF receptor 1 during development. *EMBO Journal*, *31*, 1679–1691.

Murphy, J. M., Czabotar, P. E., Hildebrand, J. M., Lucet, I. S., Zhang, J. G., Alvarez-Diaz, S., et al. (2013). The pseudokinase MLKL mediates necroptosis via a molecular switch mechanism. *Immunity*, *39*, 443–453.

Oberst, A., Dillon, C. P., Weinlich, R., McCormick, L. L., Fitzgerald, P., Pop, C., et al. (2011). Catalytic activity of the caspase-8-FLIP(L) complex inhibits RIPK3-dependent necrosis. *Nature*, *471*, 363–367.

Panayotova-Dimitrova, D., Feoktistova, M., Ploesser, M., Kellert, B., Hupe, M., Horn, S., et al. (2013). cFLIP regulates skin homeostasis and protects against TNF-induced keratinocyte apoptosis. *Cell Reports*, *5*(2), 397–408.

Reubold, T. F., & Eschenburg, S. (2012). A molecular view on signal transduction by the apoptosome. *Cellular Signalling*, *24*, 1420–1425.

Sun, L., Wang, H., Wang, Z., He, S., Chen, S., Liao, D., et al. (2012). Mixed lineage kinase domain-like protein mediates necrosis signaling downstream of RIP3 kinase. *Cell*, *148*, 213–227.

Tenev, T., Bianchi, K., Darding, M., Broemer, M., Langlais, C., Wallberg, F., et al. (2011). The Ripoptosome, a signaling platform that assembles in response to genotoxic stress and loss of IAPs. *Molecular Cell*, *43*, 432–448.

Varfolomeev, E., Blankenship, J. W., Wayson, S. M., Fedorova, A. V., Kayagaki, N., Garg, P., et al. (2007). IAP antagonists induce autoubiquitination of c-IAPs, NF-kappaB activation, and TNFalpha-dependent apoptosis. *Cell*, *131*, 669–681.

Varfolomeev, E. E., Schuchmann, M., Luria, V., Chiannilkulchai, N., Beckmann, J. S., Mett, I. L., et al. (1998). Targeted disruption of the mouse Caspase 8 gene ablates cell death induction by the TNF receptors, Fas/Apo1, and DR3 and is lethal prenatally. *Immunity, 9*, 267–276.

Vince, J. E., Wong, W. W., Khan, N., Feltham, R., Chau, D., Ahmed, A. U., et al. (2007). IAP antagonists target cIAP1 to induce TNFalpha-dependent apoptosis. *Cell, 131*, 682–693.

Wagner, L., Marschall, V., Karl, S., Cristofanon, S., Zobel, K., Deshayes, K., et al. (2013). Smac mimetic sensitizes glioblastoma cells to Temozolomide-induced apoptosis in a RIP1- and NF-κB-dependent manner. *Oncogene, 32*, 988–997.

Wallberg, F., Tenev, T., & Meier, P. (2013). Time-lapse imaging of necrosis. *Methods in Molecular Biology, 1004*, 17–29.

Weinlich, R., Oberst, A., Dillon, C. P., Janke, L. J., Kang, T. B., Milasta, S., et al. (2013). Protective roles for caspase-8 and cFLIP in adult homeostasis. *Cell Reports, 5*(2), 340–348.

Wittkopf, N., Gunther, C., Martini, E., He, G., Amann, K., He, Y. W., et al. (2013). Cellular FLICE-like inhibitory protein secures intestinal epithelial cell survival and immune homeostasis by regulating caspase-8. *Gastroenterology, 145*, 1369–1379.

Yeh, W. C., Itie, A., Elia, A. J., Ng, M., Shu, H. B., Wakeham, A., et al. (2000). Requirement for Casper (c-FLIP) in regulation of death receptor-induced apoptosis and embryonic development. *Immunity, 12*, 633–642.

Zhang, J., Cado, D., Chen, A., Kabra, N. H., & Winoto, A. (1998). Fas-mediated apoptosis and activation-induced T-cell proliferation are defective in mice lacking FADD/Mort1. *Nature, 392*, 296–300.

Zhang, H., Zhou, X., McQuade, T., Li, J., Chan, F. K., & Zhang, J. (2011). Functional complementation between FADD and RIP1 in embryos and lymphocytes. *Nature, 471*, 373–376.

Zhao, J., Jitkaew, S., Cai, Z., Choksi, S., Li, Q., Luo, J., et al. (2012). Mixed lineage kinase domain-like is a key receptor interacting protein 3 downstream component of TNF-induced necrosis. *Proceedings of the National Academy of Sciences of the United States of America, 109*, 5322–5327.

Zhou, Z., Han, V., & Han, J. (2012). New components of the necroptotic pathway. *Protein & Cell, 3*, 811–817.

CHAPTER FIVE

Tools and Techniques to Study Ligand–Receptor Interactions and Receptor Activation by TNF Superfamily Members

Pascal Schneider[1], Laure Willen, Cristian R. Smulski

Department of Biochemistry, University of Lausanne, Epalinges, Switzerland

[1]Corresponding author: e-mail address: pascal.schneider@unil.ch

Contents

Abstract

Ligands and receptors of the TNF superfamily are therapeutically relevant targets in a wide range of human diseases. This chapter describes assays based on ELISA, immunoprecipitation, FACS, and reporter cell lines to monitor interactions of tagged receptors and ligands in both soluble and membrane-bound forms using unified detection techniques. A reporter cell assay that is sensitive to ligand oligomerization can identify ligands with high probability of being active on endogenous receptors. Several assays are also suitable to measure the activity of agonist or antagonist antibodies, or to detect interactions with proteoglycans. Finally, self-interaction of membrane-bound receptors can be evidenced using a FRET-based assay.

This panel of methods provides a large degree of flexibility to address questions related to the specificity, activation, or inhibition of TNF–TNF receptor interactions in independent assay systems, but does not substitute for further tests in physiologically relevant conditions.

Methods in Enzymology, Volume 545
ISSN 0076-6879
http://dx.doi.org/10.1016/B978-0-12-801430-1.00005-6

ABBREVIATIONS

GPI glycosyl-phosphatidylinositol
THD TNF homology domain

1. INTRODUCTION

The TNF and TNFR superfamilies of ligands and receptors regulate many aspects of the development, homeostasis, activation, effector function, inhibition, and death of immune cells (Aggarwal, 2003; Grewal, 2009; Strasser, Jost, & Nagata, 2009). Some family members also control bone turnover, muscle regeneration, mammary gland biology, and the development of skin-derived appendages (Bhatnagar & Kumar, 2012; Hanada, Hanada, Sigl, Schramek, & Penninger, 2011; Mikkola, 2008). Mutations of TNF/TNFR members are the cause of several genetic defects (Lobito, Gabriel, Medema, & Kimberley, 2011). Also, imbalanced or miss-localized expression of TNF ligands and receptors occur in a number of pathologies such as autoimmunity or bone erosion. Several TNF and TNFR members are either good candidates or proven targets for therapeutic intervention (Grewal, 2009). For example, TNF is targeted in rheumatoid arthritis, RANKL in diseases with bone involvement, and BAFF in systemic lupus erythematosus.

TNF family ligands are characterized by a C-terminal, extracellular TNF homology domain (THD), whereas the signature of TNF receptors lies into one to several cysteine-rich domains (CRD) also located in the extracellular portion of the protein (Bodmer, Schneider, & Tschopp, 2002; Fig. 5.1A and B). THDs assemble as homotrimers usually able to bind and recruit three individual receptors by THD–CRD interactions. Several receptor–ligand complexes have been characterized by crystallographic studies (reviewed in Schneider, 2009). Some TNF family ligands such as APRIL and EDA interact with proteoglycans in addition to their cognate receptors. TNF family ligands are synthesized as type II transmembrane proteins. They can be active as such, and/or can be processed by proteolytic cleavage to soluble forms. Some receptors transmit signals in response to trimeric ligands, whereas others are only activated by membrane-bound ligands, or by at least two ligand trimers in close proximity, which could mimic a membrane-bound ligand (Holler et al., 2003). It is not fully understood why receptors differ in their requirement for ligand multimerization: different

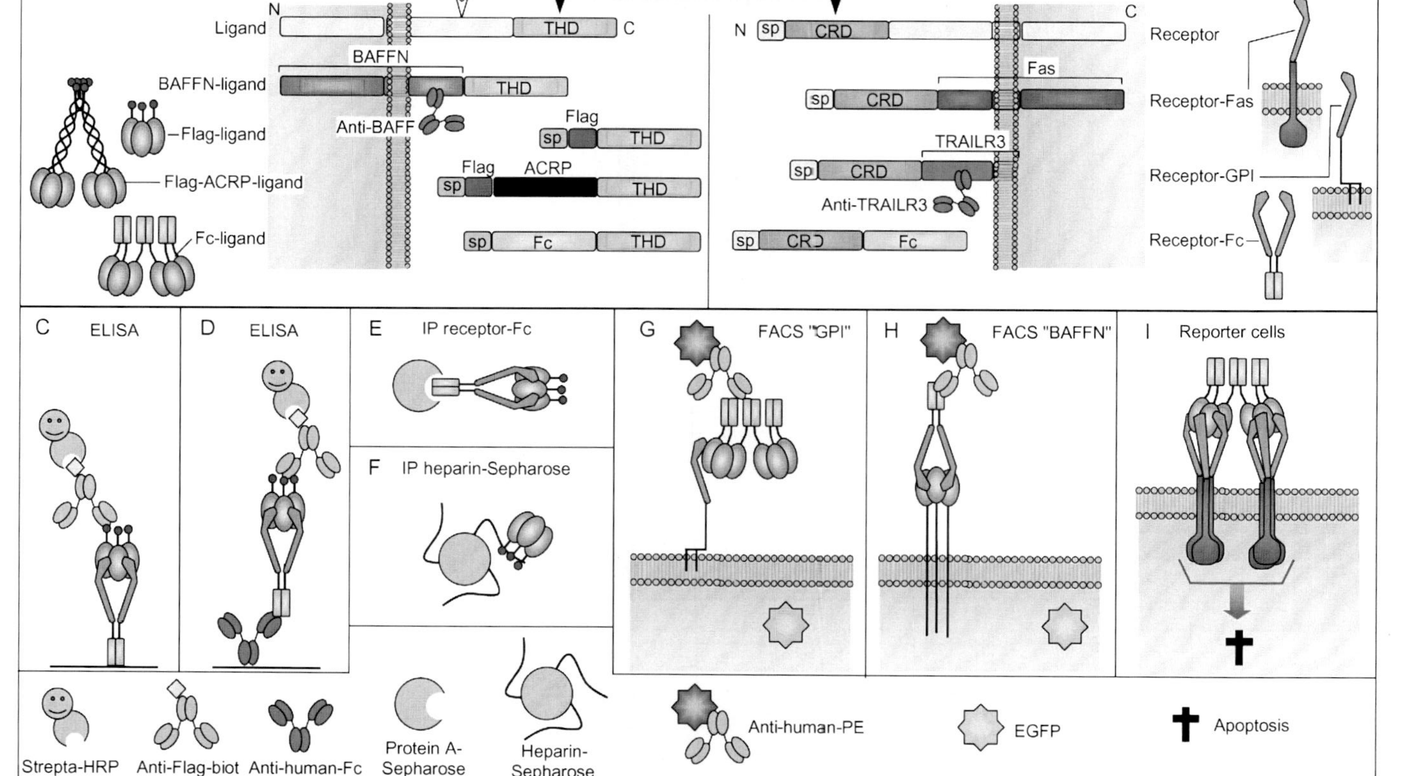

Figure 5.1 Reagents and assay principles for measuring receptor–ligand interactions. (A) Ligand constructs. BAFFN, N-terminal sequence of BAFF; sp, signal peptide; ACRP, collagen domain of ACRP30 protein. (B) Receptor constructs. CRD, cysteine-rich domains; Sp, signal peptide. (C and D) ELISA-based assays to detect receptor–ligand interactions using purified or nonpurified receptors-Fc, respectively. (E) Coimmunoprecipitation assay to monitor receptor–ligand interactions. (F) Precipitation assay to detect interactions of ligands with proteoglycans or heparin. (G) FACS-based assay with GPI-anchored receptors. (H) FACS-based assay with membrane-bound BAFFN-ligands. (I) Cell-based reporter assay in which apoptosis is induced by engagement of receptor-Fas chimeric receptors by multimerized ligands. (See the color plate.)

receptors may recruit signaling complexes of distinct composition and stoichiometries. Alternatively, the lower affinity of some ligands may be compensated by a multimerization-induced increase of avidity. In any case, activation of the receptor-Fas by FasL is sensitive to the multimerization state of the ligand.

Outside of their CRDs, TNFRs are quite diverse in sequence. Most receptors are type I transmembrane proteins (with a signal peptide), but some are type III proteins (without signal peptide: TACI, BCMA, BAFFR, and XEDAR), a few are soluble decoy receptors (OPG and DcR3) and one is a glycolipid-anchored receptor (TRAILR3). The intracellular domains of transmembrane receptors have variable lengths (from 22 amino acid residues in Fn14 1 to 381 in RANK) and can differ in their functions, although some conserved motives such as the death domain or consensus TRAF-binding sequences are shared by several receptors.

In this chapter, we describe a panel of methods that has been used or developed in our laboratory over the years to monitor receptor–ligand interactions within the TNF–TNFR families. These methods are generally based on recombinant proteins that are detected with anti-tag reagents to circumvent the need for specific anti-receptor or anti-ligand antibodies and to allow comparison of interactions between themselves.

2. METHODS

2.1. Tagged ligands and receptors for interaction and functional studies

2.1.1 Tagged ligands

Flag-ligands. Mammalian expression vectors for Flag-tagged ligands typically contain a first cassette coding for the hemaglutinin signal peptide (MAIIYLILLFTAVRG), the Flag tag (DYKDDDDK), and a short linker (GPGQVQLQ) followed by a second cassette coding for the C-terminal, THD of the ligands (Fig. 5.1A). For reasons that were not investigated in detail, expression of some VSV-tagged and HA-tagged ligands was less robust than that of Flag-tagged ligands. Flag-tagged TNF family ligands can sometimes be successfully expressed in bacteria (Schneider, 2000), but do not always yield soluble proteins.

Flag-ACRP-ligands. The collagen domain of the adipocyte complement-related protein 30 (ACRP30) (EDDVTTTEELAPALVPPPKGTCAG WMAGIPGHPGHNGTPGRDGRDGTPGEKGEKGDAGLLGPKGET GDVGMTGAEGPRGFPGTPGRKGEPGEAA) is inserted after the linker in the Signal-Flag-Linker cassette of Flag-ligands. The trimeric THD is multimerized by the collagen domain, thus increasing its receptor binding avidity (Holler et al., 2003; Fig. 5.1A).

Fc-ligands: Fc-ligands are obtained by replacing the Signal-Flag-Linker cassette of Flag-ligands with a cassette coding for (a) the hemaglutinin signal peptide, (b) the Fc portion of human IgG1 (amino acids 105–330 of UniProt entry P01857, excluding the Stop codon), and (c) a linker sequence (RSPQPQPKPQPKPEPEGS). Fc-ligands form hexamers comprising two trimeric ligands and three dimeric Fc domains, or higher order (multiples of six) oligomers (Holler et al., 2003; Fig. 5.1A).

BAFFN-ligands: BAFFN-ligands are obtained by replacing the Signal-Flag-Linker cassette of Flag-ligand with a cDNA coding for amino acids 1–132 of human BAFF. BAFFN-ligands can be stained by flow cytometry with an antibody recognizing the stalk of BAFF (Fig. 5.1A).

2.1.2 Tagged receptors

Receptors-Fc. cDNAs coding for the extracellular domains of receptors are fused to a cassette encoding the Fc-portion of human IgG1 (amino acids 105–330 of UniProt entry P01857, including the Stop codon) (Fig. 5.1B). For type I or secreted receptors, the natural signal peptide is used to drive secretion. For Type III proteins, a signal peptide is added in place of the initiating methionine, for example, the signal peptide of hemaglutinin (MAIIYLILLFTAVRG) or that of mouse immunoglobulin kappa light chain (METDTLLLWVLLLWVPGSTG). The entire extracellular domain of the receptor is included but can be shortened in case of proteolytic cleavage between the receptor and Fc-portions.

Receptors-GPI. Glycosyl-phosphatidylinositol (GPI)-anchored receptors are obtained by replacing the Fc cassette with a cassette coding for the C-terminal portion of TRAIL-R3 (amino acid 157–259, of which amino acid residues 237–259 are replaced by the GPI anchor in the mature protein). GPI-anchored receptors can be stained by flow cytometry with an antibody recognizing the C-terminal portion of TRAIL-R3 (Fig. 5.1B).

Receptors-Fas. Receptor-Fas chimeric proteins are obtained by replacing the Fc cassette with a cassette coding for the C-terminal portion of Fas (amino acids 169–335, comprising the transmembrane and intracellular domains) (Fig. 5.1B).

2.1.3 Purification and storage of ligands and receptors

Flag-tagged ligands can be purified from supernatant of stable mammalian cell clones or from bacterial extracts by anti-Flag affinity chromatography. Receptors-Fc and Fc-ligands fusion proteins are conveniently purified by

Protein A affinity chromatography. All of these proteins can also be used as crude conditioned supernatants of producing cells, usually transiently transfected 293T cells grown for 7 days in serum-free Opti-MEM medium. Such supernatants normally contain proteins of interest at concentrations of 0.2–10 μg/ml and can be stored active for over 10 years at −70 °C.

2.2. The measure of ligand–receptor interactions by ELISA

Ligand–receptors interactions can be measured in an ELISA-like assay (Fig. 5.1C and D). With this assay, the signal is quantitative, many measures can be performed in parallel (e.g., for a titration), and it is possible to circumvent the need for purified proteins (Fig. 5.1D). In addition, this assay responds to untagged competitors that bind either the ligand or the receptor (Fig. 5.2A). Competition assays are particularly informative when using ligands with altered affinity for the receptor (e.g., point mutants with partially reduced activity). Indeed, whereas some mutants behave normally in the direct ELISA, they can be significantly impaired in their ability to compete with the wild-type ligand for receptor binding.

2.2.1 Measure of ligand–receptor interactions with crude or purified tagged proteins

2.2.1.1 Materials

96-Wells Maxisorp Nunc Immunoplates (e.g., Nunc, 439454). Receptor-Fc fusion proteins (purified or in serum-free Opti-MEM supernatants). Flag-tagged ligands (purified or in Opti-MEM). PBS. Blocking buffer (PBS, 4% powdered skimmed milk, 0.5% Tween-20). Incubation buffer (PBS, 0.4% skimmed milk, 0.05% Tween-20). Wash buffer (PBS 0.05% Tween 20). Mouse anti-human IgG Fc(gamma) fragment-specific (Jackson Immuno-Research (JIR) 209-005-139 or 209-005-098), or goat anti-human IgG Fc (JIR 109-005-098). Biotinylated anti-Flag M2 mAb (Sigma F9291). Horseradish peroxidase (HRP)-conjugated streptavidin (JIR 016-030-084). HRP-conjugated donkey anti-human IgG (H+L) (JIR 709-036-149). *ortho*-Phenylenediamine (OPD) tablets (Sigma P9187). Dissolve one tablet of each in 20 ml of water shortly before use. 2 *M* HCl.

2.2.1.2 Method

Note: For purified receptors-Fc, coat the proteins at 1 μg/ml in PBS, block, and continue at Step 5.

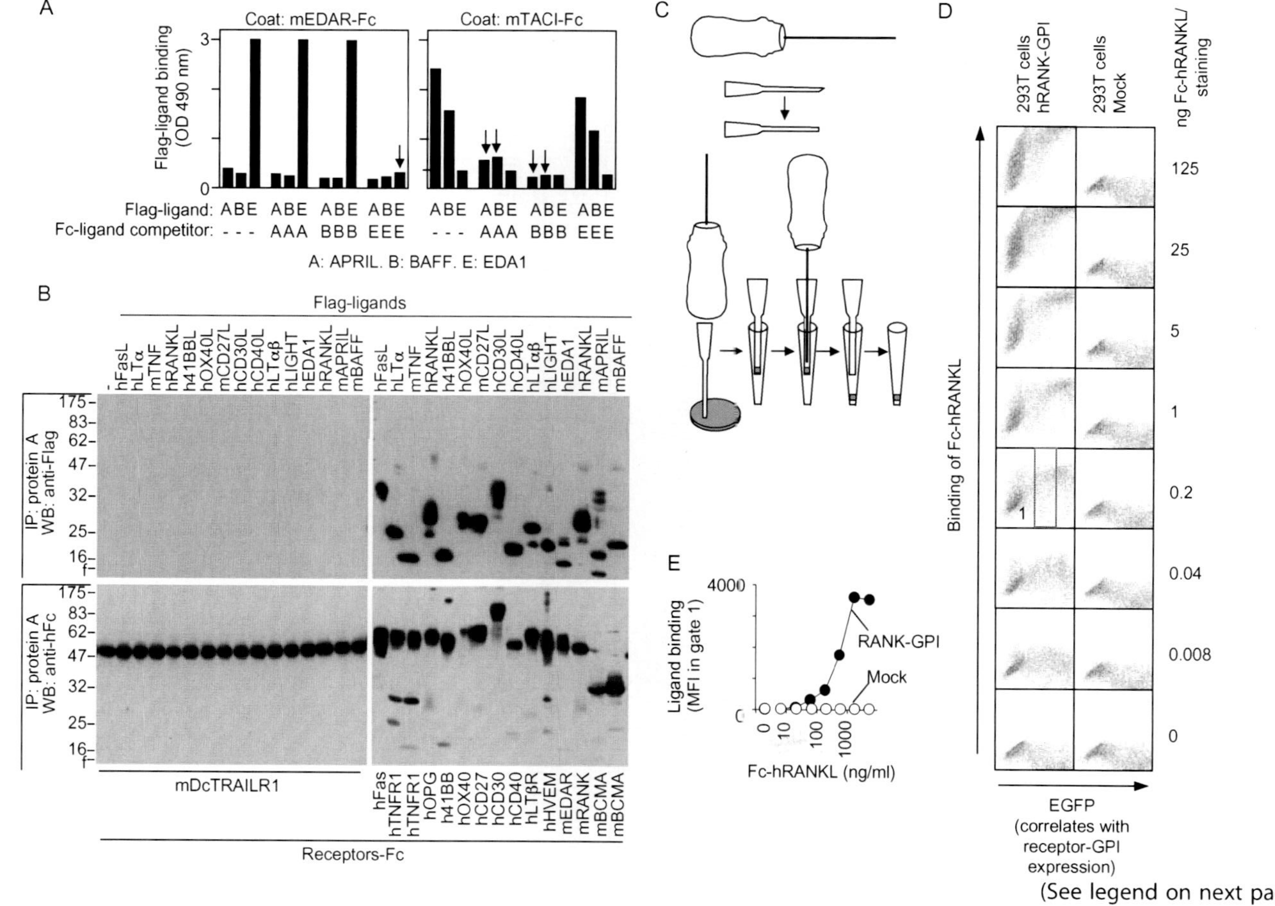

(See legend on next page)

1. Coat 96-wells immunoplate(s) with 100 μl/well of mouse (or goat) anti-human IgG at 5 μg/ml in PBS (2 h at 37 °C, or overnight at room temperature).
2. Empty the plate. Block with 300–350 μl of block buffer for 1 h at 37 °C.
3. Wash the plate 3 × with wash buffer (empty the plate, hit it upside down on an absorbing paper to remove residual liquid, fill the plate by dipping it into a tray containing wash buffer, and repeat these steps as necessary).
4. Add 20 μl of receptor-Fc in Opti-MEM (or 50 μl if receptor is known to be poorly expressed). Bring the volume to 100 μl with incubation buffer. Incubate for 1 h at 37 °C. Wash 3 ×.
5. Add 20 μl of Flag-ligand in Opti-MEM (or 50 μl for poorly expressed ligands. If the interaction is too weak, increasing the concentration will not help, but the use of Flag-ACRP30-ligands generally yields much better results. Purified ligands can also be used, for example, as two-fold titrations starting at 1 μg/ml). Do not add ligands in receptor coating controls. Instead, add 100 μl of incubation buffer. 1 h at 37 °C. Wash 3 ×.
6. Add 100 μl of M2-biot at 0.5 μg/ml in incubation buffer. In receptor coating controls, add incubation buffer only. 30 min to 1 h at 37 °C. Wash 3 ×.
7. Add 100 μl of HRP-conjugated streptavidin (1/4000 in incubation buffer). In receptor coating controls, add 100 μl of HRP-conjugated anti-human (1/8000). 30 min to 1 h at 37 °C. Wash 3 ×.
8. Add 100 μl of OPD solution. Wait as necessary. Stop by adding 50 μl of 2 *M* HCl (add in the same order as the OPD solution to ensure equal reaction time between all wells). Remove any bubble before reading at 490 nm in an ELISA plate reader. Store the plate at −20 °C until data has been processed (color remains stable at −20 °C but not at room temperature).

Figure 5.2 Detection of ligand–receptor interactions by ELISA, by coimmunoprecipitation and by FACS. (A) ELISA plates were coated with the indicated receptor-Fc and revealed with the indicated Flag-ligands, in the presence or absence of the indicated Fc-ligands as competitors. Arrows indicate conditions in which competition did occur. (B) A collection of Flag-ligands were immunoprecipitated with an irrelevant receptor (mDcTRAILR1) or with their cognate receptors, as indicated. Western blot was revealed with anti-Flag (blots at the top), then with anti-human Fc (blots at the bottom). Molecular mass markers are in kDa. (C) Mini-column preparation for immunoprecipitations. A blunted gauge 18 needle is used to punch a frit. The needle is introduced in a tip, and the frit is pushed down with a metal wire mounted on a handle. (D) 293T cells transfected with RANK-GPI and EGFP (or cells transfected with EGFP only) were stained with serial dilutions of Fc-RANKL. (E) Quantification of the mean fluorescence in gate 1 of panel (D), which represents the binding of Fc-RANKL to RANK-GPI. (See the color plate.)

2.2.2 *Measure interactions of untagged proteins or inhibitors by competition in the receptor-Fc and Flag-ligand ELISA*

Any reagent interfering with the binding of a ligand to its receptor is expected to decrease the signal in the receptor–ligand ELISA (Fig. 5.2A). This is best achieved when the ligand is used at non-saturating concentrations, and when the competitor is preincubated with its binding partner prior to running the assay. A specificity control with an irrelevant receptor–ligand pair is required to control for potential artifacts.

2.2.2.1 Method

1. Using method Section 2.2.1.2, titrate the Flag-tagged ligand to determine the dilution that gives a fair but non-saturating signal. If the inhibitor or competitor is expected to target the receptor, it may be useful to titrate the amount of receptor coated too. The less receptor is coated, the less inhibitor or competitor will be required to produce a visible effect.

2. Coat the ELISA plate with the receptor-Fc of interest (directly or via Fc-mediated capture), and block the plate.

3a. Add inhibitors or competitors targeting the receptor into the plate and incubate for 1 h or more at room temperature or at 37 °C. Without washing, add the Flag-tagged ligand (e.g., the ligand at 10-fold the desired concentration in one-tenth of the volume).

3b. For inhibitors or competitors targeting the ligand, incubate the Flag-tagged ligand with the inhibitor in a separate, preblocked plate and incubate for 1 h or more. Transfer the content of the preincubation plate into the receptor-Fc-coated plate of point 2. Incubate for 1 h.

4. Wash the plate 3 × and proceed with points 6–8 of method 2.2.1.2.

Notes: Care should be taken when using Fc-containing competitors (e.g., receptor-Fc or Fc-ligand) as they may interfere with the anti-human capture antibody coated in the plate. In this case, it is advised to coat purified receptor-Fc directly onto the plate. Receptor–ligand interactions are pH-sensitive, and care should be taken to ensure that a neutral pH is maintained throughout the procedure.

2.3. The measure of ligand–receptor interactions by immunoprecipitation

Immunoprecipitations of Flag-tagged ligands with Fc-tagged receptors followed by Western blot analysis provide qualitative informations on receptor–ligand interactions together with size information on the denatured ligands and receptors (Figs. 5.1E and 5.2B). Immunoprecipitations with immobilized heparin is also suitable to detect the binding of ligands to proteoglycans (Fig. 5.1F). As the later interaction is salt-sensitive, elution can be performed with high-salt buffers.

2.3.1 Reagents

Flag-tagged ligands and receptors-Fc, purified, or in Opti-MEM supernatants. Protein A-Sepharose beads (GE Healthcare). Heparin-Sepharose (GE Healthcare). PBS. PBS supplemented with 0.8 *M* NaCl. 50 m*M* Citrate–NaOH pH 2.7. 1 *M* Tris–HCl pH 9. Reagents for SDS–PAGE and Western blot. Anti-Flag M2 mAb (Sigma F3165 or F1804. Both work equally well in the author's experience). HRP-conjugated donkey anti-human IgG (JIR 709-036-149). HRP-conjugated goat anti-mouse IgG (JIR 115-036-166).

2.3.2 Method

2.3.2.1 Precipitations with receptors-Fc

1. Mix a receptor-Fc (about 0.5–1 μg, purified, or in Opti-MEM) with a Flag-tagged ligand (about 0.2 μg) and adjust the volume to about 1 ml with PBS. Add 10 μl of a 50% slurry of Protein A-Sepharose beads in PBS. Incubate for 1 h at 4 °C on a rotating wheel.
2. Centrifuge tubes for 1 min at 5000 rpm in a tabletop centrifuge. Pipette beads with about 200 μl buffer at the bottom of the tube with a wide-opening tip (obtained by cutting the extremity of a tip) and transfer them into 200 μl mini columns (made in tips plugged with a frit according to Fig. 5.2C).
3. Drain buffer (e.g., push it out with a syringe) and wash beads twice with 200 μl of PBS, then remove all liquid. Elute beads with 15 μl of 0.1 *M* citrate–NaOH pH 2.7. Add 5 μl of 1 *M* Tris–HCl pH 9 and 10 μl of 3 × reducing SDS–PAGE sample buffer. Heat-denature samples and analyze by Western blot.
4. Reveal Western blots with anti-Flag M2 (1 μg/ml, e.g., Sigma F3165) followed by HRP-conjugated anti-mouse (1/2000). Quench the membrane with 0.5% H_2O_2, 0.1% NaN_3, wash with PBS, and reveal again with HRP-conjugated anti-human (1/5000).

2.3.2.2 Precipitations with heparin-Sepharose

Same as above, except that (a) Protein A-Sepharose beads are replaced by heparin-Sepharose beads (50% slurry in PBS, freshly prepared), (b) receptors-Fc are omitted, and (c) the elution is performed with 15 μl of PBS supplemented with 0.8 *M* NaCl.

Remarks: Elution of heparin-Sepharose beads with high salt is preferable to elution by boiling in SDS–PAGE sample buffer. Indeed, heparin-Sepharose nonspecifically binds small amounts of any ligands irrespective of the presence or absence of proteoglycan-binding regions. Nonspecifically bound proteins can be eluted in denaturing buffer but not by high-salt elution. To avoid artifactual migrations of high-salt containing samples by SDS–PAGE, add similar amounts of salt in all samples, including standards and empty wells.

2.4. The measure of ligand–receptor interactions by FACS

2.4.1 Interactions of tagged ligands with GPI-anchored receptors

Flow cytometry is convenient to detect ligand–receptor interactions (Bossen et al., 2006). Receptors-GPI are cotransfected in 293T cells with an EGFP tracer. GPI-anchored receptors easily reach the cell surface and alleviate issues of intracellular retention or toxicity caused by intracellular domains. Receptors-GPI are stained with tagged soluble ligands (Fc-ligand or Flag-ligands) (Figs. 5.1G and 5.2D and E), or with a monoclonal antibody directed against the C-terminal portion of TRAIL-R3 to enable the comparison of surface expression of various receptors if required (Fig. 5.1B).

2.4.1.1 Reagents

Expression plasmids for receptors-GPI (Bossen et al., 2006). An EGFP expression plasmid with a CMV promoter (the EGFP containing *Bam*HI/*Not*I fragment of pEGFP-N2, Clonetech #6081-1, was cloned into pCDNA3.1/zeo, Invitrogen #V86020. Any EGFP expression plasmid should also be suitable). Fc-ligand or Flag-ligand (purified or in conditioned medium). Heparin (Liquemin 5000 IU/ml; Roche Pharma). Rat IgG2a monoclonal antibody anti-TRAILR3 (LEIA = mAb572. EnzoLifeSciences ALX-804-136) (Bossen et al., 2006). Anti-Flag M2 (e.g., Sigma F3165) or biotinylated anti-Flag M2 (Sigma F9291). Phycoerythrin (PE)-coupled streptavidin (eBiosciences); PE-coupled goat anti-human IgG (Southern Biotech Associate 2040-09); PE-coupled goat anti-rat IgG (H + L) (Southern Biotech Associate 3010-09); PE-coupled goat anti-mouse IgG (H + L) (Caltag M30004-1). FACS buffer (PBS, 5% FCS). 0.5 ml cylindrical, round-bottomed polystyrene tubes that fit into 96-well plates (e.g., Milian, 080048).

2.4.1.2 Method

Cell transfection and preparation

1. Cotransfect 293T cells in 9 cm diameter dishes in DMEM 10% FCS with plasmids for receptor-GPI (7 μg) and EGFP (1 μg). After transfection, wash cells, and grow them in DMEM 10% FCS for 24 h.
2. Detach cells by pipetting and distribute about 350 μl/well ($\sim 2 \times 10^5$ transfected 293T cells/well) in a round-bottom 96-well plate. Spin for 5 min at 1000 rpm. Aspirate supernatant.
3. Stainings are performed in a total volume of 50 μl (25 μl is also possible) with typically 25 ng purified ligand (for titrations 125, 25, 5, 1, 0.2, and 0 ng) or 10 μl ligands in Opti-MEM (for titrations 25, 5, 1, 0.2, 0.04, and 0 μl) in the presence of 0.1 μl heparin and FACS buffer. Resuspend cells by pipetting and incubate for 20 min on ice.
4. Add 200 μl of FACS buffer. Spin for 5 min at 1000 rpm. Aspirate supernatants.

For Fc-ligands

5a. Add 50 μl of PE-coupled goat anti-human IgG at 1/500 in FACS buffer.

5b. Add 50 μl of anti-Flag M2 at 1/500 in FACS buffer. Incubate and wash according to point 6a.

5c. Add 50 μl of biotinylated anti-Flag M2 at 1/500 in FACS buffer. Perform point 6a.

5d. Add 50 μl of rat anti-TRAIL-R3 LEIA at 40 μg/ml in FACS buffer. Perform point 6a.

For Flag-ligands using unconjugated anti-Flag (human cells only)

6a. Resuspend cells by pipetting. Incubate for 20 min on ice. Add 200 μl of FACS buffer. Spin for 5 min at 1000 rpm. Aspirate supernatants.

6b. Add 50 μl of PE-coupled goat anti-mouse at 1/100 in FACS buffer. Perform point 6a.

For Flag-ligands using biotinylated anti-Flag (for any cell)

6c. Add 50 μl of PE-coupled streptavidin at 1/500 in FACS buffer. Perform point 6a.

For anti-GPI

6d. Add 50 μl of PE-coupled goat anti-rat-PE 1/100 in FACS buffer. Perform point 6a.

Final steps

7. Add 200 μl of FACS buffer. Resuspend and transfer cells in 0.5 ml tubes arranged in a round-bottomed 96-well plate.
8. Acquire data on any flow cytometry apparatus able to measure EGFP and PE (e.g., a FACScan using the CellQuest program; Becton Dickinson) and analyze them with suitable analysis software (e.g., FlowJo).

Additional remarks

A. Some TNF family ligands bind to proteoglycans (e.g., APRIL, Ingold et al., 2005; EDA, Swee et al., 2009). To detect the interaction with proteoglycans, transfect 293T cells with EGFP alone, then stain cells with the ligand in the presence or absence of heparin. The difference of staining between both conditions indicates that the ligand interacts with proteoglycans. Alternatively, 293T cells can first be grown for up to 4 days in the presence or absence of 50 m*M* of the sulfation inhibitor sodium chlorate. Decreased staining in chlorate-treated cells is indicative of a ligand–proteoglycan interaction. Cationic transfection reagents (e.g., PolyFect) that use proteoglycans for plasmid delivery do not work in chlorate-treated cells.

B. The procedure is easily adapted for anti-receptor antibodies.

C. For each staining, prepare cells for compensation: untransfected 293T cells (double negative), unstained EGFP-transfected 293T cells (green only), and untransfected 293T cells stained with the proteoglycan-binding ligand Fc-muAPRIL (containing the proteoglycan-binding sequence immediately preceding the THD) in the absence of heparin, followed by anti-human PE (red only), or any procedure resulting in red staining only.

2.4.2 Interactions of tagged receptors with BAFFN-fusion ligands

BAFFN-ligands are cotransfected with EGFP and stained with soluble receptors-Fc (Fig. 5.1H) or with an anti-BAFF antibody recognizing the extracellular, membrane-proximal region of BAFF present in all constructs (Fig. 5.1A). Transfected human BAFF expresses well at the cell surface. However, other TNF family ligands, for example, mouse BAFF or human TNF, do not, and thus expression as BAFFN-fusion proteins enhances their surface expression (Bossen et al., 2011).

2.4.2.1 Reagents

Soluble receptors-Fc (purified or in Opti-MEM). Rat IgG2a monoclonal antibody anti-BAFF (Buffy1, EnzoLifeSciences). Other reagents are as described in Section 2.4.1.1.

2.4.2.2 Method

The method is the same as that of Section 2.4.1.2, with the following modifications:

1. Cotransfect 293T cells with EGFP and BAFFN-ligand expression plasmids at a 1:7 ratio.
2. Stain cells with receptors-Fc followed by anti-human PE to detect specific interactions, or with Buffy1 mAb at 40 μg/ml in FACS buffer followed by anti-rat-PE to detect surface expression of the BAFFN-ligand chimer.

2.5. The measure of ligand activity using reporter cells

The ELISA-, immunoprecipitation-, and FACS-based assays monitor interactions. Assays based on reporter cells monitor the ability of a receptor to induce a biological response, alone, or upon ligand stimulation. This response can either reflect the natural signaling pathway of the receptor (e.g., NF-κB activation), or the activation of a reporter response (e.g., Fas-dependent apoptosis induced by receptor-Fas chimeric proteins).

2.5.1 Generation of receptor: Fas-expressing reporter cell lines

Cells expressing the death receptor Fas die upon stimulation with FasL (Schneider et al., 1998). The Jurkat T cell line is particularly sensitive to FasL. When these cells (or the Fas-negative variant Jurkat JOM2 clone 6) are modified to express receptor-Fas chimeric protein, for example, BCMA-Fas (Bossen et al., 2008), EDAR-Fas (Swee et al., 2009), or OX40-Fas (Fig. 5.3A), they die in response to ligands (BAFF, EDA, or OX40L) that are usually not inducing apoptosis (Fig. 5.1I). Fas-dependent apoptosis is sensitive to the multimerization status of FasL. For example, crosslinking of trimeric, Flag-tagged FasL with an anti-Flag antibody increases the apoptotic response in FasL-sensitive Jurkat cells by greater than three orders of magnitude (Schneider et al., 1998). This property is usually conserved in cells expressing receptor-Fas fusion proteins and can be used to distinguish different states of ligand multimerization (Bossen et al., 2008; Schneider et al., 1998; Swee et al., 2009).

2.5.1.1 Reagents

Puromycin (EnzoLifeSciences, ALX-380-028). Polybrene (Sigma H9268). Expression vector for receptor-Fas fusion protein in the retroviral vector pMSCV and accessory plasmids (Bossen et al., 2008).

2.5.1.2 Method

Preparation of retrovirus

1. Day 1 morning: Dilute confluent 293T cells 1/8 in DMEM 10% FCS.
2. Day 2 morning: Transfect 293T cells.

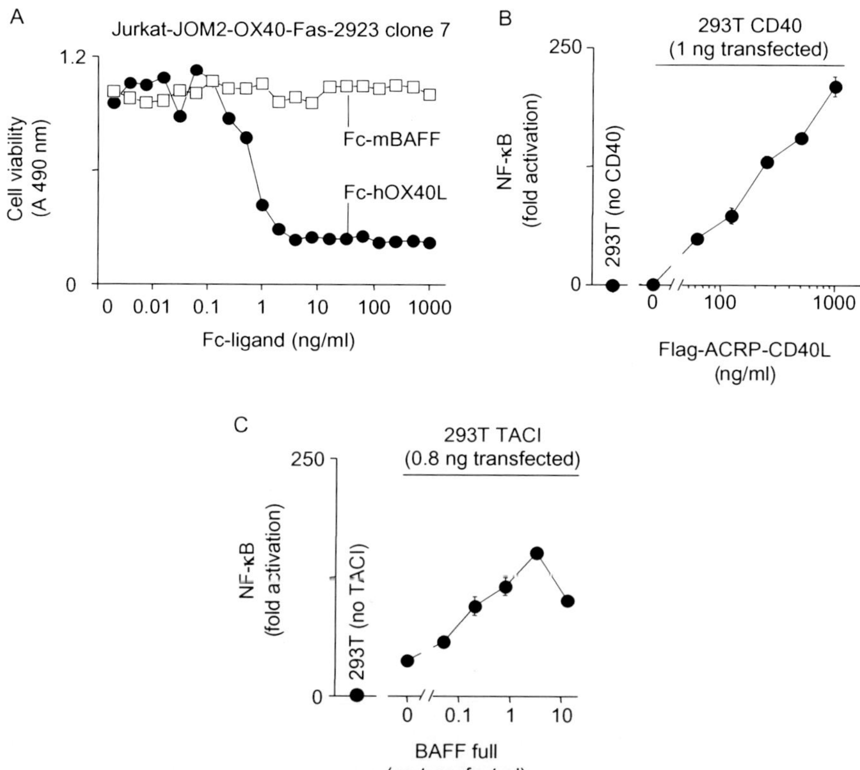

Figure 5.3 Reporter assays of ligand–receptor interactions. (A) A clone of Jurkat-JOM2 cells expressing a human OX40-Fas chimeric protein was exposed to titrated amounts of Fc-hOX40L or to Fc-mBAFF as a control. Cell viability was monitored with the PMS/MTS assay. (B) 293T cells (10^4 in 100 μl) were transfected with 1 ng of CD40 expression vector and stimulated with the indicated concentrations of recombinant Flag-ACRP-CD40L. NF-κB luciferase reporter activity was monitored. (C) 293T cells were cotransfected with a fixed (low) amount of TACI and variable amount of full-length BAFF to stimulate NF-κB. Note that TACI alone induces significant NF-κB activation in the absence of ligand.

2a. Mix 10 μg of pMSCV plasmid of interest, 10 μg of Hit60 MoMuLV gag-pol expression vector, 1.5 μg of pCG VSV envelope protein (G) expression vector, 50 μl of 2.5 *M* $CaCl_2$, and 450 μl of water.

2b. Add 500 μl of 2 × HeBS (280 m*M* NaCl, 50 m*M* HEPES, 1.5 m*M* Na_2HPO_4 pH 7.05) while vortexing. Add to cells 1 min later, mix by swirling, and put back in the incubator.

3. Day 2 late afternoon (7–8 h posttransfection): Wash cells with PBS, add 4 ml of DMEM 10% FCS (optionally with 10 m*M* sodium butyrate) and leave overnight.

4. Day 3 morning: Wash cells and add 6 ml of RPMI (not DMEM) containing 10% FCS.

5. Day 4 morning: Infection.

5a. Collect supernatant of transfected 293T cells, filter at 0.45 μm, add polybrene to a final concentration of 8 μg/ml and mix.

5b. Centrifuge 1–2 × 10^6 Jurkat JOM2 clone 6 cells (or Jurkat wt cells) for 5 min at 1250 rpm. Discard supernatant. Add 2.5 ml of virus-containing supernatants to the pellet, and seed the mix in a 12-well plate. Also prepare nontransduced Jurkat-JOM2 clone 6 cells as a control for selection.

6. Day 5: Harvest cells by centrifugation for 5 min at 1250 rpm. Resuspend in 4 ml of RPMI 10% FCS and seed in two wells of a 12-well plate.

7. Day 8: Cells transduced with control EGFP-pMSCV plasmid should be green under the fluorescence microscope. Inoculate transduced cells at a four-fold dilution to get four cultures of 2 ml in RPMI 10% FCS containing 2, 1.2, 0.8, or 0.6 μg/ml puromycin.

8. Day 11: Check cultures. Untransduced cells should be dead, cells transduced with empty pMSCV should be mostly alive, and very little to many live cells should be seen in cells transduced with receptor-Fas expression vectors.

9. Let cells grow until non-adherent spherical clusters of live cells emerge. There are then two possible options: either pick individual cell clusters under the microscope or seed them in 96-well plates in 200 μl/well of RPMI 10% FCS (no puromycin), or count cells and seed them in 96-well plates in 200 μl/well of RPMI 10% FCS at 1, 3, or 10 cells/well. Let clones grow. Spot wells with a single clone. If required, change supernatant from time to time, or pipette cells up and down to disrupt clusters and speed up growth.

10. When clones are confluent, for each clone inoculate two wells of a 96-well plate with about 40% of the cells, and add medium back to the 20% remaining cells in the parental plate. Treat clones in the test plate overnight with either the cognate or a control Fc-ligand at 200 ng/ml. Under the microscope, spot clones that selectively die in response to the cognate ligand, and also perform a phenazine methosulfate (PMS)/3-(4,5-dimethylthiazol-2-yl)-5-(3-carboxymethoxyphenyl)-2-(4-sulfophenyl)-2*H*-tetrazolium (MTS) cell viability test (see Section 2.5.2).

11. Gently amplify clones of interest (they usually do not like strong dilutions at this stage) to about 2 ml cultures. Perform a cytotoxic assay with the Fc-ligand of interest at 1 μg/ml and nineteen two-fold

dilutions (as described in Section 2.5.2.1). Select, amplify, and freeze the best clone.

Remarks: Transduction efficiency decreases with increasing size of the insert Cells transduced with the empty pMSCV are more resistant to puromycin than cells transduced with pMSCV-receptor-Fas. It is therefore recommended to select cells over a range of puromycin concentrations. In general, it is easier to produce EGFP-positive control clones than clones expressing receptor-Fas, probably due to the intrinsic toxic properties of receptor-Fas fusion proteins.

2.5.2 Monitoring ligand activity by apoptosis induction in reporter cell lines

Cell viability can be determined by measuring the reducing activity of dehydrogenases in metabolically active cells. In a cell viability assay, PMS reduced by dehydrogenases transfers electrons to MTS whose formazan reduction product absorbs at 490 nm. The quantity of formazan generated is proportional to the number of live cells in the culture.

2.5.2.1 Reagents

Reporter cells in exponential growing phase. Ligand of interest (purified or in Opti-MEM). 96-Well cell culture plates, flat-bottomed. MTS (Promega G1111) at 2 mg/ml in PBS. PMS (Sigma P9625) at 0.9 mg/ml in PBS in light-protected tubes.

2.5.2.2 Method

1. In a 96-well plate, perform titrations of relevant or control ligands at two-fold the final desired concentration in 50 μl. For example, perform 20 two-fold dilutions starting with the ligand at 2 μg/ml (to give a 1 μg/ml final concentration). For ligands in Opti-MEM, use 50 μl in the first well. The last well of the titration should contain no ligand.
2. Add 50 μl of cell suspension in medium to all wells (for Jurkat or Jurkat JOM2 clone 6-based reporter cells, use 20,000–50,000 cells/well. Fill any empty well of the plate with water. Wrap the plate in an aluminum foil and incubate it for about 16 h at 37 °C, 5% CO_2.
3. If desired, check cell morphology by microscopy. Add 20 μl per well of a freshly prepared PMS/MTS (20:1, v/v) solution. Incubate the plate at 37 °C, 5% CO_2 for several hours until color develops, then read absorbance at 490 nm with an ELISA reader.

2.5.3 Monitoring inhibitors or activators of ligands and receptors in reporter cell lines

The cell viability assay with reporter cell lines expressing receptor-Fas chimeric receptors can readily be adapted to characterize, for example (a) anti-receptor antibodies with agonist activity (the antibody coated in the plate or added in solution kills the reporter cell), (b) anti-receptor antibodies with antagonist activity (reporter cells preincubated with the antibody become resistant to the ligand), or (c) blocking antiligand antibodies (ligand preincubated with the antibody become unable to kill reporter cells).

2.5.4 Monitoring ligand activity with NF-κB reporter cells

Several TNF receptors family members activate the classical NF-κB signaling pathway. This pathway is easy to monitor with NF-κB-driven transcription of a luciferase reporter gene. In contrast to reporter cells expressing receptor-Fas fusion receptors, NF-κB signaling is relevant to the normal signaling of many receptors. Whether a ligand–receptor interaction of interest can be monitored with NF-κB reporter assays must be determined on a case-by-case basis. Some receptors like BCMA or CD40 only activate NF-κB in response to ligands (Fig. 5.3B), while receptors like TACI or EDAR tend to spontaneously activate NF-κB in the absence of ligand, making ligand-dependent responses more difficult (Fig. 5.3C) or even impossible to measure. In this assay, ligands can be either cotransfected (Fig. 5.3C) or provided exogenously as a recombinant protein (Fig. 5.3B).

2.5.4.1 Reagents

Plasmids: Expression plasmids for full-length receptors and ligands of the TNF family, empty control plasmid, EGFP-encoding plasmid, NF-κB luciferase reporter plasmid, and control renilla luciferase expression plasmid. Purified recombinant TNF family ligands. *Transfection reagent*: Lipofectamine 2000 (Invitrogen 11668-019) or PolyFect (Qiagen 301-105).

Dual luciferase reporter assay system (Promega E1910). Home-made solutions whose efficacy approaches that of the kit can also be made (quantities given for 25 wells): firely luciferase substrate: mix 400 μl buffer (6 m*M* ATP [Sigma A2383], 15 m*M* $MgSO_4$, 30 m*M* Tricine pH 7.8, 300 m*M* 2-mercaptoethanol) with 16 μl of 100 m*M* D-luciferin (Biosynth L8240) in water. Stop and glow solution: mix 400 μl buffer (PBS, 0.1 mg/ml $CaCl_2$, 1 mg/ml D-Glucose, 1 μg/ml aprotinin) with 4 μl of 50 m*M* luciferase inhibitor (Biosynth C-7004) in ethanol and with 4 μl of 10 m*M* coelenterazine H (Calbiochem 119113) in ethanol.

2.5.4.2 Method

1. Day 1 morning: Plate 2×10^4 293T cells per well in 100 μl DMEM 10% FCS in a 96-well plate.
2. Day 2 morning: Prepare plasmids for transfection as follows:

2a. Prepare a master mix containing 7 ng/well of plasmid renilla luciferase, 7 ng/well of plasmid NF-κB luciferase, 7 ng/well of plasmid EGFP, and 14 ng/well of empty plasmid.

2b. Prepare a plasmid mix with the desired quantities of ligands (0.05–32 ng/well) and/or receptors (1–12.8 ng/well), and empty plasmid (add to 35 ng/well).

2c. Mix both plasmid master mixes in equal proportion to get 70 ng/well of plasmids.

2d. Mix 4 μl of the final plasmid mix with 20.5 μl of DMEM without FCS and without antibiotics and 2.2 μl of Polyfect. Vortex and incubate for 10 min at room temperature. Add 130 μl of DMEM, 10% FCS to get a total volume of 156.7 μl. Add 24 μl of this mix containing 70 ng plasmid to cells.

3. Day 3: Transfection efficiency is evaluated by looking at GFP-positive cells under an inverted fluorescence microscope. If transfection is judged sufficient, cells are washed once with 200 μl of PBS, then lysed in 50 μl of passive lysis buffer for 15 min at room temperature on a rotatory agitator and, if required, frozen until use. 25 μl of lysate is transferred to black 96-wells plates (PerkinElmer 6005270), 15 μl/well of firefly luciferase substrate is added and luminescence is measured with a TopCount luminometer (Packard). 15 μl/well of freshly made stop and glow solution is added, and the plate measured again. Luciferase signals are normalized to renilla signals.

2.6. The measure of ligand-independent receptor interactions by Förster resonance energy transfer

Förster resonance energy transfer (FRET) is a noninvasive technique to study protein–protein interactions. FRET is based on the transfer of energy from an excited donor (ECFP) to an acceptor (EYFP) that fluoresces at a wavelength different from the donor's emission wavelength. This phenomenon takes place when the distance between donor and acceptor is less than 5–10 nm and the emission spectrum of the donor overlaps with the excitation spectrum of the acceptor, as is the case for ECFP and EYFP proteins (Selvin, 2000).

Flow cytometry is one of the techniques by which FRET can be measured. Here, we exemplify how ligand-independent CD40–CD40 self-interactions (Smulski et al., 2013) can be measured by FRET with suitable positive and negative controls. Cotransfection of equal amounts of ECFP and EYFP is used to establish the background FRET levels, whereas transfection of an ECFP–EYFP fusion protein serves to define FRET-positive cells (Fig. 5.4A and B). It must be stressed that the fluorescence signal recorded in the FRET channel consists of FRET signals but also of direct emission signals of both EYFP and ECFP. Careful removal of background signals in the FRET channel is achieved using a FRET gate (in Fig. 5.4B) whose shape is function of the EYFP to ECFP ratio. Signals obtained in the FRET channel with the experimental sample can, therefore, only be interpreted as *bona fide* FRET for cells expressing equal amount of the EYFP- and ECFP-fusion partners (i.e., for cells falling into gate 1 of Fig. 5.4A). If the ECFP to EYFP ratio is changed without adaptation of the FRET gate (which is difficult to achieve in practice), some background fluorescence will be collected in this gate at higher EYFP to ECFP ratio, while real FRET signals will not be recorded in this gate at lower EYFP to ECFP ratio, yielding false-positive and false-negative

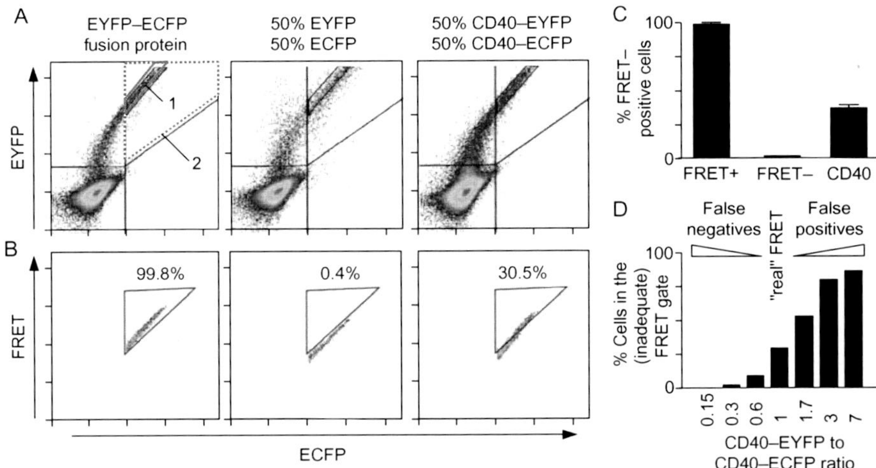

Figure 5.4 Ligand-independent interaction of CD40 as measured by FRET. 293T cells were transfected with the indicated plasmids at the indicated ratio. The total amount of plasmid was always constant. (A) Scattergrams showing expression of both ECFP and EYFP in all transfections. (B) FRET signal of cells in gate 1 of panel A. (C) Graphic representation of data from panel (B). Mean ± SEM of 10 independent experiments. (D) 293T cells were transfected with different CD40–EYFP to CD40–ECFP ratios. All cells falling in gate 2 were analyzed. The percentage of cells falling in the FRET gate increases with increasing EYFP to ECFP ratios. This is mainly due to the fact that the FRET gate established with positive and negative controls at a one-to-one ratio is not valid for other ECFP to EYFP ratios. (See the color plate.)

signals, respectively. Thus, differential expression of one of the interaction partners (e.g., a receptor with or without a point mutation) can strongly affect the percentage of "FRET-positive" cells even if no changes occurred in the interaction. The dramatic effect of imbalanced EYFP to ECFP ratio on "FRET" signals is illustrated in Fig. 5.4D.

2.6.1 Method

1. Seed 10^5 293T cells in 500 μl of DMEM 10% FCS in 24-well plates.
2. 24 h later, using Polyfect (Qiagen), transfect cells with a total amount of 0.4 μg plasmids encoding either the ECFP–EYFP fusion protein, the ECFP and EYFP proteins, or CD40–ECFP and CD40–EYFP fusion receptors (containing full-length CD40 with fluorescent proteins fused at the C-terminus).
3. 16–20 h posttransfection, analyze cells by flow cytometry using a LSRII (BD Biosciences) instrument. EYFP signal is recorded using the 488 nm laser with a 530/30 filter, ECFP signal is recorded using the 405 nm laser with a 450/50 filter and FRET signal is recorded using the 405 nm laser with a 585/42 filter.
4. For data analysis (e.g., with FlowJo), use cells transfected with the ECFP–EYFP fusion protein to set the gate of cells with one-to-one ratio of ECFP to EYFP expression. Then, set the gate of FRET-positive cells using the ECFP–EYFP and ECFP plus EYFP transfections as positive and negative controls, respectively, according to Banning et al. (2010).

3. CONCLUSIONS

The panel of methods described in this chapter is suitable to monitor receptor–ligand interactions within the TNF family and to characterize inhibitors or competitors of these interactions. Because of the important roles of TNF family members in human health and disease, this is of valuable interest. The use of tagged, recombinant, or chimeric proteins greatly facilitates comparisons and controls in independent experimental settings. However, these methods may be sensitive to false-negatives if one of the binding partners is not or only partially active as a recombinant protein. In addition, these methods mostly provide qualitative data about interaction strengths and are not performed at the endogenous level. Quantitative methods, such as surface plasmon resonance, and physiologically relevant assays at the level of the endogenous proteins in primary cells or live organisms remain essential to tackle the relevance of these interactions. The reporter cell assay described in this chapter can be extremely sensitive and may in certain cases serve as a bioassay for the detection of active endogenous ligands. Because of its sensitivity to ligand oligomerization, recombinant ligands that are effective in this test are likely to be active on their physiological target too (Gaide & Schneider, 2003; Swee et al., 2009). Regarding future developments, assays to monitor receptor-intrinsic signaling pathways such as alternative NF-κB *in vitro*, and cell-based assays equally sensitive to trimeric and oligomeric TNF family ligands would be useful complements to this collection of methods.

ACKNOWLEDGMENTS

The authors thank Olivier Micheau (University of Dijon, France) for the gift of Fas-deficient Jurkat JOM2 cells. This work was supported by grants from the Swiss National Science Foundation (to P. S.).

REFERENCES

Aggarwal, B. B. (2003). Signalling pathways of the TNF superfamily: A double-edged sword. *Nature Reviews. Immunology*, *3*, 745–756.

Banning, C., Votteler, J., Hoffmann, D., Koppensteiner, H., Warmer, M., Reimer, R., et al. (2010). A flow cytometry-based FRET assay to identify and analyse protein–protein interactions in living cells. *PLoS One*, *5*, e9344.

Bhatnagar, S., & Kumar, A. (2012). The TWEAK-Fn14 system: Breaking the silence of cytokine-induced skeletal muscle wasting. *Current Molecular Medicine*, *12*, 3–13.

Bodmer, J. L., Schneider, P., & Tschopp, J. (2002). The molecular architecture of the TNF superfamily. *Trends in Biochemical Sciences*, *27*, 19–26.

Bossen, C., Cachero, T. G., Tardivel, A., Ingold, K., Willen, L., Dobles, M., et al. (2008). TACI, unlike BAFF-R, is solely activated by oligomeric BAFF and APRIL to support survival of activated B cells and plasmablasts. *Blood*, *111*, 1004–1012.

Bossen, C., Ingold, K., Tardivel, A., Bodmer, J. L., Gaide, O., Hertig, S., et al. (2006). Interactions of tumor necrosis factor (TNF) and TNF receptor family members in the mouse and human. *Journal of Biological Chemistry*, *281*, 13964–13971.

Bossen, C., Tardivel, A., Willen, L., Fletcher, C. A., Perroud, M., Beermann, F., et al. (2011). Mutation of the BAFF furin cleavage site impairs B-cell homeostasis and antibody responses. *European Journal of Immunology*, *41*, 787–797.

Gaide, O., & Schneider, P. (2003). Permanent correction of an inherited ectodermal dysplasia with recombinant EDA. *Nature Medicine*, *9*, 614–618.

Grewal, I. S. (2009). Overview of TNF superfamily: A chest full of potential therapeutic targets. *Advances in Experimental Medicine and Biology*, *647*, 1–7.

Hanada, R., Hanada, T., Sigl, V., Schramek, D., & Penninger, J. M. (2011). RANKL/RANK-beyond bones. *Journal of Molecular Medicine (Berlin, Germany)*, *89*, 647–656.

Holler, N., Tardivel, A., Kovacsovics-Bankowski, M., Hertig, S., Gaide, O., Martinon, F., et al. (2003). Two adjacent trimeric Fas ligands are required for Fas signaling and formation of a death-inducing signaling complex. *Molecular and Cellular Biology*, *23*, 1428–1440.

Ingold, K., Zumsteg, A., Tardivel, A., Huard, B., Steiner, Q. G., Cachero, T. G., et al. (2005). Identification of proteoglycans as the APRIL-specific binding partners. *Journal of Experimental Medicine*, *201*, 1375–1383.

Lobito, A. A., Gabriel, T. L., Medema, J. P., & Kimberley, F. C. (2011). Disease causing mutations in the TNF and TNFR superfamilies: Focus on molecular mechanisms driving disease. *Trends in Molecular Medicine, 17*, 494–505.

Mikkola, M. L. (2008). TNF superfamily in skin appendage development. *Cytokine & Growth Factor Reviews, 19*, 219–230.

Schneider, P. (2000). Production of recombinant TRAIL and TRAIL receptor: Fc chimeric proteins. *Methods in Enzymology, 322*, 322–345.

Schneider, P. (2009). The beautiful structures of BAFF, APRIL and their receptors. BLyS ligands and receptors. In M. Cancro (Ed.), *BLyS ligands and receptors* (pp. 1–18): New York: Humana Press.

Schneider, P., Holler, N., Bodmer, J. L., Hahne, M., Frei, K., Fontana, A., et al. (1998). Conversion of membrane-bound Fas(CD95) ligand to its soluble form is associated with downregulation of its proapoptotic activity and loss of liver toxicity. *Journal of Experimental Medicine, 187*, 1205–1213.

Selvin, P. R. (2000). The renaissance of fluorescence resonance energy transfer. *Nature Structural Biology, 7*, 730–734.

Smulski, C. R., Beyrath, J., Decossas, M., Chekkat, N., Wolff, P., Estieu-Gionnet, K., et al. (2013). Cysteine-rich domain 1 of CD40 mediates receptor self-assembly. *Journal of Biological Chemistry, 288*, 10914–10922.

Strasser, A., Jost, P. J., & Nagata, S. (2009). The many roles of FAS receptor signaling in the immune system. *Immunity, 30*, 180–192.

Swee, L. K., Ingold-Salamin, K., Tardivel, A., Willen, L., Gaide, O., Favre, M., et al. (2009). Biological activity of ectodysplasin A is conditioned by its collagen and heparan sulfate proteoglycan-binding domains. *Journal of Biological Chemistry, 284*, 27567–27576.

CHAPTER SIX

Necrotic Cell Death in *Caenorhabditis elegans*

Vassiliki Nikoletopoulou, Nektarios Tavernarakis[1]
Institute of Molecular Biology and Biotechnology, Foundation for Research and Technology—Hellas, Heraklion, Greece
[1]Corresponding author: e-mail address: tavernarakis@imbb.forth.gr

Contents

Abstract

Similar to other organisms, necrotic cell death in the nematode *Caenorhabditis elegans* is manifested as the catastrophic collapse of cellular homeostasis, in response to overwhelming stress that is inflicted either in the form of extreme environmental stimuli or by intrinsic insults such as the expression of proteins carrying deleterious mutations. Remarkably, necrotic cell death in *C. elegans* and pathological cell death in humans share multiple fundamental features and mechanistic aspects. Therefore, mechanisms mediating necrosis are also conserved across the evolutionary spectrum and render the worm a versatile tool, with the capacity to facilitate studies of human pathologies. Here, we overview necrotic paradigms that have been characterized in the nematode and

Methods in Enzymology, Volume 545
ISSN 0076-6879
http://dx.doi.org/10.1016/B978-0-12-801430-1.00006-8

outline the cellular and molecular mechanisms that mediate this mode of cell demise. In addition, we discuss experimental approaches that utilize *C. elegans* to elucidate the molecular underpinnings of devastating human disorders that entail necrosis.

1. INTRODUCTION

1.1. Characteristics of necrotic cells

Early studies in the field of cell death described two major forms of cellular demise, apoptosis and necrosis, and contrasted them as being diametrically different in every aspect examined (Walker, Harmon, Gobe, & Kerr, 1988). Apoptosis, also known as caspase-dependent programmed cell death (PCD), was described as a controlled cell death process, proposed to function as a tissue homeostatic mechanism that is complementary and opposite to cell division (Kerr, Wyllie, & Currie, 1972). Necrosis was classically contrasted to apoptosis not only on grounds of context and mechanistic regulation or lack thereof but also based on notable morphological differences. The apoptotic cell profile is characterized by cell rounding, detachment from the basal membrane or cell culture substrate, chromatin condensation and nuclear fragmentation, blebbing of the plasma membrane, and shedding of vacuoles known as apoptotic bodies (Galluzzi et al., 2007). Necrotic cells were initially characterized in a negative fashion, exhibiting neither an apoptotic morphological profile nor an extensive vacuolization characteristic of autophagic cell death. However, specific morphological features were soon attributed to necrotic cells. These included an increasingly translucent cytoplasm, osmotic swelling of most organelles, increased cell volume, and finally rupture of the plasma membrane. The morphological profiles of apoptotic, necrotic, and autophagic cells are shown in Fig. 6.1. Notably, unlike apoptosis, necrosis does not feature major nuclear modifications but only minor ultrastructural changes. Moreover, necrotic cells do not fragment into distinct corpses as their apoptotic counterparts do (Galluzzi et al., 2007).

At the organismal level, a recent study demonstrated that necrotic death is accompanied by a burst of intense blue fluorescence immediately after the worms succumb to the necrotic stimuli. Such death fluorescence marks an anterior to posterior wave of intestinal cell death that is accompanied by cytosolic acidosis. This wave is propagated via the innexin INX-16, likely by calcium influx. Notably, inhibition of systemic necrosis can delay stress-induced death. Initially present in intestinal lysosome-related organelles (gut granules), the fluorescent substance was identified as anthranilic

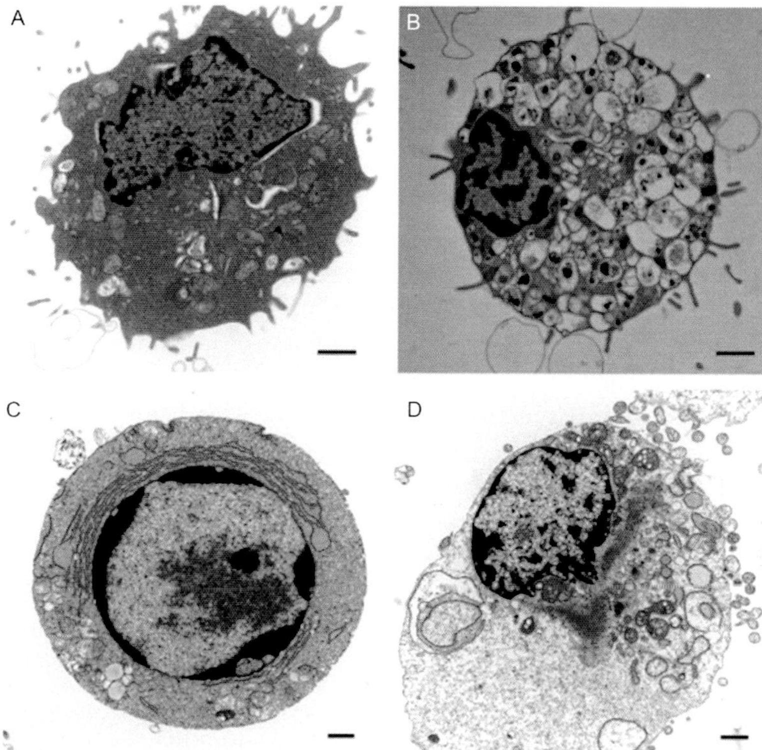

Figure 6.1 Morphological features of autophagic, apoptotic, and necrotic cells. (A) Normal, (B) autophagic, (C) apoptotic, (D) and necrotic cells. *Reprinted from Edinger and Thompson (2004), copyright (2004), with permission from Elsevier.*

acid glucosyl esters, derived from tryptophan by action of the kynurenine pathway (Coburn et al., 2013).

1.2. *Caenorhabditis elegans* as a model to study necrosis

Caenorhabditis elegans has been instrumental in deciphering both apoptotic and necrotic cellular programs. This can be largely attributed to the specific characteristics and well-described developmental stages of this nematode, which make it exceptionally well suited for the study of both normal and aberrant cell death at the cellular, genetic, and molecular level. Due to its transparency, the visualization and tracking of single cells as well as of individual nuclei is readily feasible by differential interference contrast optics, enabling researchers to follow somatic cell divisions from the fertilized egg all the way to the 959 cell adult hermaphrodite (Sulston & Horvitz,

1977; Sulston, Schierenberg, White, & Thomson, 1983). The resulting cell lineage map indicated early on that in certain lineages, particular divisions generate cells which are destined to die at specific times and locations that remain faithfully invariant from one animal to another. Exactly 131 somatic cells die every time the fertilized egg normally develops into the adult animal, by an apoptotic PCD process.

1.3. The apoptotic machinery in *C. elegans*

Genetic and molecular studies performed in *C. elegans* provided a fundamental insight into the mechanisms underlying this cell death process. In the 131 cells destined to die during development, the level of EGL-1, a BH3 domain protein, is increased. EGL-1 interacts with a protein complex composed of CED-9 (similar to the mammalian B-cell lymphoma protein 2) and CED-4 (similar to the mammalian apoptotic protease-activating factor 1), releasing CED-4 which in turn activates CED-3 (similar to human caspases) (Hengartner, 2000). In *C. elegans*, four caspase-related genes exist: *ced-3*, *csp-1*, *csp-2*, and *csp-3* (Shaham, 1998; Yuan, Shaham, Ledoux, Ellis, & Horvitz, 1993); however, only *ced-3* seems to be required for PCD (Abraham & Shaham, 2004; Yuan et al., 1993), and only ced-3 and *csp-1* are proteolytically active (Shaham, 1998). The CSP-2 caspase lacks key active-site residues, and *csp-3* encodes only a C-terminal caspase domain, entirely lacking the active site (Shaham, 1998). As it turns out, the genetic encoding for the regulation and execution of developmental apoptosis has been remarkably conserved between *C. elegans* and mammals.

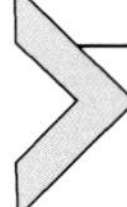

2. NECROTIC CELL DEATH PARADIGMS DURING *C. ELEGANS* DEVELOPMENT

The identification of the caspase CED-3 as a key regulator of apoptosis has been a key contribution of *C. elegans* to the cell death field, as caspases also play crucial roles in the execution of PCD across many species. However, as it turns out, not quite all cell death events during *C. elegans* development follow the typical apoptotic pathway that involves CED-4 and CED-3. Below, we elaborate on some well-studied examples of developmental death in the nematode that follow a necrotic pathway.

2.1. Death of the linker cell

The *C. elegans* linker cell has been described as an example of death that occurs during development following a mode that is independent of

CED-4 and CED-3 (Horvitz, Sternberg, Greenwald, Fixsen, & Ellis, 1983). The linker cell is born during the second larval stage (L2) in the central region of the animal and follows a stereotypical path of migration. As the cell migrates, it leads the extension of the male gonad behind it (Kimble & Hirsh, 1979; Sulston, Albertson, & Thomson, 1980), and upon completion of its migratory route, it is positioned between the gonad (vas deferens) and the cloacal tube, serving as an exit channel for sperm in the adult. It is generally thought that the death and removal of the linker cell around the L4/adult transition facilitates the fusion between the vas deferens and cloaca, to connect the male reproductive system to the exterior.

Following up on early observations that the programmed death of the linker cell persists even in *ced-3* mutant animals, the fate of this cell was thoroughly studied by following a GFP-marked linker cell in animals harboring mutations in core genes of the apoptotic machinery, such as *ced-3* and *ced-4*, as well as in engulfment genes. These studies demonstrated that the linker cell dies in a cell autonomous manner that, unlike it was postulated by previous reports (Sulston et al., 1980), does not require extrinsic signals from engulfing or other cells. Moreover, they showed that this death event is independent of any known apoptotic genes, in line with the lack of apoptotic morphological features, such as chromatin condensation. Instead, there was a noted presence of swollen and degraded mitochondria within large multilayered membrane-bound structures, as well as small electron-translucent "empty" membrane-bound cytoplasmic structures that resembled vacuoles typically seen during necrotic cell death in *C. elegans* (Hall et al., 1997) (Fig. 6.2). Although linker cell death does not satisfy all classical criteria of necrotic death, it is even further away from classical apoptotic paradigms. Possibly, the death of the linker cells falls under the characteristics of more recently described programmed necrosis processes, also known as necroptosis. However, additional experiments would be required to test this hypothesis and to further characterize the precise mode of death of the linker cell.

2.2. Death of mis-specified uterine–vulval (uv1) cells

A robust example of a necrotic event during development is the demise of mis-specified uterine–vulval (uv1) cells that have an important role in egg laying. Egg laying in *C. elegans* requires a connection between the lumens of the uterus in the somatic gonad and the vulva in the extragonadal epithelium, facilitated by cell–cell interactions between gonadal and vulval cells. Two specialized cell types of the ventral uterine π lineage are integral

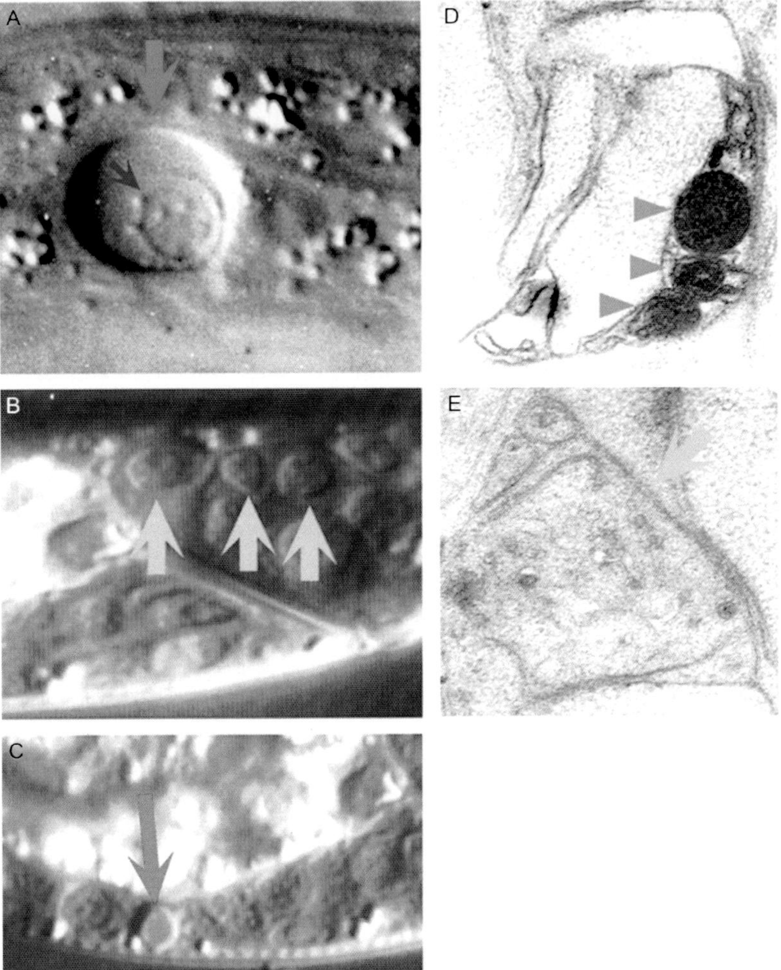

Figure 6.2 Necrotic cell death in *C. elegans*. The most prominent morphological characteristic of necrosis is the outstretched swelling of the cell to several times its normal diameter, which is manifested by a hollow, vacuole-like appearance under the optic microscope. For example, a dying PVM (posterior ventral microtubule) touch receptor is shown in (A) by a red arrow. This neuron is expressing a toxic variant of the degenerin MEC-4 (mechanosensory) protein that induces necrosis. The nucleus follows the cellular expansion (A; blue arrow). Healthy cells are indicated by green arrows for comparison (B). In sharp contrast, apoptosis, which normally occurs during nematode development, generates retractile cell corpses, compact in size, with a characteristic button-like appearance (C; red arrow). Under the electron microscope, the same degenerating neuron exhibits dark, electron-dense formations, most likely originating from plasma membrane-internalized material, arranged in onion-like concentric circles (D; arrowheads). At later stages of degeneration, the cytoplasm of the dying cell appears extensively depredated and fragmented. A normal neuron is shown in (E) by a green arrow. *Reprinted from Syntichaki and Tavernarakis (2002), copyright (2002), with permission from Nature Publishing Group.* (See the color plate.)

components of the uterine–vulval connection. These are the syncytial uterine seam (utse) cell, which overlies the vulval lumen, and the four uterine–vulval (uv1) cells, which directly contact the most dorsal vulval cell vulF (Newman, White, & Sternberg, 1996). The temporal and spatial specification of both these cell types largely relies on a specific signaling axis, where an inductive LIN-3 epidermal growth factor (EGF) signal derived from a single gonadal cell called the anchor cell activates the LET-23 EGF receptor on the receiving vulval precursor cells (Aroian, Koga, Mendel, Ohshima, & Sternberg, 1990; Hill & Sternberg, 1992). Mutations in genes of the LIN-3/LET-23/Ras signaling pathway compromise uv1 fate specification. A key study (Huang & Hanna-Rose, 2006) described the isolation of the *cog-3(ku212)* mutant, which uncouples gonadogenesis from its normal progression relative to the development of the vulva and shares phenotypes with heterochronic mutations that disturb the temporal coordination of vulval and uterine development. In *cog-3(ku212)* mutants, the entire uterus, including the pre-uv1 cells, is generated at a later stage of vulval development than is normal. Notably, the delayed pre-uv1 cells subsequently die by necrosis, leading to the absence of uv1 cells in the adult stage. Moreover, the study investigated if a LIN-3/LET-23/Ras signaling defect underlies the necrosis of uv1 defect in *cog-3(ku212)* mutants, by analyzing *cog-3(ku212)* double mutants with a gain-of-function allele of *let-23*. The results indicated that the *let-23(gf)* mutation rescued the mis-specification and death phenotype of uv1 cells, suggesting that the necrotic program is recruited during development in response to uncoordinated spatiotemporal development.

A recent study revealed the involvement of the *ku212* allele in uv1 cell necrosis, which maps to the *pnc-1* gene locus, encoding a nicotinamidase (van der Horst, Schavemaker, Pellis-van Berkel, & Burgering, 2007; Vrablik, Huang, Lange, & Hanna-Rose, 2009). Nicotinamidases are the first enzymes of the NAD^+ salvage pathway in invertebrates, using nicotinamide (NAM) as a substrate (Magni, Amici, Emanuelli, Raffaelli, & Ruggieri, 1999). Administration of high levels of NAM causes uv1 cells to die by necrosis at high frequency in wild-type animals. Thus, instead of compromised EGF signaling, the necrotic death of uv1 cells in *pnc-1* mutants may result from accumulation of the substrate NAM. In addition, the gonad-defective and uv1 cell death phenotypes are separable in *pnc-1* mutants. Constitutively active LET-23/EGF receptor prevents NAM-induced uv1 necrotic cell death, suggesting that EGF signaling may provide a survival cue that rescues uv1 cells from NAM-induced necrosis (reviewed in Vlachos & Tavernarakis, 2010).

3. NONDEVELOPMENTAL NECROTIC DEATH

In the adult nematode, necrotic cell death can be triggered by a wide variety of both extrinsic and intrinsic signals (Walker et al., 1988). Several well-defined conditions are known to trigger necrotic cell death in *C. elegans* and will be discussed below. The best-characterized case is the gain-of-function mutations in several ion channel genes, which result in an ionic imbalance and inflict a necrotic pattern of death on neurons. Other stimuli include extreme heat, hypo-osmotic shock, and bacterial infections. Cell demise in all these paradigms is accompanied by characteristic morphological features of necrosis, starting with the appearance of a distorted nucleus and cell body during the early phase of death. Gradually, the cell swells to several times its normal diameter and small, tightly wrapped membrane whorls form, originating from the plasma membrane and coalescing into large, electron-dense membranous structures (Hall et al., 1997). Interestingly enough, these membranous inclusions also represent characteristic hallmarks in mammalian neurodegenerative disorders, such as in neuronal ceroid lipofuscinosis (Batten's disease; the *mnd* mouse) as well as in the wobbler mouse, a model of amyotrophic lateral sclerosis (Blondet, Carpentier, Ait-Ikhlef, Murawsky, & Rieger, 2002; Cooper, Messer, Feng, Chua-Couzens, & Mobley, 1999).

3.1. Cell death induced by ionic imbalance

The most extensively characterized paradigm of non-PCD in adult *C. elegans* animals is the necrosis of cells expressing aberrant ion channels harboring unusual gain-of-function mutations (Syntichaki & Tavernarakis, 2003).

3.1.1 Degenerins

Dominant mutations in *deg-1* (degenerin; *deg-1(d)*) induce death of a group of interneurons of the nematode posterior touch sensory circuit (Chalfie & Wolinsky, 1990). Similarly, dominant mutations in the *mec-4* gene (mechanosensory; *mec-4(d)*) induce degeneration of six touch receptor neurons required for the sensation of gentle touch to the body (Syntichaki & Tavernarakis, 2004).

deg-1 and *mec-4* encode proteins that are very similar in sequence and were the first identified members of the *C. elegans* "degenerin" family, so named because several members can mutate to forms that induce cell degeneration (Chalfie, Driscoll, & Huang, 1993). Degenerins bear sequence similarity to mammalian epithelial sodium channels (ENaCs). The time of degeneration onset correlates with the initiation of degenerin gene expression, and the severity of cell death is analogous to the dose of the toxic allele (Hall et al., 1997). Expression of mammalian homologous proteins, carrying amino acid substitutions analogous to those of toxic degenerins, leads to degeneration of cells in a manner reminiscent of necrotic cell death in *C. elegans*. Additional members of the degenerin family are *mec-10*, which can be engineered to encode toxic degeneration-inducing substitutions, *unc-8*, which can mutate to a semi-dominant form that induces swelling and dysfunction of ventral nerve cord and *unc-105*, which appears to be expressed in muscle and can mutate to a semi-dominant form that induces muscle hypercontraction (Syntichaki & Tavernarakis, 2004). Thus, a unifying feature of degenerin family members is that specific gain-of-function mutations have deleterious consequences for the cells in which they are expressed, which, at least in neurons, culminate into a necrotic cell death event.

C. elegans degenerins share sequence similarity with *Drosophila* ripped pocket and pickpocket, with subunits of the vertebrate amiloride-sensitive ENaC and with other neuronally expressed ion channels. Together, these proteins define the DEG/ENaC protein superfamily (Tavernarakis & Driscoll, 2001). Although mutant degenerins can kill different groups of neurons depending on their expression patterns, the morphological features of the cell death that they induce are the same and resemble those of mammalian cells undergoing necrotic cell death. The pattern of necrotic cell death inflicted by degenerins is not a peculiarity of this gene class. For example, *C. elegans deg-3*, whose product is related to the vertebrate α-7 nicotinic acetylcholine receptor (nAChr) and together with the related protein DES-2 forms a very efficient calcium channel, can mutate to induce necrotic cell death similar to that induced by degenerins (Treinin, Gillo, Liebman, & Chalfie, 1998). In addition, mutant-activated forms of the heterotrimeric G-protein α subunit ($G\alpha_s$ Q208L), from both *C. elegans* and rat, cause swelling and degeneration of many cell types when expressed in *C. elegans* (Berger, Hart, & Kaplan, 1998; Korswagen, Park, Ohshima, & Plasterk, 1997).

3.1.2 Other ion channels

In addition to degenerins, gain-of-function mutations in other ion channel genes such as deg-3 lead to vacuolar degeneration of various types of *C. elegans* neurons. deg-3 encodes an acetylcholine receptor ion channel, related to the vertebrate nAChr that participates in the formation of a channel highly permeable to Ca^{2+} (Treinin & Chalfie, 1995). Moreover, expression of a constitutively active form of a heterotrimeric G-protein α subunit $G\alpha_s$ results in degeneration of a specific subset of neurons. Genetic suppressor analysis identified an adenyl cyclase as a downstream effector of $G\alpha_s$-induced neurodegeneration, indicating that cAMP signaling is critical for degeneration (Berger et al., 1998; Korswagen, van der Linden, & Plasterk, 1998).

Ionic imbalance and subsequent necrotic cell death induced by aberrant ion channel function in *C. elegans* is mechanistically and morphologically similar to excitotoxicity in vertebrates. Excitotoxic cell death is prevalent during stroke, where the energy required for sustaining ionic gradients and the resting potential of neurons is lost. Because membrane potential collapses, massive amounts of the excitatory neurotransmitter glutamate are released at synaptic clefts (Kauppinen, Enkvist, Holopainen, & Akerman, 1988; Kauppinen, McMahon, & Nicholls, 1988). Energy depletion also prevents reuptake of glutamate by dedicated transporters leading to accumulation of glutamate at synapses, hyperexcitation, and eventually necrotic death of downstream synaptic target neurons. Excitotoxicity is critically dependent on Ca^{2+} influx through glutamate-gated receptor ion channels (reviewed in Kourtis & Tavernarakis, 2007).

Malfunction of glutamate transporters and the resulting accumulation of glutamate are known to trigger excitotoxicity in several neurodegenerative diseases (Cleveland & Rothstein, 2001). However, the details on the cascade of events leading to neurodegeneration remain unclear. The molecular components of glutamatergic synapses assembled in *C. elegans* are highly conserved from nematodes to humans. A recent study describes a novel paradigm for nematode excitotoxicity, by investigating the *in vivo* effects of multiple mediators of glutamate-induced neuronal necrosis (Mano & Driscoll, 2009). Combined Δ*glt-3* glutamate transporter-null mutations and expression of a constitutively active form of the α subunit of the G-protein $G\alpha_s$ induces extensive neurodegeneration in head interneurons. Δ*glt-3*-dependent neurodegeneration acts through Ca^{2+}-permeable Glu receptors of the α-amino-3-hydroxyl-5-methyl-4-isoxazolepropionic acid subtype, requires calreticulin function, and is modulated by calcineurin and type-9 adenylyl cyclase (AC9). This glutamate-dependent toxicity defines a novel necrotic death paradigm in *C. elegans* that shares many basic features with excitotoxicity in mammalian neurons and may potentially be operative also in higher organisms.

3.2. Heat-induced necrotic death

Climate change has brought about a dramatic increase in the cases of heat stroke and related pathologies in humans. Causing core body temperature to reach over 40 °C, heat stroke inflicts immediate devastating tissue damage and inflammatory response that can be fatal, as well as long-term defects. To gain insight into the molecular mechanisms of heat cytotoxicity and to circumvent the confounding influence of secondary physiological and inflammatory responses, our laboratory developed and characterized a genetically tractable model of heat stroke in *C. elegans*. Widespread cell death across several tissues could be observed in animals exposed to hyperthermia, which in the nematode was simulated by a short exposure to 39 °C (Kourtis, Nikoletopoulou, & Tavernarakis, 2012). Dying cells displayed morphological features characteristic of necrosis, expressed markers of necrotic death, and became permeable to propidium iodide. Moreover, depletion of proteins required for necrosis strongly facilitated survival after heat stroke. In contrast, loss of key mediators and core components of the apoptotic or autophagic machineries did not suppress heat-stroke-induced cell death. Thus, heat stroke compromises viability by triggering extensive necrotic cell death and represents a newly added necrotic cell paradigm in the nematode.

Notably, we also observed that preconditioning animals at an intermediate, nonlethal temperature markedly enhanced their capacity to withstand a subsequent heat stroke. This protective effect is in line with the previously described phenomenon of hormesis (Calabrese, 2004), where preexposure to mild stress elicits increased resistance to subsequent severe stress. It is also worth noting that in addition to heat stroke, heat preconditioning conferred resistance against a wide range of necrotic death insults, including in particular ionic imbalance paradigms (discussed earlier), overexpression of aggregation-prone proteins (such as α-synuclein), and hypoxic conditions. In the case of hormesis by heat preconditioning, we found that cytoprotection is orchestrated at the molecular level by the hermetic induction of a single sHSP, HSP-16.1. sHSPs assemble into oligomeric complexes and serve as molecular chaperones, efficiently binding denatured proteins and/or preventing irreversible protein aggregation and insolubilization (Van Montfort, Slingsby, & Vierling, 2001). HSP-16.1 localizes in the Golgi, where it functions together with the PMR-1 pump to prevent cytoplasmic Ca^{2+} overload under extreme stress. We propose that HSP-16.1 contributes to stabilize and protect the stress-labile PMR-1 pump, allowing for efficient clearance of Ca^{2+} from the cytoplasm, after necrotic insult (Fig. 6.3).

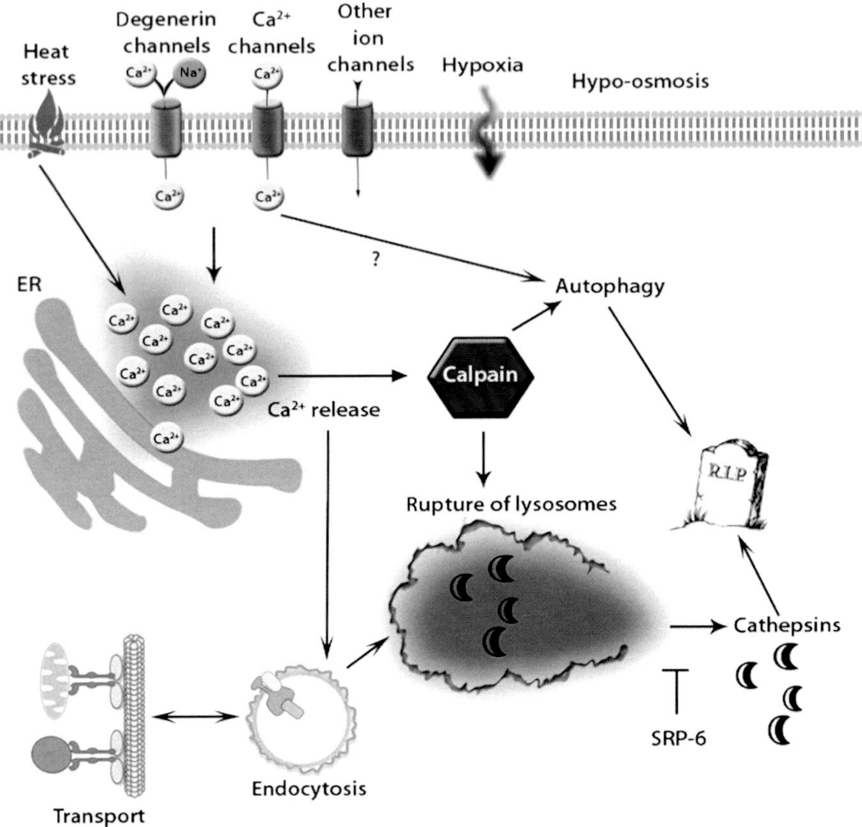

Figure 6.3 Necrotic cell death mechanisms. Various types of extrinsic necrotic insults converge on increased intracellular Ca^{2+} levels, caused by increased influx from extracellular pools through plasma membrane channels, or by Ca^{2+} efflux from intracellular stores, such as the endoplasmic reticulum (ER). Ca^{2+} then activates calpain proteases in the cytoplasm that attack lysosomal membrane proteins to compromise lysosomal integrity. Rupture of the lysosomes follows, and release of hydrolytic enzymes such as cathepsin proteases. In addition, autophagy is induced during necrosis, either directly by Ca^{2+} or via calpains and also contributes to cellular destruction. Moreover, both clathrin-mediated endocytosis and intracellular transport are required for necrotic death and are induced by necrosis-triggering insults. (See the color plate.)

Importantly, mammalian PMR-1 is selectively impaired during ischemic or reperfusion brain injury (Gidday, 2006; Lehotsky, Kaplan, Murin, & Raeymaekers, 2002; Pavlikova et al., 2009). Given the strong evolutionary conservation of the proteins involved, this mechanism is probably relevant to related human pathologies. Relevant to that, we also demonstrated that heat stroke induces widespread necrotic death in mammalian neurons, which can be largely prevented by heat preconditioning. Moreover, hormesis in mammalian neurons in response to heat preconditioning also requires the function of PMR-1 and is mediated by the same molecular players as in the nematode.

3.3. Bacterial infection-induced necrosis

Infection of *C. elegans* with different bacterial pathogens has been shown to induce necrotic death of intestinal cells as part of a pathogen-shared response to infection (Wong, Bazopoulou, Pujol, Tavernarakis, & Ewbank, 2007). Using whole-genome microarrays representing 20,334 genes, this study analyzed the transcriptional response of *C. elegans* to four bacterial pathogens (*Serratia marcescens*; *Enterococcus faecalis*; *Erwinia carotovora*; *Photorhabdus luminescens*). Different bacteria provoked pathogen-specific signatures within the host, including genes that encode potential pathogen-recognition and antimicrobial proteins. Additionally, variance analysis also revealed a robust signature that was commonly elicited by the pathogens. This involved 22 genes associated with proteolysis, necrotic cell death, and stress responses. Necrosis aggravated the pathogenesis and accelerated the death of the host. At later stages of infection, necrotic vacuoles are also observed in epidermal and gonadal cells. Moreover, mutations in genes required for necrosis ameliorated the consequences of infection, suggesting that necrosis is an integral part of host–pathogen interaction that contributes to the pathology associated with infection in *C. elegans*. These results are the first indication that necrosis is important for disease susceptibility in *C. elegans*. Moreover, given the striking similarities between the innate immune systems of invertebrates and vertebrates, as well as the fact that necrosis has been implicated in infections of human tissues, these findings invite the possibility to employ *C. elegans* as a model for the study of innate immunity in humans in response to infections.

3.4. Hypo-osmotic shock-induced cell death

Lysosomal integrity and lysosomal proteolytic mechanisms are key factors modulating necrotic cell death in the nematode. Serpins are extracellular or intracellular regulators of proteolytic pathways and inhibitors of multiple peptidases (Silverman et al., 2001). One of the functions of intracellular serpins is the inhibition of lysosomal cysteine peptidases. SRP-6 is such an intracellular serpin in *C. elegans*. *srp-6* null mutants experiencing hypo-osmotic conditions die rapidly and display marked increase of necrotic cell death of the intestinal epithelium (Luke et al., 2007). Ca^{2+} release from endoplasmic reticulum (ER) stores, together with other factors, induces calpain-mediated lysosomal rupture and massive release of lysosomal peptidases into the cytoplasm that mediate necrotic cell death. In addition to hypo-osmotic conditions, *srp-6* null mutants are susceptible to other stressors such as thermal and oxidative stress, hypoxia, and channel hyperactivity. SRP-6 appears to protect cells from lysosomal rupture and also ameliorates the deleterious consequences of lysosomal rupture triggered by various stressors. The protective function of SRP-6 may be adaptive by enhancing the degradation of misfolded proteins or by aiding cytoskeletal rearrangements through altering lysosomal membrane permeability and allowing the leakage of small amounts of peptidases. In the absence of SRP-6, the uncontrolled release of these peptidases leads to necrotic cell death.

4. EXECUTION OF NECROSIS

Intracellular calcium overload through different sources is considered as one of the leading steps in the necrotic pathway. Calcium may enter the cell through voltage-gated channels and the Na^{+}/Ca^{2+} exchanger and mutations that increase sodium influx augment calcium entry through these paths. The main intracellular compartment for calcium storage is the ER (Mattson et al., 2000; Paschen, 2001; Paschen & Frandsen, 2001), where calcium is sequestered by the sarcoendoplasmic reticulum Ca^{2+}-ATPase and is released back to the cytoplasm by ryanodine and inositol-1,4,5-triphosphate receptors. In *C. elegans*, extensive genetic screens for suppressors of *mec-4*(*d*)-induced necrosis have identified genes required for the execution of necrotic cell death. Two of these genes encode the calcium-binding chaperones calreticulin and calnexin, which were found to regulate intracellular calcium levels and to be required for necrotic cell death (Xu, Tavernarakis, & Driscoll, 2001). Moreover, treatment of animals with thapsigargin, a drug that induces release of calcium from the ER to the cytoplasm, triggers necrotic cell death, whereas pharmacological treatments or genetic mutations that inhibit calcium release from the ER have a strong protective effect against necrotic cell death.

Genetic studies in *C. elegans* have also shown that in addition to calcium homeostasis, intracellular pH is also an important modulator of necrotic cell death. Cytoplasmic acidification occurs during necrosis, whereas the vacuolar H^{+}-ATPase, which is a pump that acidifies lysosomes and other intracellular organelles, is required downstream of cytoplasmic calcium overload to promote necrotic cell death (Syntichaki, Samara, & Tavernarakis, 2005). In line with this, reduced vacuolar H^+-ATPase activity or alkalization of acidic endosomal/lysosomal compartments by weak bases has a neuroprotective role against necrosis. Acidic conditions are required for full activity of cathepsins, aspartyl proteases that are primarily confined to lysosomes and other acidic endosomal compartments (Ishidoh & Kominami, 2002).

Lysosomal as well as cytoplasmic proteases have been implicated as downstream effectors of cellular destruction in necrosis. Calpains are cytoplasmic, papain-like cysteine proteases that depend on calcium for their activity. Under normal conditions, calpains function to mediate essential signaling and metabolic processes. However, during the course of necrotic cell death, these proteases localize onto lysosomal membranes and may compromise lysosomal integrity, thereby causing leakage of their acidic contents, including lysosomal proteases, into the cytoplasm (Yamashima, 2004). In primates, calpains rapidly localize to lysosomal membranes after the onset of ischemic episodes (Yamashima, 2000). In *C. elegans*, two specific calpains—TRA-3 and CLP-1—and two lysosomal cathepsin proteases—ASP-3 and ASP-4—are required for neurodegeneration (Syntichaki, Xu, Driscoll, & Tavernarakis, 2002). It is likely that ensuing cytoplasmic acidification, activation of the lysosomal, low-pH-dependent cathepsins and hydrolases contributes to cell demise. Mutations that interfere with lysosomal biogenesis and function influence necrotic cell death. For example, necrosis is exacerbated in mutants that accumulate abnormally large lysosomes, whereas impairment of lysosomal biogenesis protects from cell death (Artal-Sanz, Samara, Syntichaki, & Tavernarakis, 2006). Interestingly, lysosomes appear to coalesce around the nucleus and dramatically enlarge during early and intermediate stages of necrosis. In advanced stages of cell death, GFP-labeled lysosomal membranes fade, as lysosomes rupture.

In a recent study from our laboratory, we utilized well-characterized necrosis models in *C. elegans* to dissect the involvement of clathrin-mediated endocytosis and intracellular trafficking by kinesin motor proteins in cellular destruction during necrotic death (Troulinaki & Tavernarakis, 2012). Our findings revealed for the first time that both clathrin-mediated endocytosis and intracellular trafficking are required for the execution of necrosis in the nematode. Downregulation of endocytosis or kinesin-mediated trafficking by interfering with key proteins regulating these processes, including SNT-1, endophilin (UNC-57), AP180 (UNC-11), synaptojanin (UNC-26), heavy chain of kinesin 1 (UNC-116), and the monomeric kinesin UNC-104, significantly suppresses neurodegeneration induced by hyperactive ion channels without affecting the expression, the localization, or the function of the toxic insults.

Moreover, using the same well-defined necrotic cell paradigm, we assayed animals that were deficient for both autophagy and endocytosis and observed significant synergistic protection against degeneration. These results suggest that autophagy and endocytosis function in parallel to contribute to necrotic cell death (Troulinaki & Tavernarakis, 2012).

5. *C. ELEGANS* AS A MODEL FOR HUMAN DISEASES ENTAILING NECROSIS

Nematode genes and major signaling pathways show significant conservation during evolution and more than 50% of the *C. elegans* genes have counterparts in humans. In addition to its contribution in elucidating developmental processes, the worm has also served as a platform to model many human pathological conditions such as neurodegenerative disorders, cancer, aging, and associated diseases (Baumeister & Ge, 2002; Lee, Goedert, & Trojanowski, 2001; Poulin, Nandakumar, & Ahringer, 2004). Systematic mapping of gene interactions and signaling pathways implicated in human disease using *C. elegans* has provided better understanding of complex pathologies (Bussey, Andrews, & Boone, 2006). The ability to produce "humanized" worms, which express human genes not present in the *C. elegans* genome, has further enhanced the experimental value of the nematode by allowing the dissection of molecular mechanisms relevant to human disorders. In addition, the ease of drug testing coupled with the efficiency of genetic screens in worms has made *C. elegans* a favorable tool for the identification and validation of novel drugs and drug targets, aiming to battle human pathological conditions (Kaletta & Hengartner, 2006). Here, we overview *C. elegans* models of human diseases that entail necrosis, focusing on hypoxia, Parkinson's disease, and tauopathies. Clearly, this list is only indicative of the applications of *C. elegans* in understanding complex human pathologies that involve necrotic death, and many more such diseases that are not mentioned here have been usefully modeled in the nematode.

5.1. Hypoxia

In humans, oxygen deprivation induces cell death in pathological conditions such as stroke and heart attack. In *C. elegans*, hypoxia inflicts necrotic death in a variety of cell types (Scott, Avidan, & Crowder, 2002). Interestingly, mutations in the *daf-2* gene, which encodes the *C. elegans* insulin/IGF receptor tyrosine kinase, confer resistance against hypoxic cell death.

DAF-2 is also known to regulate aging and dauer formation in *C. elegans* (Libina, Berman, & Kenyon, 2003). Related to this, many human neurodegenerative disorders show a late-onset pathogenesis, indicating that aging may alter the vulnerability of cells to various insults. However, while hypoxia resistance in *C. elegans* appears to be modulated by insulin signaling, other *daf-2* mutations that affect longevity and stress resistance do not affect hypoxic death. Selective expression of wild-type *daf-2* in neurons and muscles restores hypoxic death in *daf-2* hypoxia-resistant mutants, demonstrating a role of the insulin/IGF receptor in the protection of myocytes and neurons from hypoxic injury. Na^+-activated potassium (KNa) channels have been identified in cardiomyocytes and neurons as mediators of the protective mechanisms against hypoxic death (Bader, Bernheim, & Bertrand, 1985; Kameyama et al., 1984). In *C. elegans*, a KNa ion channel is encoded by the *slo-2* gene. *slo-2* mutants are hypersensitive to hypoxic death, suggesting that SLO-2 protects against hypoxia effects. Thus, molecular characterization of KNa channels may allow the development of specific agonists and antagonists, in an effort to combat hypoxia-caused pathologies (Yuan et al., 2003).

A recent study reported that SLO-2 channels, SLO-2a and a novel N-terminal variant isoform, SLO-2b, are activated by Ca^{2+} and voltage, but in contrast to previous reports, they do not exhibit Cl^- sensitivity. In contrast to SLO-1, SLO-2 loss-of-function mutants confer resistance to hypoxia in *C. elegans* (Zhang et al., 2013).

5.2. Parkinson's disease

α-Synuclein is a small protein expressed primarily at presynaptic terminals in the central nervous system. It is enriched at presynaptic terminals, where it promotes the assembly of the SNARE machinery, and is proposed to play a role in neurotransmitter release and in regulating membrane stability and neuronal plasticity (Kalia, Kalia, McLean, Lozano, & Lang, 2013; Recchia et al., 2004). Inclusions of α-synuclein represent a hallmark feature of pathology in both sporadic and familial cases of Parkinson's disease, and constitute the main component of Lewy bodies found in degenerating dopamine neurons (Spillantini et al., 1997). Mutations in the α-synuclein gene or multiplications of the α-synuclein locus have also been associated with some autosomal-dominant familial cases of Parkinson's disease (Chartier-Harlin et al., 2004; Polymeropoulos et al., 1997; Singleton et al., 2003). The A53T and A30P mutations have been shown to cause rare cases of autosomal-dominant heritable early-onset PD.

C. elegans models of wild-type or mutated human α-synuclein overexpression have been established, either pan-neuronally or specifically in dopaminergic neurons (Cao, Gelwix, Caldwell, & Caldwell, 2005; Cooper et al., 2006; Kuwahara et al., 2006; Lakso et al., 2003; Qiao et al., 2008), and result in significant motor deficits. No inclusion bodies or α-synuclein aggregation is observed, and intracellular inclusions are rarely observed in these transgenic animals. Overexpression of wild-type or mutant human α-synuclein specifically in worm dopaminergic neurons causes their degeneration, which becomes more pronounced as animal age (Cao et al., 2005; Cooper et al., 2006; Kuwahara et al., 2006).

One of the mechanisms implicated in the pathogenesis of Parkinson's disease is mitochondrial dysfunction (Schapira, 2008). Autosomal-dominant mutations in the leucine-rich repeat kinase 2 (LRRK2) have been associated with both familial and late-onset cases of PD, with G2019S being a prominent such mutation. *C elegans* engineered to express the human LRRK2 (G2019S) mutant form shows extensive loss of dopaminergic neurons (Saha et al., 2009), by increasing their vulnerability to mitochondrial stress. Expression of the wild-type LRRK2 has a milder effect on neuron loss. Similarly, loss-of-function mutations in the *lrk-1* gene, encoding the worm ortholog of LRRK2, also sensitize dopaminergic neurons to mitochondrial stress.

C. elegans models of α-synuclein-induced dopaminergic neurodegeneration have been used as a platform to identify suppressors of dopaminergic neuron loss with some success. For example, specific overexpression of human torsinA or the worm homolog TOR-2 protects dopamine neurons in these models (Cao et al., 2005). In addition, overexpression of the human lysosomal enzyme cathepsin D has a similar neuroprotective effect (Qiao et al., 2008). Several other molecules involved in autophagy, lysosomal function, trafficking, and G-protein signaling have also been identified in RNAi suppressor screenings (Hamamichi et al., 2008).

Moreover, a recent study (Ruan, Harrington, Caldwell, Caldwell, & Standaert, 2010) reported that the overexpression of hVPS41, which is the human ortholog of *vps-41*, could prevent dopamine neuron degeneration induced by α-synuclein overexpression in *C. elegans*. Furthermore, the neuroprotective effect of hVPS41 can enhance the clearance of misfolded and aggregated α-synuclein through the AP-3 (heterotetrameric adaptor protein complex) interaction domain and clathrin heavy-chain repeat domain (Harrington, Yacoubian, Slone, Caldwell, & Caldwell, 2012). These data reveal the critical role of lysosomal trafficking in maintaining cellular homeostasis in the presence of toxic proteins.

ATP13A2 (also known as PARK9) encodes a lysosomal protein that is linked with PD. Overexpression of ATP13A2 can rescue the loss of dopamine neuron caused by α-synuclein overexpression in *C. elegans*, while knockdown of the ATP13A2 ortholog in worms can enhance α-synuclein misfolding and toxicity (Gitler et al., 2009). A recent study performed a screen to identify the ATP13A2 interacting partners responsible for modifying α-synuclein aggregation and toxicity to dopamine neurons in *C. elegans* (Usenovic et al., 2012). The modifiers of α-syn misfolding and neurotoxicity belong to groups responsible for ER and Golgi transport (YIF1A), clathrin-mediated vesicular transport (AAK1), and lysosomal fusion and degradation of aggregated proteins (HDAC6), indicating the importance of these processes in α-synuclein-mediated toxicity.

Another study (Su et al., 2010) performed a high-throughput chemical screen in yeast and identified strong suppressors of α-synuclein toxicity. The compounds identified were also verified in *C. elegans*, where they also rescue α-synuclein-mediated dopaminergic neuron loss in the worm model of PD (Su et al., 2010). These results suggest that it is possible to develop novel therapeutic strategies to simultaneously target the multiple pathological features of PD.

5.3. Tau toxicity: Modeling Alzheimer's disease in *C. elegans*

Several neurodegenerative diseases (including, in particular, Alzheimer's disease, frontotemporal dementia and Parkinsonism linked to chromosome 17, FTDP-17) are characterized by neurofibrillary tangles consisting of hyperphosphorylated forms of the microtubule-associated protein *Tau*, encoded by the *mapt* gene (Lee et al., 2001). Although the exact role of *tau* in the pathogenesis of these diseases is not clear, the identification of autosomal-dominant mutations in the *mapt* gene indicates a crucial role for the altered *tau* protein in the neurodegenerative process (Hutton et al., 1998; Poorkaj et al., 1998; Spillantini, Crowther, Kamphorst, Heutink, & van Swieten, 1998).

Two studies have investigated the effects of expressing different forms of human *tau* (wild-type *tau* or *tau* carrying FTDP-17 mutations) in the nervous system of *C. elegans*. These studies included pan-neuronal expression, under the control of the *aex-3* promoter (Kraemer et al., 2003), or specifically in touch receptor neurons of *C. elegans*, under the control of the *mec-7* promoter (Miyasaka et al., 2005). In the first study, expression of either wild-type or FTDP-17 human *tau* resulted in reduced lifespan, behavioral abnormalities, progressive uncoordinated movement, accumulation of insoluble phosphorylated *tau*, defective cholinergic neurotransmission, and age-dependent axonal and neuronal degeneration. This degenerative phenotype was more severe in lines expressing FTDP-17 mutant tau compared to those expressing wild-type tau. Morphologically, neurodegeneration was manifested by axonal vacuolar clearing, collapsed membrane structure, and membranous infoldings and whorls (which are characteristic of necrotic cell death), with associated amorphous *tau* accumulations and abnormal *tau*-positive aggregates, without, however, tau filaments observed (Kraemer et al., 2003).

The second study analyzed transgenic worms expressing wild-type or mutant (P301L and R406W) tau specifically in the touch (mechanosensory) neurons. Whereas worms expressing wild-type tau showed a small decrease in the touch response across their lifespan, worms expressing mutant tau displayed a large and progressive decrease. Loss of touch neurons function was accompanied by prominent neuritic abnormalities and microtubular loss. A substantial fraction of degenerating neurons developed tau accumulation in the cell body and neuronal processes. Notably, this neuronal dysfunction was not related to the apoptotic process because little recovery from touch abnormality was observed in *ced-3* or *ced-4*-deficient backgrounds.

A recent study employed *C. elegans* to investigate the relationship between tau aggregation and toxicity by comparing transgenic worms expressing pro- or antiaggregative tau species in their nervous system. The findings indicated that animals expressing the highly amyloidogenic tau species showed accelerated aggregation and pathology manifested by severely impaired motility, impaired axonal transport of mitochondria, and evident neuronal dysfunction. By contrast, control animals expressing the antiaggregant combination had rather mild phenotype (Fatouros et al., 2012). Furthermore, this study used the transgenic worms to screen for compounds that act as inhibitors of tau aggregation. Treatment of the proaggregant transgenic strains with a novel tau aggregation inhibitor, a compound belonging to the aminothienopyridazine class, ameliorated the motility phenotype, reflected also by a reduced extent of the progressive accumulation of neuronal morphological abnormalities.

C. elegans was also recently used to identify factors that are required for tau-mediated neurotoxicity. *Sut-2* was identified as such as gene, as recessive loss-of-function mutations in the *sut-2* locus suppress the tau aggregation and neurodegenerative changes caused by human tau expression in worms (Guthrie, Schellenberg, & Kraemer, 2009). The *sut-2* gene encodes a novel subtype of CCCH zinc finger protein conserved across animal phyla, sharing significant identity with the mammalian SUT-2 (MSUT-2). Notably, the involvement of sut-2 in tau-mediated toxicity was also verified in mammalian cells, both *in vitro* and in postmortem human tissues (Guthrie, Greenup, Leverenz, & Kraemer, 2011). Specifically, RNAi knockdown of MSUT-2 in cultured human cells overexpressing tau causes a marked decrease in tau aggregation. Both cell culture and postmortem tissue studies suggest that MSUT-2 levels may influence neuronal vulnerability to tau toxicity and aggregation. Thus, neuroprotective strategies targeting MSUT-2 may be of therapeutic interest for tauopathy disorders.

6. CONCLUDING REMARKS

In this chapter, we have attempted to provide a comprehensive overview of the necrotic cell death paradigms that have been established in *C. elegans* (see Table 6.1) and also to convey our current understanding

Table 6.1 Triggers and paradigms of necrotic death in *C. elegans*

Death initiator	Type of insult	Dying cells	References
mec-4(u231), referred to as *mec-4(d)*	Hyperactive degenerin ion channel	Touch receptor neurons	Driscoll and Chalfie (1991)
mec-10(A673V), referred to as *mec-10(d)*	Hyperactive degenerin ion channel	Touch receptor neurons	Huang and Chalfie (1994)
deg-1(u38), referred to as *deg-1(d)*	Hyperactive degenerin ion channel	Some polymodal neurons and specific interneurons	Chalfie and Wolinsky (1990)
unc-8(n491)	Hyperactive degenerin ion channel	Motor neurons	Shreffler, Magardino, Shekdar, & Wolinsky, (1995), Tavernarakis Shreffler, Wang, & Driscoll, (1997)
pnc-1(ku212) or *cog-3(ku212)* (as was initially named)	Excess nicotinamide levels	Uterine vulval 1 cells	Huang and Hanna-Rose (2006), Vrablik et al. (2009)

Continued

Table 6.1 Triggers and paradigms of necrotic death in *C. elegans*—cont'd

Death initiator	Type of insult	Dying cells	References
Nicotinamide	Excess nicotinamide levels	Uterine vulval 1 cells	Vrablik et al. (2009)
deg-3(u662), referred to as *deg-3(d)*	Hyperactive nicotinic acetylcholine receptor	Subset of sensory neurons and interneurons	Treinin and Chalfie (1995)
gsa-1(Q208L) and $G\alpha_s$(Q227L), referred to as α_s(*gf*)	Constitutively active GTP-binding protein $G\alpha_s$	Motor neurons, interneurons, head and tail ganglia neurons, and pharyngeal neurons or epithelial cells (unidentified)	Korswagen et al. (1997), Berger et al. (1998)
Thapsigargin	Elevation of intracellular Ca^{2+} levels	Random cells (including neuronal)	Xu et al. (2001)
Δ*glt-3*;α_s(*gf*)	Glutamate-dependent toxicity	Head neurons	Mano and Driscoll (2009)
Erwinia carotovora, Photorhabdus luminescens	Pathogen infection	Intestinal, epidermal, and gonadal cells	Wong et al. (2007)
Hypoxic treatment	Oxygen/energy limitation	Pharynx, gonad primordium, body wall muscles, and unidentified cells	Scott et al. (2002)
α-Synuclein	Stress induction	Dopaminergic neurons	Lakso et al. (2003), Cao et al. (2005), Cooper et al. (2006), Kuwahara et al. (2006), Qiao et al. (2008)
LRRK2 (leucine-rich repeat kinase 2)	Stress induction	Dopaminergic neurons	Saha et al. (2009)
Tau protein	Stress induction	Several neurons (including motor neurons)	Kraemer et al. (2003)

Summary of the necrotic stimuli and the cell populations they affect, as discussed in this chapter.

of the molecular mechanisms involved. The rich repertoire of necrotic cell death events that occur in *C. elegans* both during development, as well as, in the adult renders the nematode a particularly attractive platform for dissecting the mechanisms of pathological cell death in humans, which is typically mediated by necrotic processes.

The similarity of necrotic cell death triggered by hyperactive ion channels in *C. elegans* to excitotoxic cell death and neurodegeneration in mammals, both in terms of morphological characteristics and mechanistic aspects, reflects the extensive evolutionary conservation of necrosis-relevant genes between *C. elegans* and mammals. Moreover, conservation of the mechanisms that protect *C. elegans* and mammalian cells from necrotic death inflicted by diverse stimuli, as exhibited, for example, by the hormetic induction of HSF16.1 upon heat preconditioning, provides new prospects for employing the nematode in the battle against degeneration. Concomitantly, modeling of human degenerative disorders, such as Parkinson's disease and others, in *C. elegans* has already accelerated the pace of the molecular dissection of the underlying mechanisms and holds promise for the development and testing of innovative intervention strategies.

ACKNOWLEDGMENTS

Work in the authors' laboratory is funded by grants from the European Research Council (ERC), the European Commission Framework Programmes, and the Greek Ministry of Education. V. N. is supported by an EMBO long-term postdoctoral fellowship.

REFERENCES

Abraham, M. C., & Shaham, S. (2004). Death without caspases, caspases without death. *Trends in Cell Biology*, *14*(4), 184–193. http://dx.doi.org/10.1016/j.tcb.2004.03.002.

Aroian, R. V., Koga, M., Mendel, J. E., Ohshima, Y., & Sternberg, P. W. (1990). The let-23 gene necessary for *Caenorhabditis elegans* vulval induction encodes a tyrosine kinase of the EGF receptor subfamily. *Nature*, *348*(6303), 693–699. http://dx.doi.org/10.1038/348693a0.

Artal-Sanz, M., Samara, C., Syntichaki, P., & Tavernarakis, N. (2006). Lysosomal biogenesis and function is critical for necrotic cell death in *Caenorhabditis elegans*. *Journal of Cell Biology*, *173*(2), 231–239. http://dx.doi.org/10.1083/jcb.200511103.

Bader, C. R., Bernheim, L., & Bertrand, D. (1985). Sodium-activated potassium current in cultured avian neurones. *Nature*, *317*(6037), 540–542.

Baumeister, R., & Ge, L. (2002). The worm in us—*Caenorhabditis elegans* as a model of human disease. *Trends in Biotechnology*, *20*(4), 147–148.

Berger, A. J., Hart, A. C., & Kaplan, J. M. (1998). G alphas-induced neurodegeneration in *Caenorhabditis elegans*. *Journal of Neuroscience*, *18*(8), 2871–2880.

Blondet, B., Carpentier, G., Ait-Ikhlef, A., Murawsky, M., & Rieger, F. (2002). Motoneuron morphological alterations before and after the onset of the disease in the wobbler mouse. *Brain Research*, *930*(1–2), 53–57.

Bussey, H., Andrews, B., & Boone, C. (2006). From worm genetic networks to complex human diseases. *Nature Genetics*, *38*(8), 862–863. http://dx.doi.org/10.1038/ng0806-862.

Calabrese, E. J. (2004). Hormesis: A revolution in toxicology, risk assessment and medicine. *EMBO Reports*, *5*(Suppl. 1), S37–S40. http://dx.doi.org/10.1038/sj.embor.7400222.

Cao, S., Gelwix, C. C., Caldwell, K. A., & Caldwell, G. A. (2005). Torsin-mediated protection from cellular stress in the dopaminergic neurons of *Caenorhabditis elegans*. *Journal of Neuroscience*, *25*(15), 3801–3812. http://dx.doi.org/10.1523/JNEUROSCI.5157-04.2005.

Chalfie, M., Driscoll, M., & Huang, M. (1993). Degenerin similarities. *Nature*, *361*(6412), 504. http://dx.doi.org/10.1038/361504a0.

Chalfie, M., & Wolinsky, E. (1990). The identification and suppression of inherited neurodegeneration in *Caenorhabditis elegans*. *Nature*, *345*(6274), 410–416. http://dx.doi.org/10.1038/345410a0.

Chartier-Harlin, M. C., Kachergus, J., Roumier, C., Mouroux, V., Douay, X., Lincoln, S., et al. (2004). Alpha-synuclein locus duplication as a cause of familial Parkinson's disease. *Lancet*, *364*(9440), 1167–1169. http://dx.doi.org/10.1016/S0140-6736(04)17103-1.

Cleveland, D. W., & Rothstein, J. D. (2001). From Charcot to Lou Gehrig: Deciphering selective motor neuron death in ALS. *Nature Reviews. Neuroscience*, *2*(11), 806–819. http://dx.doi.org/10.1038/35097565.

Coburn, C., Allman, E., Mahanti, P., Benedetto, A., Cabreiro, F., Pincus, Z., et al. (2013). Anthranilate fluorescence marks a calcium-propagated necrotic wave that promotes organismal death in *C. elegans*. *PLoS Biology*, *11*(7), e1001613. http://dx.doi.org/10.1371/journal.pbio.1001613.

Cooper, A. A., Gitler, A. D., Cashikar, A., Haynes, C. M., Hill, K. J., Bhullar, B., et al. (2006). Alpha-synuclein blocks ER-Golgi traffic and Rab1 rescues neuron loss in Parkinson's models. *Science*, *313*(5785), 324–328. http://dx.doi.org/10.1126/science.1129462.

Cooper, J. D., Messer, A., Feng, A. K., Chua-Couzens, J., & Mobley, W. C. (1999). Apparent loss and hypertrophy of interneurons in a mouse model of neuronal ceroid lipofuscinosis: Evidence for partial response to insulin-like growth factor-1 treatment. *Journal of Neuroscience*, *19*(7), 2556–2567.

Driscoll, M., & Chalfie, M. (1991). The mec-4 gene is a member of a family of *Caenorhabditis elegans* genes that can mutate to induce neuronal degeneration. *Nature*, *349*(6310), 588–593.

Edinger, A. L., & Thompson, C. B. (2004). Death by design: Apoptosis, necrosis and autophagy. *Current Opinion in Cell Biology*, *16*(6), 663–669. http://dx.doi.org/10.1016/j.ceb.2004.09.011.

Fatouros, C., Pir, G. J., Biernat, J., Koushika, S. P., Mandelkow, E., Mandelkow, E. M., et al. (2012). Inhibition of tau aggregation in a novel *Caenorhabditis elegans* model of tauopathy mitigates proteotoxicity. *Human Molecular Genetics*, *21*(16), 3587–3603. http://dx.doi.org/10.1093/hmg/dds190.

Galluzzi, L., Maiuri, M. C., Vitale, I., Zischka, H., Castedo, M., Zitvogel, L., et al. (2007). Cell death modalities: Classification and pathophysiological implications. *Cell Death and Differentiation*, *14*(7), 1237–1243. http://dx.doi.org/10.1038/sj.cdd.4402148.

Gidday, J. M. (2006). Cerebral preconditioning and ischaemic tolerance. *Nature Reviews. Neuroscience*, 7(6), 437–448. http://dx.doi.org/10.1038/nrn1927.

Gitler, A. D., Chesi, A., Geddie, M. L., Strathearn, K. E., Hamamichi, S., Hill, K. J., et al. (2009). Alpha-synuclein is part of a diverse and highly conserved interaction network that includes PARK9 and manganese toxicity. *Nature Genetics*, *41*(3), 308–315. http://dx.doi.org/10.1038/ng.300.

Guthrie, C. R., Greenup, L., Leverenz, J. B., & Kraemer, B. C. (2011). MSUT2 is a determinant of susceptibility to tau neurotoxicity. *Human Molecular Genetics*, *20*(10), 1989–1999. http://dx.doi.org/10.1093/hmg/ddr079.

Guthrie, C. R., Schellenberg, G. D., & Kraemer, B. C. (2009). SUT-2 potentiates tau-induced neurotoxicity in *Caenorhabditis elegans*. *Human Molecular Genetics*, *18*(10), 1825–1838. http://dx.doi.org/10.1093/hmg/ddp099.

Hall, D. H., Gu, G., Garcia-Anoveros, J., Gong, L., Chalfie, M., & Driscoll, M. (1997). Neuropathology of degenerative cell death in *Caenorhabditis elegans*. *Journal of Neuroscience*, *17*(3), 1033–1045.

Hamamichi, S., Rivas, R. N., Knight, A. L., Cao, S., Caldwell, K. A., & Caldwell, G. A. (2008). Hypothesis-based RNAi screening identifies neuroprotective genes in a Parkinson's disease model. *Proceedings of the National Academy of Sciences of the United States of America*, *105*(2), 728–733. http://dx.doi.org/10.1073/pnas.0711018105.

Harrington, A. J., Yacoubian, T. A., Slone, S. R., Caldwell, K. A., & Caldwell, G. A. (2012). Functional analysis of VPS41-mediated neuroprotection in *Caenorhabditis elegans* and mammalian models of Parkinson's disease. *Journal of Neuroscience*, *32*(6), 2142–2153. http://dx.doi.org/10.1523/JNEUROSCI.2606-11.2012.

Hengartner, M. O. (2000). The biochemistry of apoptosis. *Nature*, *407*(6805), 770–776. http://dx.doi.org/10.1038/35037710.

Hill, R. J., & Sternberg, P. W. (1992). The gene lin-3 encodes an inductive signal for vulval development in *C. elegans*. *Nature*, *358*(6386), 470–476. http://dx.doi.org/10.1038/358470a0.

Horvitz, H. R., Sternberg, P. W., Greenwald, I. S., Fixsen, W., & Ellis, H. M. (1983). Mutations that affect neural cell lineages and cell fates during the development of the nematode *Caenorhabditis elegans*. *Cold Spring Harbor Symposia on Quantitative Biology*, *48*(Pt. 2), 453–463.

Huang, M., & Chalfie, M. (1994). Gene interactions affecting mechanosensory transduction in *Caenorhabditis elegans*. *Nature*, *367*(6462), 467–470.

Huang, L., & Hanna-Rose, W. (2006). EGF signaling overcomes a uterine cell death associated with temporal mis-coordination of organogenesis within the *C. elegans* egg-laying apparatus. *Developmental Biology*, *300*(2), 599–611. http://dx.doi.org/10.1016/j.ydbio.2006.08.024.

Hutton, M., Lendon, C. L., Rizzu, P., Baker, M., Froelich, S., Houlden, H., et al. (1998). Association of missense and 5'-splice-site mutations in tau with the inherited dementia FTDP-17. *Nature*, *393*(6686), 702–705. http://dx.doi.org/10.1038/31508.

Ishidoh, K., & Kominami, E. (2002). Processing and activation of lysosomal proteinases. *Biological Chemistry*, *383*(12), 1827–1831. http://dx.doi.org/10.1515/BC.2002.206.

Kaletta, T., & Hengartner, M. O. (2006). Finding function in novel targets: *C. elegans* as a model organism. *Nature Reviews. Drug Discovery*, *5*(5), 387–398. http://dx.doi.org/10.1038/nrd2031.

Kalia, L. V., Kalia, S. K., McLean, P. J., Lozano, A. M., & Lang, A. E. (2013). alpha-Synuclein oligomers and clinical implications for Parkinson disease. *Annals of Neurology*, *73*(2), 155–169. http://dx.doi.org/10.1002/ana.23746.

Kameyama, M., Kakei, M., Sato, R., Shibasaki, T., Matsuda, H., & Irisawa, H. (1984). Intracellular Na+ activates a K+ channel in mammalian cardiac cells. *Nature*, *309*(5966), 354–356.

Kauppinen, R. A., Enkvist, K., Holopainen, I., & Akerman, K. E. (1988). Glucose deprivation depolarizes plasma membrane of cultured astrocytes and collapses transmembrane potassium and glutamate gradients. *Neuroscience*, *26*(1), 283–289.

Kauppinen, R. A., McMahon, H. T., & Nicholls, D. G. (1988). Ca2+-dependent and Ca2+-independent glutamate release, energy status and cytosolic free Ca2+

concentration in isolated nerve terminals following metabolic inhibition: Possible relevance to hypoglycaemia and anoxia. *Neuroscience*, *27*(1), 175–182.

Kerr, J. F., Wyllie, A. H., & Currie, A. R. (1972). Apoptosis: A basic biological phenomenon with wide-ranging implications in tissue kinetics. *British Journal of Cancer*, *26*(4), 239–257.

Kimble, J., & Hirsh, D. (1979). The postembryonic cell lineages of the hermaphrodite and male gonads in *Caenorhabditis elegans*. *Developmental Biology*, *70*(2), 396–417.

Korswagen, H. C., Park, J. H., Ohshima, Y., & Plasterk, R. H. (1997). An activating mutation in a *Caenorhabditis elegans* Gs protein induces neural degeneration. *Genes and Development*, *11*(12), 1493–1503.

Korswagen, H. C., van der Linden, A. M., & Plasterk, R. H. (1998). G protein hyperactivation of the *Caenorhabditis elegans* adenylyl cyclase SGS-1 induces neuronal degeneration. *EMBO Journal*, *17*(17), 5059–5065. http://dx.doi.org/10.1093/emboj/17.17.5059.

Kourtis, N., Nikoletopoulou, V., & Tavernarakis, N. (2012). Small heat-shock proteins protect from heat-stroke-associated neurodegeneration. *Nature*, *490*(7419), 213–218. http://dx.doi.org/10.1038/nature11417.

Kourtis, N., & Tavernarakis, N. (2007). Non-developmentally programmed cell death in *Caenorhabditis elegans*. *Seminars in Cancer Biology*, *17*(2), 122–133. http://dx.doi.org/10.1016/j.semcancer.2006.11.004.

Kraemer, B. C., Zhang, B., Leverenz, J. B., Thomas, J. H., Trojanowski, J. Q., & Schellenberg, G. D. (2003). Neurodegeneration and defective neurotransmission in a *Caenorhabditis elegans* model of tauopathy. *Proceedings of the National Academy of Sciences of the United States of America*, *100*(17), 9980–9985. http://dx.doi.org/10.1073/pnas.1533448100.

Kuwahara, T., Koyama, A., Gengyo-Ando, K., Masuda, M., Kowa, H., Tsunoda, M., et al. (2006). Familial Parkinson mutant alpha-synuclein causes dopamine neuron dysfunction in transgenic *Caenorhabditis elegans*. *Journal of Biological Chemistry*, *281*(1), 334–340. http://dx.doi.org/10.1074/jbc.M504860200.

Lakso, M., Vartiainen, S., Moilanen, A. M., Sirvio, J., Thomas, J. H., Nass, R., et al. (2003). Dopaminergic neuronal loss and motor deficits in *Caenorhabditis elegans* overexpressing human alpha-synuclein. *Journal of Neurochemistry*, *86*(1), 165–172.

Lee, V. M., Goedert, M., & Trojanowski, J. Q. (2001). Neurodegenerative tauopathies. *Annual Review of Neuroscience*, *24*, 1121–1159. http://dx.doi.org/10.1146/annurev.neuro.24.1.1121.

Lehotsky, J., Kaplan, P., Murin, R., & Raeymaekers, L. (2002). The role of plasma membrane Ca2+ pumps (PMCAs) in pathologies of mammalian cells. *Frontiers in Bioscience*, 7, d53–d84.

Libina, N., Berman, J. R., & Kenyon, C. (2003). Tissue-specific activities of *C. elegans* DAF-16 in the regulation of lifespan. *Cell*, *115*(4), 489–502.

Luke, C. J., Pak, S. C., Askew, Y. S., Naviglia, T. L., Askew, D. J., Nobar, S. M., et al. (2007). An intracellular serpin regulates necrosis by inhibiting the induction and sequelae of lysosomal injury. *Cell*, *130*(6), 1108–1119. http://dx.doi.org/10.1016/j.cell.2007.07.013.

Magni, G., Amici, A., Emanuelli, M., Raffaelli, N., & Ruggieri, S. (1999). Enzymology of NAD+ synthesis. *Advances in Enzymology and Related Areas of Molecular Biology*, *73*, 135–182, xi.

Mano, I., & Driscoll, M. (2009). *Caenorhabditis elegans* glutamate transporter deletion induces AMPA-receptor/adenylyl cyclase 9-dependent excitotoxicity. *Journal of Neurochemistry*, *108*(6), 1373–1384. http://dx.doi.org/10.1111/j.1471-4159.2008.05804.x.

Mattson, M. P., LaFerla, F. M., Chan, S. L., Leissring, M. A., Shepel, P. N., & Geiger, J. D. (2000). Calcium signaling in the ER: Its role in neuronal plasticity and neurodegenerative disorders. *Trends in Neurosciences*, *23*(5), 222–229.

Miyasaka, T., Ding, Z., Gengyo-Ando, K., Oue, M., Yamaguchi, H., Mitani, S., et al. (2005). Progressive neurodegeneration in *C. elegans* model of tauopathy. *Neurobiology of Disease, 20*(2), 372–383. http://dx.doi.org/10.1016/j.nbd.2005.03.017.

Newman, A. P., White, J. G., & Sternberg, P. W. (1996). Morphogenesis of the *C. elegans* hermaphrodite uterus. *Development, 122*(11), 3617–3626.

Paschen, W. (2001). Dependence of vital cell function on endoplasmic reticulum calcium levels: Implications for the mechanisms underlying neuronal cell injury in different pathological states. *Cell Calcium, 29*(1), 1–11. http://dx.doi.org/10.1054/ceca.2000.0162.

Paschen, W., & Frandsen, A. (2001). Endoplasmic reticulum dysfunction—A common denominator for cell injury in acute and degenerative diseases of the brain? *Journal of Neurochemistry, 79*(4), 719–725.

Pavlikova, M., Tatarkova, Z., Sivonova, M., Kaplan, P., Krizanova, O., & Lehotsky, J. (2009). Alterations induced by ischemic preconditioning on secretory pathways Ca2+-ATPase (SPCA) gene expression and oxidative damage after global cerebral ischemia/reperfusion in rats. *Cellular and Molecular Neurobiology, 29*(6–7), 909–916. http://dx.doi.org/10.1007/s10571-009-9374-6.

Polymeropoulos, M. H., Lavedan, C., Leroy, E., Ide, S. E., Dehejia, A., Dutra, A., et al. (1997). Mutation in the alpha-synuclein gene identified in families with Parkinson's disease. *Science, 276*(5321), 2045–2047.

Poorkaj, P., Bird, T. D., Wijsman, E., Nemens, E., Garruto, R. M., Anderson, L., et al. (1998). Tau is a candidate gene for chromosome 17 frontotemporal dementia. *Annals of Neurology, 43*(6), 815–825. http://dx.doi.org/10.1002/ana.410430617.

Poulin, G., Nandakumar, R., & Ahringer, J. (2004). Genome-wide RNAi screens in *Caenorhabditis elegans*: Impact on cancer research. *Oncogene, 23*(51), 8340–8345. http://dx.doi.org/10.1038/sj.onc.1208010.

Qiao, L., Hamamichi, S., Caldwell, K. A., Caldwell, G. A., Yacoubian, T. A., Wilson, S., et al. (2008). Lysosomal enzyme cathepsin D protects against alpha-synuclein aggregation and toxicity. *Molecular Brain, 1*, 17. http://dx.doi.org/10.1186/1756-6606-1-17.

Recchia, A., Debetto, P., Negro, A., Guidolin, D., Skaper, S. D., & Giusti, P. (2004). Alpha-synuclein and Parkinson's disease. *FASEB Journal, 18*(6), 617–626. http://dx.doi.org/10.1096/fj.03-0338rev.

Ruan, Q., Harrington, A. J., Caldwell, K. A., Caldwell, G. A., & Standaert, D. G. (2010). VPS41, a protein involved in lysosomal trafficking, is protective in *Caenorhabditis elegans* and mammalian cellular models of Parkinson's disease. *Neurobiology of Disease, 37*(2), 330–338. http://dx.doi.org/10.1016/j.nbd.2009.10.011.

Saha, S., Guillily, M. D., Ferree, A., Lanceta, J., Chan, D., Ghosh, J., et al. (2009). LRRK2 modulates vulnerability to mitochondrial dysfunction in *Caenorhabditis elegans*. *Journal of Neuroscience, 29*(29), 9210–9218. http://dx.doi.org/10.1523/JNEUROSCI.2281-09.2009.

Schapira, A. H. (2008). Mitochondria in the aetiology and pathogenesis of Parkinson's disease. *Lancet Neurology, 7*(1), 97–109. http://dx.doi.org/10.1016/S1474-4422(07)70327-7.

Scott, B. A., Avidan, M. S., & Crowder, C. M. (2002). Regulation of hypoxic death in *C. elegans* by the insulin/IGF receptor homolog DAF-2. *Science, 296*(5577), 2388–2391. http://dx.doi.org/10.1126/science.1072302.

Shaham, S. (1998). Identification of multiple *Caenorhabditis elegans* caspases and their potential roles in proteolytic cascades. *Journal of Biological Chemistry, 273*(52), 35109–35117.

Shreffler, W., Magardino, T., Shekdar, K., & Wolinsky, E. (1995). The unc-8 and sup-40 genes regulate ion channel function in *Caenorhabditis elegans* motorneurons. *Genetics, 139*(3), 1261–1272.

Silverman, G. A., Bird, P. I., Carrell, R. W., Church, F. C., Coughlin, P. B., Gettins, P. G., et al. (2001). The serpins are an expanding superfamily of structurally similar but functionally diverse proteins. Evolution, mechanism of inhibition, novel functions, and a revised nomenclature. *Journal of Biological Chemistry*, *276*(36), 33293–33296. http://dx.doi.org/10.1074/jbc.R100016200.

Singleton, A. B., Farrer, M., Johnson, J., Singleton, A., Hague, S., Kachergus, J., et al. (2003). alpha-Synuclein locus triplication causes Parkinson's disease. *Science*, *302*(5646), 841. http://dx.doi.org/10.1126/science.1090278.

Spillantini, M. G., Crowther, R. A., Kamphorst, W., Heutink, P., & van Swieten, J. C. (1998). Tau pathology in two Dutch families with mutations in the microtubule-binding region of tau. *American Journal of Pathology*, *153*(5), 1359–1363. http://dx.doi.org/10.1016/S0002-9440(10)65721-5.

Spillantini, M. G., Schmidt, M. L., Lee, V. M., Trojanowski, J. Q., Jakes, R., & Goedert, M. (1997). Alpha-synuclein in Lewy bodies. *Nature*, *388*(6645), 839–840. http://dx.doi.org/10.1038/42166.

Su, L. J., Auluck, P. K., Outeiro, T. F., Yeger-Lotem, E., Kritzer, J. A., Tardiff, D. F., et al. (2010). Compounds from an unbiased chemical screen reverse both ER-to-Golgi trafficking defects and mitochondrial dysfunction in Parkinson's disease models. *Disease Models & Mechanisms*, *3*(3–4), 194–208. http://dx.doi.org/10.1242/dmm.004267.

Sulston, J. E., Albertson, D. G., & Thomson, J. N. (1980). The *Caenorhabditis elegans* male: Postembryonic development of nongonadal structures. *Developmental Biology*, *78*(2), 542–576.

Sulston, J. E., & Horvitz, H. R. (1977). Post-embryonic cell lineages of the nematode, *Caenorhabditis elegans*. *Developmental Biology*, *56*(1), 110–156.

Sulston, J. E., Schierenberg, E., White, J. G., & Thomson, J. N. (1983). The embryonic cell lineage of the nematode *Caenorhabditis elegans*. *Developmental Biology*, *100*(1), 64–119.

Syntichaki, P., Samara, C., & Tavernarakis, N. (2005). The vacuolar H+-ATPase mediates intracellular acidification required for neurodegeneration in *C. elegans*. *Current Biology*, *15*(13), 1249–1254. http://dx.doi.org/10.1016/j.cub.2005.05.057.

Syntichaki, P., & Tavernarakis, N. (2002). Death by necrosis. Uncontrollable catastrophe, or is there order behind the chaos? *EMBO Reports*, *3*(7), 604–609. http://dx.doi.org/10.1093/embo-reports/kvf138.

Syntichaki, P., & Tavernarakis, N. (2003). The biochemistry of neuronal necrosis: Rogue biology? *Nature Reviews. Neuroscience*, *4*(8), 672–684. http://dx.doi.org/10.1038/nrn1174.

Syntichaki, P., & Tavernarakis, N. (2004). Genetic models of mechanotransduction: The nematode *Caenorhabditis elegans*. *Physiological Reviews*, *84*(4), 1097–1153. http://dx.doi.org/10.1152/physrev.00043.2003.

Syntichaki, P., Xu, K., Driscoll, M., & Tavernarakis, N. (2002). Specific aspartyl and calpain proteases are required for neurodegeneration in *C. elegans*. *Nature*, *419*(6910), 939–944. http://dx.doi.org/10.1038/nature01108.

Tavernarakis, N., & Driscoll, M. (2001). Degenerins. At the core of the metazoan mechanotransducer? *Annals of the New York Academy of Sciences*, *940*, 28–41.

Tavernarakis, N., Shreffler, W., Wang, S., & Driscoll, M. (1997). unc-8, a DEG/ENaC family member, encodes a subunit of a candidate mechanically gated channel that modulates *C. elegans* locomotion. *Neuron*, *18*(1), 107–109.

Treinin, M., & Chalfie, M. (1995). A mutated acetylcholine receptor subunit causes neuronal degeneration in *C. elegans*. *Neuron*, *14*(4), 871–877.

Treinin, M., Gillo, B., Liebman, L., & Chalfie, M. (1998). Two functionally dependent acetylcholine subunits are encoded in a single *Caenorhabditis elegans* operon. *Proceedings of the National Academy of Sciences of the United States of America*, *95*(26), 15492–15495.

Troulinaki, K., & Tavernarakis, N. (2012). Endocytosis and intracellular trafficking contribute to necrotic neurodegeneration in *C. elegans*. *EMBO Journal*, *31*(3), 654–666. http://dx.doi.org/10.1038/emboj.2011.447.

Usenovic, M., Knight, A. L., Ray, A., Wong, V., Brown, K. R., Caldwell, G. A., et al. (2012). Identification of novel ATP13A2 interactors and their role in alpha-synuclein misfolding and toxicity. *Human Molecular Genetics, 21*(17), 3785–3794. http://dx.doi.org/10.1093/hmg/dds206.

van der Horst, A., Schavemaker, J. M., Pellis-van Berkel, W., & Burgering, B. M. (2007). The *Caenorhabditis elegans* nicotinamidase PNC-1 enhances survival. *Mechanisms of Ageing and Development, 128*(4), 346–349. http://dx.doi.org/10.1016/j.mad.2007.01.004.

Van Montfort, R., Slingsby, C., & Vierling, E. (2001). Structure and function of the small heat shock protein/alpha-crystallin family of molecular chaperones. *Advances in Protein Chemistry, 59*, 105–156.

Vlachos, M., & Tavernarakis, N. (2010). Non-apoptotic cell death in *Caenorhabditis elegans*. *Developmental Dynamics, 239*(5), 1337–1351. http://dx.doi.org/10.1002/dvdy.22230.

Vrablik, T. L., Huang, L., Lange, S. E., & Hanna-Rose, W. (2009). Nicotinamidase modulation of NAD+ biosynthesis and nicotinamide levels separately affect reproductive development and cell survival in *C. elegans*. *Development, 136*(21), 3637–3646. http://dx.doi.org/10.1242/dev.028431.

Walker, N. I., Harmon, B. V., Gobe, G. C., & Kerr, J. F. (1988). Patterns of cell death. *Methods and Achievements in Experimental Pathology, 13*, 18–54.

Wong, D., Bazopoulou, D., Pujol, N., Tavernarakis, N., & Ewbank, J. J. (2007). Genome-wide investigation reveals pathogen-specific and shared signatures in the response of *Caenorhabditis elegans* to infection. *Genome Biology, 8*(9), R194. http://dx.doi.org/10.1186/gb-2007-8-9-r194.

Xu, K., Tavernarakis, N., & Driscoll, M. (2001). Necrotic cell death in *C. elegans* requires the function of calreticulin and regulators of Ca(2+) release from the endoplasmic reticulum. *Neuron, 31*(6), 957–971.

Yamashima, T. (2000). Implication of cysteine proteases calpain, cathepsin and caspase in ischemic neuronal death of primates. *Progress in Neurobiology, 62*(3), 273–295.

Yamashima, T. (2004). Ca2+-dependent proteases in ischemic neuronal death: A conserved 'calpain-cathepsin cascade' from nematodes to primates. *Cell Calcium, 36*(3–4), 285–293. http://dx.doi.org/10.1016/j.ceca.2004.03.001.

Yuan, A., Santi, C. M., Wei, A., Wang, Z. W., Pollak, K., Nonet, M., et al. (2003). The sodium-activated potassium channel is encoded by a member of the Slo gene family. *Neuron, 37*(5), 765–773.

Yuan, J., Shaham, S., Ledoux, S., Ellis, H. M., & Horvitz, H. R. (1993). The *C. elegans* cell death gene ced-3 encodes a protein similar to mammalian interleukin-1 beta-converting enzyme. *Cell, 75*(4), 641–652.

Zhang, Z., Tang, Q. Y., Alaimo, J. T., Davies, A. G., Bettinger, J. C., & Logothetis, D. E. (2013). SLO-2 isoforms with unique Ca(2+)- and voltage-dependence characteristics confer sensitivity to hypoxia in *C. elegans*. *Channels (Austin)*, 7(3), 194–205. http://dx.doi.org/10.4161/chan.24492.

CHAPTER SEVEN

Noncanonical Cell Death in the Nematode *Caenorhabditis elegans*

Maxime J. Kinet, Shai Shaham[1]
Laboratory of Developmental Genetics, The Rockefeller University, New York, USA
[1]Corresponding author: e-mail address: shaham@rockefeller.edu

Contents

Abstract

The nematode *Caenorhabditis elegans* has served as a fruitful setting for cell death research for over three decades. A conserved pathway of four genes, *egl-1*/BH3-only, *ced-9*/Bcl-2, *ced-4*/Apaf-1, and *ced-3*/caspase, coordinates most developmental cell deaths in *C. elegans*. However, other cell death forms, programmed and pathological, have also been described in this animal. Some of these share morphological and/or molecular similarities with the canonical apoptotic pathway, while others do not. Indeed, recent studies suggest the existence of an entirely novel mode of programmed developmental cell destruction that may also be conserved beyond nematodes. Here, we review evidence for these noncanonical pathways. We propose that different cell death modalities can function as backup mechanisms for apoptosis, or as tailor-made programs that allow specific dying cells to be efficiently cleared from the animal.

Methods in Enzymology, Volume 545
ISSN 0076-6879
http://dx.doi.org/10.1016/B978-0-12-801430-1.00007-X

Highlights

- The core apoptotic pathway can be regulated at any step.
- Death processes share morphological features and possibly downstream effectors.
- Linker cell death is a novel death program possibly relevant to human biology.

1. INTRODUCTION

Cell death, programmed or otherwise, is a ubiquitous biological phenomenon. Programmed cell death is required for development, homeostasis, and the response to pathological insults in virtually all animals, from sponges to humans (Ameisen, 2002). In humans, disease processes are often accompanied by either causal or incidental cell death (Kumar, Abbas, Fausto, & Aster, 2009). Thus, a broad understanding of cell death programs may yield insights into, and possibly treatments for, human pathologies.

The nematode *Caenorhabditis elegans* has proved to be an invaluable tool for dissecting programmed cell death mechanisms. Several aspects of this organism make it well suited for cell death research. Like other nematodes, *C. elegans* has an essentially invariant cell lineage (Sulston, Schierenberg, White, & Thomson, 1983), where death features prominently as a common fate. Dying cells are easy to observe in intact, developing animals, which are small and possess a transparent cuticle. Simple genetics and animal husbandry (Brenner, 1974), efficient RNA interference (Kamath, Martinez-Campos, Zipperlen, Fraser, & Ahringer, 2001), and a fully sequenced and heavily annotated genome (C. elegans Sequencing Consortium, 1998) have enabled investigators to identify genes involved in the control and execution of developmental programmed cell death and to uncover mutations and conditions leading to pathological cellular demise.

A molecular description of apoptotic cell death emerged from studies of *C. elegans* in the 1980s and 1990s. Horvitz and colleagues identified mutants that define four core apoptotic genes: the BH3-only-like gene *egl-1*, the Bcl-2-like *ced-9*, the Apaf-1-like *ced-4*, and the caspase *ced-3* (Horvitz, Shaham, & Hengartner, 1994). Mutations in these genes abolish the death of almost all cells fated to die, and their roles in cell death are largely conserved in all metazoans examined. Most somatic *C. elegans* cells destined to die specifically induce *egl-1* transcription (Nehme & Conradt, 2008). EGL-1 protein then binds to CED-9 (Conradt & Horvitz, 1998), disrupting its interaction with CED-4 (Yan et al., 2005; Yang, Chang, & Baltimore, 1998), thereby freeing CED-4 to activate CED-3, promoting cell death (Ellis & Horvitz, 1986; Xue, Shaham, & Horvitz, 1996).

Despite the great success of these early genetic studies, which relied on tracking the survival of groups of cells, they did not initially identify programs unique to individual cells. Partially redundant pathways would have also been more difficult to detect, as mutations in individual components would likely yield only weak defects. Later genetic screens in many labs, seeking mutations affecting the deaths of individual or small groups of cells, uncovered new forms of cell death that deviate partially or entirely from the canonical molecular pathway for apoptosis. Here, we discuss these recent studies.

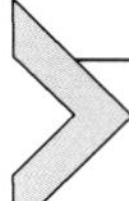

2. PATHOLOGICAL CELL DEATH INDUCED BY GENOME LESIONS AND ENVIRONMENTAL STRESS

2.1. Ion channel mutations

Genetic studies in *C. elegans* identified three proteins, MEC-4 (Driscoll & Chalfie, 1991), DEG-1 (Chalfie & Wolinsky, 1990), and UNC-8 (Shreffler, Magardino, Shekdar, & Wolinsky, 1995), whose activation by gain-of-function mutations inappropriately promotes neuronal death. Electron microscope reconstructions demonstrate that dying neurons accumulate progressively larger vacuoles and electron-dense membranous whorls, as well as what appear to be nuclear chromatin clumps. Changes in nuclear shape are also evident (Fig. 7.1A; Hall et al., 1997). Late in the process, organelle swelling and lysis can be seen.

The three affected proteins are ENaC-type cation channels, the so-called degenerins, that conduct predominantly sodium (Hong & Driscoll, 1994) but also calcium (Bianchi et al., 2004), and cell death-inducing mutations increase their open channel probability (Wang et al., 2013). Thus, abnormal ion homeostasis is likely the initiating insult that leads to cell swelling and death. Gain-of-function mutations in the nicotinic acetylcholine receptor DEG-3 (Treinin & Chalfie, 1995), another cation channel, also have similar effects.

While the mechanistic details of this pathological cell death process are still not entirely worked out, a prominent role for intracellular calcium release has been suggested. Mutants in the *C. elegans* homolog of the endoplasmic reticulum (ER) calcium-binding chaperone, calreticulin, attenuate *mec-4(gf)*-mediated neuronal cell death (Xu, Tavernarakis, & Driscoll, 2001). Similarly, mutations in calnexin, another ER calcium-binding protein, in ITR-1, the *C. elegans* ER IP_3 receptor, and in the ryanodine

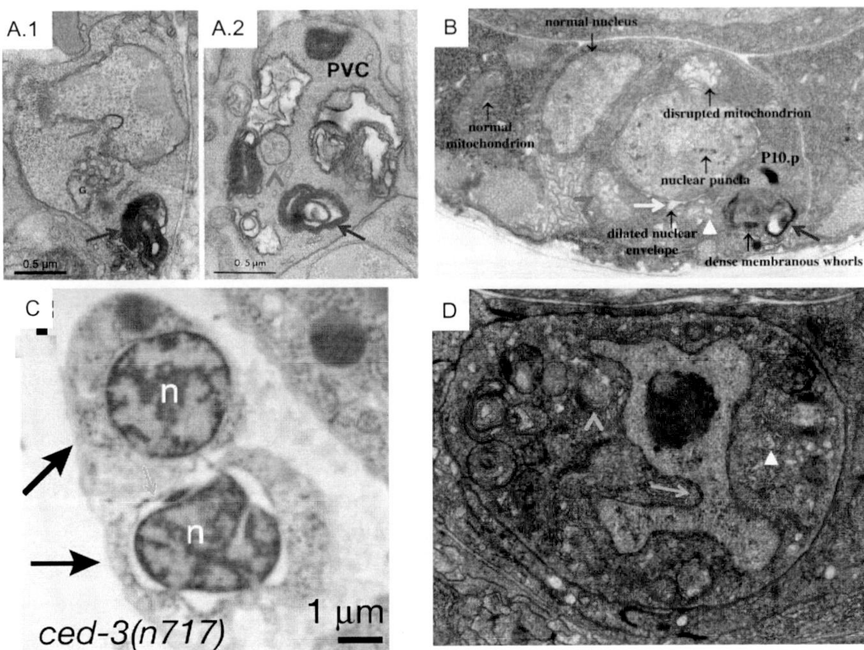

Figure 7.1 Different cell death pathways share morphological features. (A) PVM neuron (A.1) of a *mec-4*(gf) mutant and PVC neuron (A.2) of a *deg-1*(gf) mutant. (B) P10.p cell in a *lin-33*(gf) animal. (C) Shed cells (arrows) in a *ced-3*(*n717*) embryo. (D) Dying linker cell. Nuclear indentations, red arrows in A.1, D. Membranous whorls, blue arrows in A.1, A.2, B. Dilated ER, green arrowheads in B, D. Dilated nuclear envelope, yellow arrows in B, C. Dilated mitochondria, red arrowheads in A.2, B, D. Dark intranuclear structure in (D) is the linker cell nucleolus. *Reproduced with permission from (A) Hall et al. (1997), (B) Galvin, Kim, and Horvitz (2008), (C) Denning, Hatch, and Horvitz (2012), and (D) Abraham, Lu, and Shaham (2007).* (See the color plate.)

receptor ER release channel, UNC-68, also attenuate cell death (Fig. 7.2), as does the calcium chelator EGTA. Cell death can be restored in these suppressed animals by thapsigargin, which blocks the ER calcium influx pump and causes calcium release from the ER. Thapsigargin treatment also results in occasional cell death in wild-type animals, suggesting that cytosolic calcium elevation may be sufficient to promote cell death. Consistent with this idea, the *deg-3*(*gf*) mutations, which likely cause cytosolic calcium increase without the need for additional ER calcium, cannot be suppressed by mutations that block ER calcium release (Xu et al., 2001). Additionally, heat shock is also able to induce calcium-dependent necrosis, perhaps by denaturing crucial regulators of calcium homeostasis (Kourtis, Nikoletopoulou, & Tavernarakis, 2012).

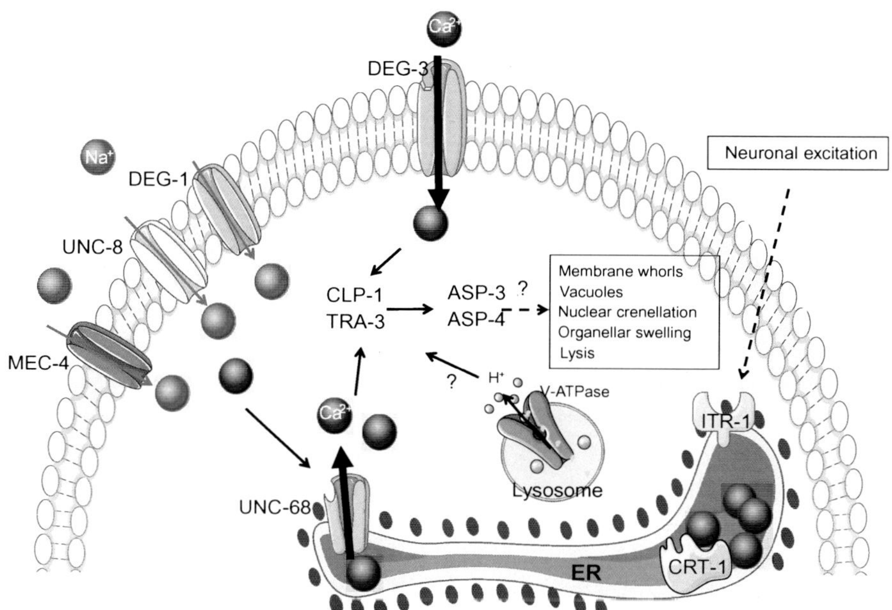

Figure 7.2 Mechanisms of ion channel mutation induced death in *C. elegans*. (See the color plate.)

While calcium has many functions in the cell, its requirement for the activation of cytosolic calpain and cathepsin proteases may play at least some role in cell degeneration (Syntichaki, Xu, Driscoll, & Tavernarakis, 2002). Overexpression of these proteases is sufficient to cause death with similar morphology, and RNAi-mediated knockdown of the calpains CLP-1 and TRA-3, or the cathepsins ASP-3 and ASP-4 inhibits cell death progression in *mec-4(gf)* mutants. While double calpain or double cathepsin knockdowns enhance cell survival in this background, reducing expression of one of each does not, suggesting that calpains and cathepsins might function in a linear pathway in which elevated cytosolic calcium activates calpains, which, in turn, promote cathepsin activation and cell demise (Fig. 7.2). However, this model has not been rigorously tested.

Several other cytoplasmic cathepsins exist in *C. elegans* that do not seem to affect activated-channel-induced neuronal death. Whether this requirement for select proteases reflects cell-type-specific expression of these proteins or substrate specificity is not clear.

Calcium may not be the only ion involved in degenerin-induced cell death. Mutations in subunits of the vacuolar-H+-ATPase (V-ATPase) ameliorate both degenerin-mediated and thapsigargin-induced death (Syntichaki, Samara, & Tavernarakis, 2005), suggesting that cytosol acidification could function downstream of calcium elevation to promote cell death (Fig. 7.2). Treating *C. elegans* with weak lysotropic bases or impairing lysosomal biogenesis can also attenuate calcium-dependent cell death, suggesting a possible role for this organelle in cytosol acidification (Artal-Sanz, Samara, Syntichaki, & Tavernarakis, 2006). How protons may affect cytosolic protease activation, if at all, is not known, but lysosomes might also contribute to cellular demise by leaking their normally sequestered acid hydrolases into the cytoplasm.

Neuronal cell death accompanied by cell swelling can also be induced in *C. elegans* by constitutive activation of the $G_{\alpha s}$ protein (Berger, Hart, & Kaplan, 1998; Korswagen, Park, Ohshima, & Plasterk, 1997), which functions through the adenylyl cyclase ACY-1 to transmit signals from metabotropic neurotransmitter receptors. This death is weakly dependent on the voltage-gated calcium channel subunit UNC-36 and on the vesicular glutamate transporter EAT-4 (Berger et al., 1998), suggesting that neuronal activity may modulate sensitivity to pathological cell death. Indeed, deletion of the *C. elegans* glutamate transporter *glt-3*, which presumably leads to higher extracellular glutamate levels, cooperates with $G_{\alpha s}$ overexpression to enhance neuronal cell death (Mano & Driscoll, 2009).

The studies of degenerative cell death in *C. elegans* neurons raise the possibility that similar processes contribute to human nervous system pathologies. For example, as in *C. elegans*, neuronal cell death induced in a mouse stroke model depends on both calcium and low pH. However, in this system, acid seems to function upstream of calcium release (Xiong et al., 2004). Glutamate-induced toxicity, thought to be an important facet of cell death induction in stroke, may also promote cell death through neuronal second messengers (Zhou, Ding, Chen, Yun, & Wang, 2013).

2.2. NAD metabolism defects

While neuronal cell death has featured prominently in studies of degenerative cell death in *C. elegans*, the degeneration of nonneuronal cells in response to specific gene mutations has also been described. In *pnc-1* mutant larvae, the uterine uv1 cells die with a vacuolated morphology through a process requiring calpains and aspartyl proteases (Huang & Hanna-Rose, 2006). Cell death seems to be a response to overabundance of nicotinamide (NAM), which PNC-1, a nicotinamidase homolog, converts to nicotinic acid. Indeed, feeding animals NAM also promotes uv1 vacuolation and death (Vrablik, Huang, Lange, & Hanna-Rose, 2009). Why uv1 cells are sensitive to NAM accumulation is not understood. One possibility is that NAM levels alter the generation and/or function of nicotinamide adenine dinucleotide (NAD), a key respiration intermediate. However, muscle cells whose energy requirements are likely much higher remain intact in *pnc-1* mutants (Vrablik et al., 2009). Boosting EGF signaling, which promotes uv1 specification, suppresses cell death, suggesting that an EGF-repressible NAD consumer, or its product, may be involved. However, if and how such an NAD consumer causes calpain and/or aspartyl protease activation is unclear.

How vacuoles accumulate within neuronal or nonneuronal *C. elegans* cells undergoing degenerative cell death is not well understood. Notably, expressing human caspase-3 in *C. elegans* body wall muscle can promote vacuole formation in these cells as well, rather than the more classical, refractile appearance induced in other cells (Chelur & Chalfie, 2007). Thus, it is possible that both apoptotic and degenerative cell death regulators in *C. elegans* engage common targets. Identification of such targets would be required to confirm this idea.

2.3. Cell differentiation mutations

In vertebrates and in *Drosophila*, cell death is often induced in response to a failure in cell fate specification or differentiation (Raymond, Murphy, O'Sullivan, Bardwell, & Zarkower, 2000), perhaps as a result of disturbances in proteostasis (Arrigo, 2005; Hetz, 2012). This may also be the case in *C. elegans*. For example, LIN-26, a Zn-finger transcription factor, normally promotes hypodermal and glial cell fate. A reduction-of-function mutation in this gene results not only in excess neuron production but also in vacuolation and death of hypodermal and glial cells (Labouesse, Sookhareea, & Horvitz, 1994). While the mechanism promoting cell demise in this case is not known, it is independent of the *ced-3* caspase (M. Labouesse, personal communication).

Developmental failure also seems to lead to cell death in animals carrying mutations in the *unc-83* and *unc-84* genes, which encode KASH and SUN domain proteins, respectively, that anchor the nucleus to the cytoskeleton (McGee, Rillo, Anderson, & Starr, 2006). In these mutants, nuclei of the P epithelial blast cells fail to migrate ventrally along with the rest of the cell body, resulting in elongated cells that eventually die in a *ced-3*-independent manner (Malone, Fixsen, Horvitz, & Han, 1999; Starr et al., 2001). Unlike hypodermal cell death in *lin-26* mutants, dying cells in *unc-83/84* mutants are not vacuolated, instead adopting a refractile appearance common to naturally dying cells in *C. elegans*. Double mutants of *unc-84* and genes that block P-cell migration do not exhibit P-cell death (Malone et al., 1999). Furthermore, failure of nuclear migration is not generally lethal to cells, since, in *unc-83* mutants, other cells exhibit nuclear migration failure without death (Starr et al., 2001). These observations suggest that the disconnect between cell migration and nuclear migration must trigger a cell-specific response that leads to death. Genetic screens for cell death suppressors could reveal the key players in this pathological process and should reveal whether it is possible to interrupt cell death without restoring nuclear migration.

While cell death in response to failed differentiation in *C. elegans* is *ced-3*-independent, and likely caspase independent, this is not the case in other animals (Böhmer, 1989; Howard et al., 1993; Kulkarni & McCulloch, 1994; Yang et al., 1999). The source of this difference is not clear, but suggests that at least in somatic cells in *C. elegans*, caspases and other apoptotic genes respond mainly to programmed stimuli. *C. elegans* germ cells can engage caspases in response to irradiation and other DNA lesions (Derry, Putzke, & Rothman, 2001), suggesting that damage responses in this tissue may be more akin to generalized responses in vertebrates.

2.4. *lin-24/lin-33* mutants

Dominant mutations in two genes, *lin-24* and *lin-33*, promote the inappropriate deaths of P*n*.p cells, daughter cells of the P cells affected by *unc-83* and *unc-84* mutations (Galvin et al., 2008). Dying cells assume a refractile, nonvacuolated appearance under differential interference contrast optics. Electron microscopy revealed that dying cells exhibit electron-dense nuclear puncta, but otherwise normal nucleoplasm, dilation of the nuclear envelope, dense membranous cytoplasmic whorls, and disrupted mitochondria (Fig. 7.1B). Some dying cells can recover and reacquire normal morphology, while others can recover but possess a small nucleus. P*n*.p cell fate is also affected in these mutants, but whether fate changes result from developmental cues missed because of injury or are independent defects is not clear.

Programmed cell death in *C. elegans* is usually an all-or-none process; however, recovery from death is also seen in animals doubly mutant for weak mutations in *ced-3* and genes promoting apoptotic cell corpse engulfment. In these animals, progeny of P*n*.p cells begin to undergo normal developmental cell death and acquire a refractile appearance, only to recover and inappropriately survive (Reddien, Cameron, & Horvitz, 2001). This phenomenon is only seen in engulfment mutant backgrounds, demonstrating that engulfment can modulate cell susceptibility to death. Indeed, one *C. elegans* cell, B.al/rapaav, always survives in engulfment-defective mutants, despite additional dependence on *ced-3* caspase activity for death (Reddien et al., 2001), and, rarely, cells in animals lacking all four *C. elegans* caspase-related genes die and are engulfed (Denning, Hatch, & Horvitz, 2013), hinting perhaps at a role for engulfment in cell death. Remarkably, cell death in dominant *lin-24/33* mutants can also be attenuated by mutations in engulfment genes (Galvin et al., 2008). While *ced-3* caspase does not seem to play a role in *lin-24/33*-mediated cell death, mutations in *egl-1*/BH3-only and *ced-4*/Apaf-1 weakly interfere with the process. The sites of action of *lin-24/33* or any of the modifying genes in the context of P*n*.p cell death are not known.

Deleting *lin-24*, *lin-33*, or both has no obvious effects on *C. elegans* development or cell survival. However, loss of function of either gene prevents P*n*.p cell death by dominant mutations in the other, suggesting that the encoded proteins may function in a complex. LIN-24 protein contains a domain similar to bacterial toxins, and both loss- and gain-of-function alleles alter this conserved domain (Galvin et al., 2008). The predicted LIN-33 protein does not resemble other known genes. Bacterial toxins homologous to LIN-24 kill eukaryotic cells by forming oligomeric pores in the plasma membrane, a mechanism shared with the membrane attack complex of the vertebrate blood complement system (Anderluh & Lakey, 2008) as well as with the perforins of cytotoxic T lymphocytes and NK cells (Chávez-Galán, Arenas-Del Angel, Zenteno, Chávez, & Lascurain, 2009). One possibility, therefore, might be that LIN-24/33-mutant proteins are inappropriately released from neighboring cells to promote P*n*.p cell death by poking holes in their membranes. Engulfment mutants might suppress death by preventing contact between P*n*.p cells and their killer neighbors. It is equally plausible that LIN-24/33 function in the dying cell to introduce membrane pores, and that engulfing cells, sensing membrane perturbations, finish off the weakened P*n*.p cells.

2.5. A latent apoptotic pathway in P*n*.p cells?

That apoptotic genes modulate *lin-24/33*-dependent P*n*.p cell death suggests that this death process may be mechanistically related to apoptosis. Support for this notion comes from studies of loss-of-function mutations in the gene *pvl-5*. Animals carrying such mutations exhibit inappropriate P*n*.p cell death, which occurs at the same developmental time as *lin-24/33*-mediated deaths. Dying cells in *pvl-5(lf)* animals can recover, and recovered cells frequently exhibit an ovoid morphology and shrunken nucleus as in *lin-24/33* mutants (Joshi & Eisenmann, 2004). Intriguingly, *pvl-5*-mediated cell death requires *ced-3* caspase and can be suppressed by *ced-9(gf)* mutations, suggesting inappropriate initiation of an apoptosis-related pathway in P*n*.p cells.

Nonetheless, *pvl-5*- and *lin-24/33*-mediated P*n*.p cell deaths are not identical. *pvl-5* mutations can cause P*n*.p cell vacuolation not reported in *lin-24/33* mutants. Moreover, the cell fate defects resulting from the two lesions are likely to be different: *lin-24/33* mutants lack the hermaphrodite vulva, which is normally generated by P*n*.p cell descendents, whereas *pvl-5* mutants exhibit a protruding vulva defect. Furthermore, while *lin-24/33*-mediated death is weakly suppressed by mutations in all core cell death genes except for *ced-3*, *pvl-5(lf)*-mediated P*n*.p cell death is suppressed by *ced-9(gf)*, but not by *egl-1* or *ced-4* loss-of-function mutations. Finally, *pvl-5* mutants suffer a small number of *ced-3*-dependent ectopic cell deaths in cells other than the P*n*.p cells, a defect not reported for *lin-24/33* mutants. The molecular identity of the *pvl-5* gene is not known, but it maps to a different chromosome from *lin-24* and *lin-33*. How *pvl-5* regulates *ced-3* function is also not understood.

In *Pristionchus pacificus* and other nematodes more distantly related to *C. elegans*, some or all P*n*.p cells that are not destined to contribute to vulva formation die by *ced-3*-dependent apoptosis (Sommer et al., 1998; Sommer & Sternberg, 1996). All P*n*.p cells have the capacity to die in *Pristionchus*, as mutants in the Hox gene *Ppa-lin-39* are vulvaless because all P*n*.p cells die. Cell death is blocked, and vulva formation is restored in animals also carrying *Ppa-ced-3* mutations. A genetic screen for *Ppa-lin-39* suppressors recovered 22 alleles of *Ppa-ced-3*, but only two alleles of other genes. This may suggest that in *Pristionchus* P*n*.p cells, *ced-3* caspase is the main cell death effector, a profile resembling that of *pvl-5*-induced cell death, although differences in gene mutability could also account for this mutational profile.

Taken together, these studies suggest that *C. elegans lin-24/33* and *pvl-5* mutations may uncover a silenced apoptotic program that is still functional in other nematodes (Fig. 7.3). However, the differential requirement for *ced-3* and *ced-4* in these death processes remains puzzling. RNAi against *icd-1*, the beta subunit of the nascent-polypeptide-associated complex (β-NAC),

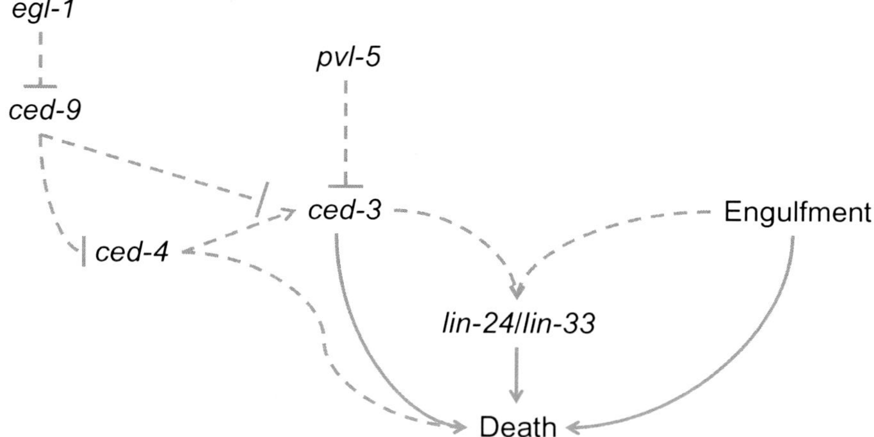

Figure 7.3 Possible P*n*.p cell death pathways in P*n*.p cells. Dashed arrows indicate tentative relevant genetic interactions.

causes widespread ectopic cell death during *C. elegans* development (Bloss, Witze, & Rothman, 2003). As in *lin-24/33* mutants, loss of *ced-4*, but not of *ced-3*, suppresses *icd-1*(*RNAi*)-mediated death, and other *C. elegans* caspases play only minor roles in this process (Denning et al., 2013). One must, therefore, entertain the possibility that *ced-4* can promote cell death in the absence of caspases.

2.6. Cell shedding in caspase mutants

In embryos lacking all four *C. elegans* caspases, six cells are shed from the anterior sensory depression and the ventral pocket (Denning et al., 2012). These cells express *egl-1*, but are still shed in *egl-1*(*lf*), *ced-9*(*gf*), and *ced-4*(*lf*) mutants as well. In wild-type embryos, shed cells die normally and are engulfed by neighboring cells. Mutations in the MELK kinase PIG-1 and in its activating kinase PAR-4/LKB1 prevent shed cell accumulation. These observations have led to the hypothesis that cell shedding is a cell death program that functions in parallel to CED-3 caspase, perhaps as a backup program, and that PIG-1 is a key activator of this program. Perhaps the strongest evidence in favor of this model is that shed corpses exhibit some features reminiscent of apoptosis (Fig. 7.1C).

Cell shedding does not appear to take place in wild-type animals and *pig-1* expression is not sufficient to promote cell death. These and other results have raised the possibility that cell shedding may not be a cell death program *per se*, but a passive result of *ced-3* caspase loss (Chien, Brinkmann,

Teuliere, & Garriga, 2013). The transcriptome of dying cells may be under reduced selective pressure, and, if so, dying cells may be poorly differentiated. Indeed, while inappropriately surviving cells in *ced-3* mutants can acquire fates of their sister cells or their progeny, differentiation reporters are weakly expressed in many of these surviving cells (Ellis & Horvitz, 1986), and fate acquisition is often incomplete (Avery & Horvitz, 1987). It is possible, therefore, that in *ced-3* mutants, these six inappropriately surviving cells have poor expression of cell adhesion proteins and are passively extruded from the animal in response to movements of adjacent cells. *pig-1* mutations could, in this model, enhance the differentiation of the "undead" cells toward their adhesive sister cell fate, thereby preventing shedding. PIG-1 has been implicated in the control of asymmetric cell division (Cordes, Frank, & Garriga, 2006; Ou, Stuurman, D'Ambrosio, & Vale, 2010) and plays important roles in cell fate specification throughout the embryo (Morton, Hoose, & Kemphues, 2012). In the context of cell shedding, *pig-1* mutants inappropriately express the α-catenin HMP-1 on the surface of would-be shed cells, and one of these cells expresses reporters specific for its sister cell progeny, the excretory cell (Denning et al., 2012).

If *pig-1* does regulate a novel cell death process, the expectation would be that its role and its targets in all dying cells be the same. Whether this is the case is unclear; however, *pig-1* mutations also affect the death and specification of the sister cell of the M4 pharyngeal neuron (Hirose & Horvitz, 2013), and this cell is not shed and remains adhesive in *ced-3* mutants (Avery & Horvitz, 1987).

3. DEVELOPMENTAL CELL DEATHS THAT DO NOT FOLLOW THE CANONICAL APOPTOTIC PATHWAY

3.1. Germline cell death

In adult *C. elegans* hermaphrodites, about half of female germ cells die by apoptosis before developing into mature oocytes (Gumienny, Lambie, Hartwieg, Horvitz, & Hengartner, 1999). Unlike dying somatic cells, whose identities are invariant, germ cells, which occupy a syncytium and appear identical, seem to die stochastically. Competence to die is imparted by ephrin and Ras/MAPK signaling, probably originating from surrounding sheath cells, resulting in germ cell exit from meiotic pachytene (Church, Guan, & Lambie, 1995; Li, Johnson, Park, Chin-Sang, & Chamberlin, 2012).

While germline cell death requires *ced-3*, *ced-4*, and *ced-9*, it is independent of *egl-1* and is not blocked by a gain-of-function mutation in *ced-9* that prevents somatic cell death (Gumienny et al., 1999). Mutations in the Pax2-related genes *egl-38* and *pax-2* promote excess germ cell death. Genetically, *egl-38* and *pax-2* seem to function upstream of *ced-9*, a model supported by the observation that EGL-38 and PAX-2 proteins bind to regulatory sequences near the *ced-9* gene. It is therefore possible that in the germline, Pax2 proteins substitute for EGL-1. Nonetheless, in the soma, *egl-1* transcription is induced in dying cells. This does not seem to be the case for *egl-38* and *pax-2* (Park, Jia, Rajakumar, & Chamberlin, 2006), suggesting that these genes may act permissively to set *ced-9* levels in germ cells. Thus, other inputs into the apoptotic pathway may control the decision to promote germ cell death. The involvement of genes acting in gonadal sheath cells in germ cell death competence (Ito, Greiss, Gartner, & Derry, 2010; Morthorst & Olsen, 2013) raises the possibility that regulation of germ cell death could have cell autonomous and nonautonomous components, which would allow the animal to make decisions about germ cell death based on both the overall state of the animal (Aballay & Ausubel, 2001; Andux & Ellis, 2008; Angelo & Van Gilst, 2009; Salinas, Maldonado, & Navarro, 2006; Sendoel, Kohler, Fellmann, Lowe, & Hengartner, 2010) and the integrity of individual germ cell genomes (Silva, Adamo, Santonicola, Martinez-Perez, & La Volpe, 2013).

3.2. Tail-spike cell death

The genetic requirements for germ cell death are mirrored, in part, in the *C. elegans* tail-spike cell. This binucleate cell, which arises by cell fusion, sends a slender posterior process that seems to serve as a scaffold for molding the *C. elegans* tail. *ced*-3 and *ced-4* are absolutely required for tail-spike cell death, but *egl-1* plays only a minor role, and a gain-of-function mutation in *ced-9* has no effect (Maurer, Chiorazzi, & Shaham, 2007). Studies of *ced-3* transcription revealed that its expression is induced in the tail-spike cell about 25 min before morphological signs of cell death are apparent. The homeodomain transcription factor PAL-1 promotes *ced-3* expression in the tail-spike cell by binding to three redundant sites upstream of the *ced-3* gene. These results suggest that transcriptional induction of *ced-3*, and not of *egl-1*, may be the key regulatory event promoting tail-spike cell death.

Additional layers of control also exist. A recent study demonstrated that tail-spike cell death requires the F-box protein DRE-1. Genetic and molecular evidence supports the idea that DRE-1 functions in a Skp/Cullin/F-box complex in parallel to EGL-1 and likely upstream of CED-9.

An attractive model is that DRE-1 substitutes for EGL-1 by inactivating CED-9 through ubiquitination and degradation, thereby creating a permissive environment for newly translated CED-3. Support for this model comes from studies of human FBX010 and BCL2, proteins similar to DRE-1 and CED-9, respectively. In a subset of B-cell lymphomas, FBX010 expression can promote BCL2 degradation. Furthermore, in these same lines, FBX010 expression promotes cell death (Chiorazzi et al., 2013).

Mutations in FBX010 are found in some patients with B-cell lymphomas, and expression of the gene is reduced in many others. Furthermore, RNAi against FBX010 in tumor cells promotes their survival (Chiorazzi et al., 2013). These results suggest that FBX010 may function as a tumor suppressor gene. Mutations in Cdx2, the human homolog of *C. elegans pal-1*, promote intestinal tumors (Barros, Freund, David, & Almeida, 2012), suggesting that this gene is a tumor suppressor as well. These observations raise the intriguing possibility that while tail-spike cell death control exhibits noncanonical features in *C. elegans*, similar regulatory mechanisms may play integral roles in controlling tumorigenesis in humans.

3.3. Sex-specific death of CEM neurons

The sexually dimorphic CEM cells survive in males, differentiating into neurons that help orchestrate the male's complex mating behavior (White et al., 2007). In hermaphrodites, which do not exhibit this behavior, the neurons die (Sulston & Horvitz, 1977). CEM cell death requires all four core cell death genes. Yet, as in the germline and tail-spike cells, CEM cell death regulation appears to require transcriptional activation of the *ced-3* caspase gene. Although *egl-1* expression is still induced in CEM neurons, this induction is not always sufficient to promote CEM death. In males carrying mutations in *unc-86*, a gene encoding a POU homeodomain transcription factor, *egl-1* expression is unaltered, but CEMs fail to die. Genetics and expression studies revealed that UNC-86 protein, LRS-1, a tRNA synthetase, and UNC-132, a novel protein, control CEM demise by promoting *ced-3* transcription (Nehme et al., 2010; Peden, Kimberly, Gengyo-Ando, Mitani, & Xue, 2007). Nonetheless, whether *ced-3* or *egl-1* transcription is the rate-determining step in CEM cell death remains unclear.

ced-3 transcription in CEMs seems to be counteracted by CEH-30, a BarH1-related transcription factor. CEH-30 functions genetically downstream of *egl-1* and *ced-9* (Peden et al., 2007; Schwartz & Horvitz, 2007). A *ceh-30* gain-of-function allele alters an intronic consensus sequence for binding by TRA-1A, a Gli-related protein that is an effector of the sex determination machinery promoting hermaphrodite identity (Hodgkin, 1987; Zarkower & Hodgkin, 1992). This observation suggests that TRA-1A normally represses *ceh-30* in hermaphrodites.

CEM neurons and the tail-spike cell survive for an extended duration after they are generated and before succumbing to cell death. Likewise, both cell types actively control transcription of *ced-3*. This correlation raises the possibility that in these long-lived cells destined to die, there is a need to replenish CED-3 protein to promote cell death. Indeed, *ced-3* transcriptional reporter studies suggest that while the gene is widely expressed, its transcription is mainly confined to early embryogenesis (Shaham, Reddien, Davies, & Horvitz, 1999), before most cell death takes place. Thus, cells that are longer lived may need to reexpress the gene to promote their demise.

3.4. The use of alternate caspases in dying cells

The *C. elegans* genome contains three caspase-encoding genes in addition to *ced-3*: *csp-1*, *csp-2*, and *csp-3* (Shaham, 1998). While CSP-1 protein has caspase activity *in vitro* (Shaham, 1998), and its overexpression can promote cell death in *C. elegans* (Denning et al., 2013), neither *csp-2* nor *csp-3* seems to encode catalytically active enzymes. CSP-2 has a catalytic cysteine, but lacks conserved residues surrounding the active site, and CSP-3 lacks the large caspase subunit and its active site.

csp-1 may play a minor role in somatic cell death. While mutants in the gene have no obvious cell death defects, enhanced cell survival is observed in conjunction with weak mutations in *ced-3* caspase (Denning et al., 2013). Enhancement is cell-specific, as only some cells destined to die, such as the sister of the pharyngeal M4 neuron, are affected. The activity of CSP-1 does not appear to be regulated by CED-4, as *ced-4* lesions do not inhibit ectopic cell death mediated by CSP-1. It seems, therefore, that if CSP-1 has a role in cell death, it may respond to different cues than CED-3.

Loss-of-function mutations in the *csp-2* and *csp-3* genes have been reported to enhance cell death in the germline (Geng et al., 2009) and soma (Geng et al., 2008) respectively, although this observation has been challenged (Denning et al., 2013). A suggested mechanism for these effects is that these caspase-related proteins bind CED-3 or CSP-1 to inhibit their activities (Geng et al., 2008, 2009). However, given the weak cell death effects of mutants in these genes, testing models regarding their activities remains challenging.

4. NONAPOPTOTIC, CASPASE-INDEPENDENT LINKER CELL DEATH

The male-specific linker cell leads the developing male gonad on a stereotyped elongation path and, upon its death, permits the lumen of the vas deferens to fuse with the cloaca to allow sperm exit (Kimble & Hirsh, 1979). Linker cell death is independent of all *C. elegans* caspases and other apoptotic cell death genes (Abraham et al., 2007; Denning et al., 2013; Ellis & Horvitz, 1986), and also seems to proceed independently of genes controlling apoptotic cell engulfment and proteases involved in other cell death forms in *C. elegans* (Abraham et al., 2007). While linker cell death was initially thought to proceed nonautonomously, through engulfment by the U.l/rp cell (Kimble & Hirsh, 1979; Sulston & Horvitz, 1977), recent studies demonstrate important cell autonomous components involved in the process (Abraham et al., 2007; Blum, Abraham, Yoshimura, Lu, & Shaham, 2012).

Consistent with the unique genetic requirements, dying linker cells are morphologically distinct from apoptotic cells in which chromatin condensation and cytoplasmic shrinkage are generally evident. Dying linker cells maintain open chromatin and display progressive nuclear envelope crenellation leading to the formation of a flower-shaped nucleus never observed outside this setting in *C. elegans*. Mitochondrial and ER swelling is also observed (Fig. 7.1D; Abraham et al., 2007).

The death of the linker cell requires both temporal and spatial cues. Mutations in the microRNA gene *let-7* and the Zn-finger transcription factor gene *lin-29* block linker cell death. These genes are components of a developmental timing program, the heterochronic pathway, that communicates the developmental stage of animals to individual cells within the animal. LIN-29 functions together with the MAB-10 transcriptional cofactor (Harris & Horvitz, 2011), and both proteins are present in the nucleus of the linker cell during its migration and death (Fig. 7.4). Mutation in the *him-4* gene, encoding a secreted immunoglobulin family member, reroutes the linker cell migration path, so that cells often end up in the head, instead of the tail. While most linker cells die on time even in the head, about 13% of *him-4* mutants exhibit linker cell survival, suggesting that local spatial cues may be important for linker cell death. Our recent studies suggest a role for Wnt signaling in transducing such a spatial cue (M. Kinet & S. Shaham, unpublished observations).

Dying linker cells morphologically resemble cells that die during normal vertebrate development. For example, approximately half of the cells

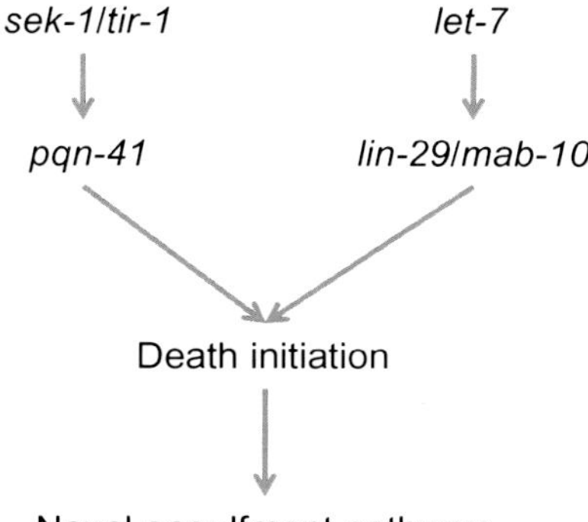

Figure 7.4 Genetics of linker cell death. Arrows indicate plausible genetic interactions.

initially present in the developing chick ciliary ganglion die during development. Electron microscopy fails to reveal apoptotic features in dying cells, but does uncover cells with swollen mitochondria and ER (Pilar & Landmesser, 1976). Nuclear crenellation can also be observed in these cells, becoming more pronounced when ganglion neurons are deprived of their target organ (Pilar & Landmesser, 1976). Developmental death of spinal motor neurons proceeds, slowed but unabated, in the absence of caspase-3 or caspase-9, and dying cells exhibit open chromatin, swollen ER and mitochondria, and cytoplasmic vacuolation (Oppenheim et al., 2001). Crenellated nuclei and swollen ER and mitochondria are also observed in dying neurons of patients with polyglutamine expansion diseases, such as Huntington's disease and some spinocerebellar ataxias, as well as in mouse models for those diseases (Blum et al., 2012). Strikingly, the gene *pqn-41*, which encodes a protein containing a 427 amino acid C-terminal domain rich in glutamines, is required for linker cell death. *pqn-41* seems to function in the same pathway as the conserved MAPKK SEK-1 and its adapter protein TIR-1 to promote linker cell death (Fig. 7.4; Blum et al., 2012). *pqn-41* is not required for other cell deaths in *C. elegans*, and ectopic expression of the rescuing PQN-41C isoform does not precociously kill the linker cell or other cells, suggesting that it must function with other components to promote linker cell death.

Recently, the TIR-1 homologs dSarm and Sarm have been implicated in distal neurite degeneration following axotomy in *Drosophila* and mice, respectively (Osterloh et al., 2012). That the same protein promotes degenerative processes in all three species is intriguing, and bolsters the possibility of a connection between linker cell death and cell death processes in humans. Further, excitotoxic injury to mouse retinal ganglion cells induced by kainate treatment also requires Sarm (Massoll, Mando, & Chintala, 2013), suggesting mechanistic commonalities between excitotoxic necrotic death and other degenerative deaths that may explain some of the observed morphological parallels. Nonetheless, it is still too early to tell whether the morphological and molecular similarities among linker cell death, normal vertebrate cell death, polyglutamine-mediated cell death, and axon degeneration represent true conservation or happenstance.

Why does the linker cell not die by apoptosis? The cell is larger than other cells that succumb to apoptosis and likely harbors extensive functional machinery required for its long migration and the concomitant morphological stages through which it must progress (Kato & Sternberg, 2009). For these reasons, the linker cell might require an alternate program to deal with its degradation. A similar idea has been invoked for the degeneration of *Drosophila* salivary glands, although in this case cell death remains caspase-dependent (Martin & Baehrecke, 2004). Alternatively, the linker cell death program may ensure that the cell will be engulfed by a specific phagocyte. Supporting this idea, the engulfment of dying linker cells is independent of genes required for apoptotic corpse engulfment (Abraham et al., 2007). Furthermore, the engulfing U.l/rp cell does not cluster CED-1::GFP at membranes making contact with the linker cell, as is the case for apoptotic cells. CED-1::GFP does surround mistargeted dying linker cells in *him-4* mutants, suggesting that the cell can be engulfed by this mechanism, and that other cells may not express the physiological engulfment program utilized by the U.l/rp cells. Mistargeted *him-4* cell corpses often persist much longer than wild-type corpses (Abraham et al., 2007), suggesting that the *ced-1*-mediated engulfment process used at these locations is not as efficient as the physiological program engulfing the dying linker cell. Supporting this notion, CED-1::GFP surrounding mistargeted linker cell corpses can be incomplete (Abraham et al., 2007), a phenomenon never seen in apoptotic corpse engulfment but reminiscent of the incomplete engulfment seen in *ced-1* mutants (Zhou, Hartwieg, & Horvitz, 2001). Corpses of cells in animals with ablated U.l/rp cells also persist in the terminal vas deferens (M. Abraham & S. Shaham, unpublished observations), appearing to block sperm exit, suggesting that terminal vas deferens cells are not competent for engulfment using either canonical or linker cell processes.

The cell-specific competence for expressing the physiological engulfment program combined with the selective efficiency between specific death and engulfment programs suggests a strategy for targeting specific cells to specific phagocytes. Such a strategy may have specific anatomical imperatives in the worm, but, in animals with cellular immune systems, it is the rule rather than the exception.

5. CONCLUSION

C. elegans has been appropriately lauded as a system for studying programmed cell death, as studies in this organism laid the foundations for understanding the conserved process of apoptosis. *C. elegans* has also been used to study cell death induced by environmental toxicants (Nass & Blakely, 2003; Nass, Hall, Miller, & Blakely, 2002), excellent studies in their own right with clear relevance to humans but outside the scope of our present discussion. Here, we have reviewed experiments, suggesting that this animal still has much to offer in the context of programmed cell death research. From the identification of a novel morphologically conserved developmental cell death program, to the characterization of different degenerative processes, *C. elegans* continues to be an exciting venue for uncovering basic mechanisms that control cell viability normally and in pathological states. Several of the seemingly disparate cell death phenomena described in this review share common threads, either morphological (Fig. 7.1) or molecular. As our understanding of cell death processes expands, additional interconnections may emerge, providing insight into what key cellular aspects must be dismantled for cells to give up the ghost.

ACKNOWLEDGMENTS

M.J.K was supported by a Medical Scientist Training Program grant from the National Institute of General Medical Sciences of the National Institutes of Health under award number T32GM07739 to the Weill Cornell/Rockefeller/Sloan-Kettering Tri-Institutional MD-PhD Program. S.S. was supported by NIH grants HD078703 and NS081490.

REFERENCES

Aballay, A., & Ausubel, F. M. (2001). Programmed cell death mediated by ced-3 and ced-4 protects Caenorhabditis elegans from Salmonella typhimurium-mediated killing. *Proceedings of the National Academy of Sciences of the United States of America, 98*, 2735–2739.

Abraham, M. C., Lu, Y., & Shaham, S. (2007). A morphologically conserved nonapoptotic program promotes linker cell death in Caenorhabditis elegans. *Developmental Cell, 12*, 73–86.

Ameisen, J. C. (2002). On the origin, evolution, and nature of programmed cell death: A timeline of four billion years. *Cell Death and Differentiation, 9*, 367–393.

Anderluh, G., & Lakey, J. H. (2008). Disparate proteins use similar architectures to damage membranes. *Trends in Biochemical Sciences, 33*, 482–490.

Andux, S., & Ellis, R. E. (2008). Apoptosis maintains oocyte quality in aging Caenorhabditis elegans females. *PLoS Genetics*, *4*, e1000295.
Angelo, G., & Van Gilst, M. R. (2009). Starvation protects germline stem cells and extends reproductive longevity in C. elegans. *Science*, *326*, 954–958.
Arrigo, A. P. (2005). In search of the molecular mechanism by which small stress proteins counteract apoptosis during cellular differentiation. *Journal of Cellular Biochemistry*, *94*, 241–246.
Artal-Sanz, M., Samara, C., Syntichaki, P., & Tavernarakis, N. (2006). Lysosomal biogenesis and function is critical for necrotic cell death in Caenorhabditis elegans. *Journal of Cell Biology*, *173*, 231–239.
Avery, L., & Horvitz, H. R. (1987). A cell that dies during wild-type C. elegans development can function as a neuron in a ced-3 mutant. *Cell*, *51*, 1071–1078.
Barros, R., Freund, J.-N., David, L., & Almeida, R. (2012). Gastric intestinal metaplasia revisited: Function and regulation of CDX2. *Trends in Molecular Medicine*, *18*, 555–563.
Berger, A. J., Hart, A. C., & Kaplan, J. M. (1998). G alphas-induced neurodegeneration in Caenorhabditis elegans. *Journal of Neuroscience*, *18*, 2871–2880.
Bianchi, L., Gerstbrein, B., Frøkjær-Jensen, C., Royal, D. C., Mukherjee, G., Royal, M. A., et al. (2004). The neurotoxic MEC-4(d) DEG/ENaC sodium channel conducts calcium: Implications for necrosis initiation. *Nature Neuroscience*, 7, 1337–1344.
Bloss, T. A., Witze, E. S., & Rothman, J. H. (2003). Suppression of CED-3-independent apoptosis by mitochondrial betaNAC in Caenorhabditis elegans. *Nature*, *424*, 1066–1071.
Blum, E. S., Abraham, M. C., Yoshimura, S., Lu, Y., & Shaham, S. (2012). Control of non-apoptotic developmental cell death in Caenorhabditis elegans by a polyglutamine-repeat protein. *Science*, *335*, 970–973.
Böhmer, R. M. (1989). Interaction of serum and colony-stimulating factor for survival of a factor-dependent hemopoietic progenitor cell line. *Journal of Cellular Physiology*, *139*, 531–537.
Brenner, S. (1974). The genetics of Caenorhabditis elegans. *Genetics*, 77, 71–94.
C. elegans Sequencing Consortium. (1998). Genome sequence of the nematode C. elegans: A platform for investigating biology. *Science*, *282*, 2012–2018.
Chalfie, M., & Wolinsky, E. (1990). The identification and suppression of inherited neurodegeneration in Caenorhabditis elegans. *Nature*, *345*, 410–416.
Chávez-Galán, L., Arenas-Del Angel, M. C., Zenteno, E., Chávez, R., & Lascurain, R. (2009). Cell death mechanisms induced by cytotoxic lymphocytes. *Cellular & Molecular Immunology*, *6*, 15–25.
Chelur, D. S., & Chalfie, M. (2007). Targeted cell killing by reconstituted caspases. *Proceedings of the National Academy of Sciences of the United States of America*, *104*, 2283–2288.
Chien, S.-C., Brinkmann, E.-M., Teuliere, J., & Garriga, G. (2013). Caenorhabditis elegans PIG-1/MELK acts in a conserved PAR-4/LKB1 polarity pathway to promote asymmetric neuroblast divisions. *Genetics*, *193*, 897–909.
Chiorazzi, M., Rui, L., Yang, Y., Ceribelli, M., Tishbi, N., Maurer, C. W., et al. (2013). Related F-box proteins control cell death in Caenorhabditis elegans and human lymphoma. *Proceedings of the National Academy of Sciences of the United States of America*, *110*, 3943–3948.
Church, D. L., Guan, K. L., & Lambie, E. J. (1995). Three genes of the MAP kinase cascade, mek-2, mpk-1/sur-1 and let-60 ras, are required for meiotic cell cycle progression in Caenorhabditis elegans. *Development*, *121*, 2525–2535.
Conradt, B., & Horvitz, H. R. (1998). The C. elegans protein EGL-1 is required for programmed cell death and interacts with the Bcl-2-like protein CED-9. *Cell*, *93*, 519–529.

Cordes, S., Frank, C. A., & Garriga, G. (2006). The C. elegans MELK ortholog PIG-1 regulates cell size asymmetry and daughter cell fate in asymmetric neuroblast divisions. *Development*, *133*, 2747–2756.

Denning, D. P., Hatch, V., & Horvitz, H. R. (2012). Programmed elimination of cells by caspase-independent cell extrusion in C. elegans. *Nature*, *488*, 226–230.

Denning, D. P., Hatch, V., & Horvitz, H. R. (2013). Both the caspase CSP-1 and a caspase-independent pathway promote programmed cell death in parallel to the canonical pathway for apoptosis in Caenorhabditis elegans. *PLoS Genetics*, *9*, e1003341.

Derry, W. B., Putzke, A. P., & Rothman, J. H. (2001). Caenorhabditis elegans p53: Role in apoptosis, meiosis, and stress resistance. *Science*, *294*, 591–595.

Driscoll, M., & Chalfie, M. (1991). The mec-4 gene is a member of a family of Caenorhabditis elegans genes that can mutate to induce neuronal degeneration. *Nature*, *349*, 588–593.

Ellis, H. M., & Horvitz, H. R. (1986). Genetic control of programmed cell death in the nematode C. elegans. *Cell*, *44*, 817–829.

Galvin, B. D., Kim, S., & Horvitz, H. R. (2008). Caenorhabditis elegans genes required for the engulfment of apoptotic corpses function in the cytotoxic cell deaths induced by mutations in lin-24 and lin-33. *Genetics*, *179*, 403–417.

Geng, X., Shi, Y., Nakagawa, A., Yoshina, S., Mitani, S., Shi, Y., et al. (2008). Inhibition of CED-3 zymogen activation and apoptosis in Caenorhabditis elegans by caspase homolog CSP-3. *Nature Structural & Molecular Biology*, *15*, 1094–1101.

Geng, X., Zhou, Q. H., Kage-Nakadai, E., Shi, Y., Yan, N., Mitani, S., et al. (2009). Caenorhabditis elegans caspase homolog CSP-2 inhibits CED-3 autoactivation and apoptosis in germ cells. *Cell Death and Differentiation*, *16*, 1385–1394.

Gumienny, T. L., Lambie, E., Hartwieg, E., Horvitz, H. R., & Hengartner, M. O. (1999). Genetic control of programmed cell death in the Caenorhabditis elegans hermaphrodite germline. *Development*, *126*, 1011–1022.

Hall, D. H., Gu, G., García-Añoveros, J., Gong, L., Chalfie, M., & Driscoll, M. (1997). Neuropathology of degenerative cell death in Caenorhabditis elegans. *Journal of Neuroscience*, *17*, 1033–1045.

Harris, D. T., & Horvitz, H. R. (2011). MAB-10/NAB acts with LIN-29/EGR to regulate terminal differentiation and the transition from larva to adult in C. elegans. *Development*, *138*, 4051–4062.

Hetz, C. (2012). The unfolded protein response: Controlling cell fate decisions under ER stress and beyond. *Nature Reviews. Molecular Cell Biology*, *13*, 89–102.

Hirose, T., & Horvitz, H. R. (2013). An Sp1 transcription factor coordinates caspase-dependent and -independent apoptotic pathways. *Nature*, *500*, 354–358.

Hodgkin, J. (1987). A genetic analysis of the sex-determining gene, tra-1, in the nematode Caenorhabditis elegans. *Genes & Development*, *1*, 731–745.

Hong, K., & Driscoll, M. (1994). A transmembrane domain of the putative channel subunit MEC-4 influences mechanotransduction and neurodegeneration in C. elegans. *Nature*, *367*, 470–473.

Horvitz, H. R., Shaham, S., & Hengartner, M. O. (1994). The genetics of programmed cell death in the nematode Caenorhabditis elegans. *Cold Spring Harbor Symposia on Quantitative Biology*, *59*, 377–385.

Howard, M. K., Burke, L. C., Mailhos, C., Pizzey, A., Gilbert, C. S., Lawson, W. D., et al. (1993). Cell cycle arrest of proliferating neuronal cells by serum deprivation can result in either apoptosis or differentiation. *Journal of Neurochemistry*, *60*, 1783–1791.

Huang, L., & Hanna-Rose, W. (2006). EGF signaling overcomes a uterine cell death associated with temporal mis-coordination of organogenesis within the C. elegans egg-laying apparatus. *Developmental Biology*, *300*, 599–611.

Ito, S., Greiss, S., Gartner, A., & Derry, W. B. (2010). Cell-nonautonomous regulation of C. elegans germ cell death by kri-1. *Current Biology*, *20*, 333–338.

Joshi, P., & Eisenmann, D. M. (2004). The Caenorhabditis elegans pvl-5 gene protects hypodermal cells from ced-3-dependent, ced-4-independent cell death. *Genetics*, *167*, 673–685.

Kamath, R. S., Martinez-Campos, M., Zipperlen, P., Fraser, A. G., & Ahringer, J. (2001). Effectiveness of specific RNA-mediated interference through ingested double-stranded RNA in Caenorhabditis elegans. *Genome Biology*, *2*(1).

Kato, M., & Sternberg, P. W. (2009). The C. elegans tailless/Tlx homolog nhr-67 regulates a stage-specific program of linker cell migration in male gonadogenesis. *Development*, *136*, 3907–3915.

Kimble, J., & Hirsh, D. (1979). The postembryonic cell lineages of the hermaphrodite and male gonads in Caenorhabditis elegans. *Developmental Biology*, *70*, 396–417.

Korswagen, H. C., Park, J. H., Ohshima, Y., & Plasterk, R. H. (1997). An activating mutation in a Caenorhabditis elegans Gs protein induces neural degeneration. *Genes & Development*, *11*, 1493–1503.

Kourtis, N., Nikoletopoulou, V., & Tavernarakis, N. (2012). Small heat-shock proteins protect from heat-stroke-associated neurodegeneration. *Nature*, *490*, 213–218.

Kulkarni, G. V., & McCulloch, C. A. (1994). Serum deprivation induces apoptotic cell death in a subset of Balb/c 3T3 fibroblasts. *Journal of Cell Science*, *107*(Pt 5), 1169–1179.

Kumar, V., Abbas, A. K., Fausto, N., & Aster, J. C. (2009). *Robbins & Cotran pathologic basis of disease*. Philadelphia: Elsevier Health Sciences.

Labouesse, M., Sookhareea, S., & Horvitz, H. R. (1994). The Caenorhabditis elegans gene lin-26 is required to specify the fates of hypodermal cells and encodes a presumptive zinc-finger transcription factor. *Development*, *120*, 2359–2368.

Li, X., Johnson, R. W., Park, D., Chin-Sang, I., & Chamberlin, H. M. (2012). Somatic gonad sheath cells and Eph receptor signaling promote germ-cell death in C. elegans. *Cell Death and Differentiation*, *19*, 1080–1089.

Malone, C. J., Fixsen, W. D., Horvitz, H. R., & Han, M. (1999). UNC-84 localizes to the nuclear envelope and is required for nuclear migration and anchoring during C. elegans development. *Development*, *126*, 3171–3181.

Mano, I., & Driscoll, M. (2009). Caenorhabditis elegans glutamate transporter deletion induces AMPA-receptor/adenylyl cyclase 9-dependent excitotoxicity. *Journal of Neurochemistry*, *108*, 1373–1384.

Martin, D. N., & Baehrecke, E. H. (2004). Caspases function in autophagic programmed cell death in Drosophila. *Development*, *131*, 275–284.

Massoll, C., Mando, W., & Chintala, S. K. (2013). Excitotoxicity upregulates SARM1 protein expression and promotes Wallerian-like degeneration of retinal ganglion cells and their axons. *Investigative Ophthalmology & Visual Science*, *54*, 2771–2780.

Maurer, C. W., Chiorazzi, M., & Shaham, S. (2007). Timing of the onset of a developmental cell death is controlled by transcriptional induction of the C. elegans ced-3 caspase-encoding gene. *Development*, *134*, 1357–1368.

McGee, M. D., Rillo, R., Anderson, A. S., & Starr, D. A. (2006). UNC-83 IS a KASH protein required for nuclear migration and is recruited to the outer nuclear membrane by a physical interaction with the SUN protein UNC-84. *Molecular Biology of the Cell*, *17*, 1790–1801.

Morthorst, T. H., & Olsen, A. (2013). Cell-nonautonomous inhibition of radiation-induced apoptosis by dynein light chain 1 in Caenorhabditis elegans. *Cell Death and Differentiation*, *4*, e799.

Morton, D. G., Hoose, W. A., & Kemphues, K. J. (2012). A genome-wide RNAi screen for enhancers of par mutants reveals new contributors to early embryonic polarity in Caenorhabditis elegans. *Genetics*, *192*, 929–942.

Nass, R., & Blakely, R. D. (2003). The Caenorhabditis elegans dopaminergic system: Opportunities for insights into dopamine transport and neurodegeneration. *Annual Review of Pharmacology and Toxicology*, *43*, 521–544.

Nass, R., Hall, D. H., Miller, D. M., & Blakely, R. D. (2002). Neurotoxin-induced degeneration of dopamine neurons in Caenorhabditis elegans. *Proceedings of the National Academy of Sciences of the United States of America*, *99*, 3264–3269.

Nehme, R., & Conradt, B. (2008). egl-1: A key activator of apoptotic cell death in C. elegans. *Oncogene*, *27*(Suppl. 1), S30–S40.

Nehme, R., Grote, P., Tomasi, T., Löser, S., Holzkamp, H., Schnabel, R., et al. (2010). Transcriptional upregulation of both egl-1 BH3-only and ced-3 caspase is required for the death of the male-specific CEM neurons. *Cell Death and Differentiation*, *17*, 1266–1276.

Oppenheim, R. W., Flavell, R. A., Vinsant, S., Prevette, D., Kuan, C. Y., & Rakic, P. (2001). Programmed cell death of developing mammalian neurons after genetic deletion of caspases. *Journal of Neuroscience*, *21*, 4752–4760.

Osterloh, J. M., Yang, J., Rooney, T. M., Fox, A. N., Adalbert, R., Powell, E. H., et al. (2012). dSarm/Sarm1 is required for activation of an injury-induced axon death pathway. *Science*, *337*, 481–484.

Ou, G., Stuurman, N., D'Ambrosio, M., & Vale, R. D. (2010). Polarized myosin produces unequal-size daughters during asymmetric cell division. *Science*, *330*, 677–680.

Park, D., Jia, H., Rajakumar, V., & Chamberlin, H. M. (2006). Pax2/5/8 proteins promote cell survival in C. elegans. *Development*, *133*, 4193–4202.

Peden, E., Kimberly, E., Gengyo-Ando, K., Mitani, S., & Xue, D. (2007). Control of sex-specific apoptosis in C. elegans by the BarH homeodomain protein CEH-30 and the transcriptional repressor UNC-37/Groucho. *Genes & Development*, *21*, 3195–3207.

Pilar, G., & Landmesser, L. (1976). Ultrastructural differences during embryonic cell death in normal and peripherally deprived ciliary ganglia. *Journal of Cell Biology*, *68*, 339–356.

Raymond, C. S., Murphy, M. W., O'Sullivan, M. G., Bardwell, V. J., & Zarkower, D. (2000). Dmrt1, a gene related to worm and fly sexual regulators, is required for mammalian testis differentiation. *Genes & Development*, *14*, 2587–2595.

Reddien, P. W., Cameron, S., & Horvitz, H. R. (2001). Phagocytosis promotes programmed cell death in C. elegans. *Nature*, *412*, 198–202.

Salinas, L. S., Maldonado, E., & Navarro, R. E. (2006). Stress-induced germ cell apoptosis by a p53 independent pathway in Caenorhabditis elegans. *Cell Death and Differentiation*, *13*, 2129–2139.

Schwartz, H. T., & Horvitz, H. R. (2007). The C. elegans protein CEH-30 protects male-specific neurons from apoptosis independently of the Bcl-2 homolog CED-9. *Genes & Development*, *21*, 3181–3194.

Sendoel, A., Kohler, I., Fellmann, C., Lowe, S. W., & Hengartner, M. O. (2010). HIF-1 antagonizes p53-mediated apoptosis through a secreted neuronal tyrosinase. *Nature*, *465*, 577–583.

Shaham, S. (1998). Identification of multiple Caenorhabditis elegans caspases and their potential roles in proteolytic cascades. *Journal of Biological Chemistry*, *273*, 35109–35117.

Shaham, S., Reddien, P. W., Davies, B., & Horvitz, H. R. (1999). Mutational analysis of the Caenorhabditis elegans cell-death gene ced-3. *Genetics*, *153*, 1655–1671.

Shreffler, W., Magardino, T., Shekdar, K., & Wolinsky, E. (1995). The unc-8 and sup-40 genes regulate ion channel function in Caenorhabditis elegans motorneurons. *Genetics*, *139*, 1261–1272.

Silva, N., Adamo, A., Santonicola, P., Martinez-Perez, E., & La Volpe, A. (2013). Pro-crossover factors regulate damage-dependent apoptosis in the Caenorhabditis elegans germ line. *Cell Death and Differentiation*, *20*, 1209–1218.

Sommer, R. J., Eizinger, A., Lee, K. Z., Jungblut, B., Bubeck, A., & Schlak, I. (1998). The Pristionchus HOX gene Ppa-lin-39 inhibits programmed cell death to specify the vulva equivalence group and is not required during vulval induction. *Development*, *125*, 3865–3873.
Sommer, R. J., & Sternberg, P. W. (1996). Apoptosis and change of competence limit the size of the vulva equivalence group in Pristionchus pacificus: A genetic analysis. *Current Biology*, *6*, 52–59.
Starr, D. A., Hermann, G. J., Malone, C. J., Fixsen, W., Priess, J. R., Horvitz, H. R., et al. (2001). unc-83 encodes a novel component of the nuclear envelope and is essential for proper nuclear migration. *Development*, *128*, 5039–5050.
Sulston, J. E., & Horvitz, H. R. (1977). Post-embryonic cell lineages of the nematode, Caenorhabditis elegans. *Developmental Biology*, *56*, 110–156.
Sulston, J. E., Schierenberg, E., White, J. G., & Thomson, J. N. (1983). The embryonic cell lineage of the nematode Caenorhabditis elegans. *Developmental Biology*, *100*, 64–119.
Syntichaki, P., Samara, C., & Tavernarakis, N. (2005). The vacuolar H+-ATPase mediates intracellular acidification required for neurodegeneration in C. elegans. *Current Biology*, *15*, 1249–1254.
Syntichaki, P., Xu, K., Driscoll, M., & Tavernarakis, N. (2002). Specific aspartyl and calpain proteases are required for neurodegeneration in C. elegans. *Nature*, *419*, 939–944.
Treinin, M., & Chalfie, M. (1995). A mutated acetylcholine receptor subunit causes neuronal degeneration in C. elegans. *Neuron*, *14*, 871–877.
Vrablik, T. L., Huang, L., Lange, S. E., & Hanna-Rose, W. (2009). Nicotinamidase modulation of NAD+ biosynthesis and nicotinamide levels separately affect reproductive development and cell survival in C. elegans. *Development*, *136*, 3637–3646.
Wang, Y., Matthewman, C., Han, L., Miller, T., Miller, D. M., & Bianchi, L. (2013). Neurotoxic unc-8 mutants encode constitutively active DEG/ENaC channels that are blocked by divalent cations. *Journal of General Physiology*, *142*, 157–169.
White, J. Q., Nicholas, T. J., Gritton, J., Truong, L., Davidson, E. R., & Jorgensen, E. M. (2007). The sensory circuitry for sexual attraction in C. elegans males. *Current Biology*, *17*, 1847–1857.
Xiong, Z.-G., Zhu, X.-M., Chu, X.-P., Minami, M., Hey, J., Wei, W.-L., et al. (2004). Neuroprotection in ischemia: Blocking calcium-permeable acid-sensing ion channels. *Cell*, *118*, 687–698.
Xu, K., Tavernarakis, N., & Driscoll, M. (2001). Necrotic cell death in C. elegans requires the function of calreticulin and regulators of Ca(2+) release from the endoplasmic reticulum. *Neuron*, *31*, 957–971.
Xue, D., Shaham, S., & Horvitz, H. R. (1996). The Caenorhabditis elegans cell-death protein CED-3 is a cysteine protease with substrate specificities similar to those of the human CPP32 protease. *Genes & Development*, *10*, 1073–1083.
Yan, N., Chai, J., Lee, E.-S., Gu, L., Liu, Q., He, J., et al. (2005). Structure of the CED-4-CED-9 complex provides insights into programmed cell death in Caenorhabditis elegans. *Nature*, *437*, 831–837.
Yang, X., Castilla, L. H., Xu, X., Li, C., Gotay, J., Weinstein, M., et al. (1999). Angiogenesis defects and mesenchymal apoptosis in mice lacking SMAD5. *Development*, *126*, 1571–1580.
Yang, X., Chang, H. Y., & Baltimore, D. (1998). Essential role of CED-4 oligomerization in CED-3 activation and apoptosis. *Science*, *281*, 1355–1357.
Zarkower, D., & Hodgkin, J. (1992). Molecular analysis of the C. elegans sex-determining gene tra-1: A gene encoding two zinc finger proteins. *Cell*, *70*, 237–249.
Zhou, X., Ding, Q., Chen, Z., Yun, H., & Wang, H. (2013). Involvement of the GluN2A and GluN2B subunits in synaptic and extrasynaptic N-methyl-D-aspartate receptor function and neuronal excitotoxicity. *Journal of Biological Chemistry*, *288*, 24151–24159.
Zhou, Z., Hartwieg, E., & Horvitz, H. R. (2001). CED-1 is a transmembrane receptor that mediates cell corpse engulfment in C. elegans. *Cell*, *104*, 43–56.

CHAPTER EIGHT

Autophagy and Cell Death in the Fly

Charles Nelson, Eric H. Baehrecke[1]

Department of Cancer Biology, University of Massachusetts Medical School, Worcester, Massachusetts, USA
[1]Corresponding author: e-mail address: eric.baehrecke@umassmed.edu

Contents

Abstract

Macroautophagy (hereafter referred to as autophagy) is a lysosome-dependent catabolic process that results in the degradation and recycling of cellular components, such as lipids, proteins, and organelles. Autophagy can function in many cellular contexts, including during infection, stress, cell survival, and cell death. During the development of the fruit fly *Drosophila melanogaster*, multiple tissues undergo a programmed cell death in which autophagy plays a key role in their destruction. Here, we describe how to analyze autophagy and its relationship to cell death in *Drosophila*.

Methods in Enzymology, Volume 545
ISSN 0076-6879
http://dx.doi.org/10.1016/B978-0-12-801430-1.00008-1

1. INTRODUCTION

1.1. *Drosophila* as a biological system for studying autophagic cell death

Programmed cell death has been studied for decades. However, most of these studies have focused on apoptotic cell death, and relatively little attention has been paid to the various forms of nonapoptotic programmed cell death, such as programmed autophagic cell death. The potentials for understanding the mechanisms of autophagic cell death cannot be understated. Our knowledge about cancer, neurodegeneration, and other age-associated diseases are likely to benefit from a clearer understanding of how autophagy functions during programmed cell death. In *Drosophila melanogaster*, at least two distinct tissues undergo programmed autophagic cell death: the larval salivary glands and the larval midgut (Lee & Baehrecke, 2001; Lee, Cooksey, & Baehrecke, 2002). Fortuitously, the powerful genetics and conservation of signaling and cellular mechanisms between *Drosophila* and humans make the fly an ideal organism in which to study the role of autophagy during programmed cell death.

Autophagy is a catabolic process utilized for the degradation of various cellular components that occurs in all animals (Mizushima & Komatsu, 2011). The initiation of autophagy begins with the formation of the preautophagic structure (PAS). This PAS acts as a nucleation point for the elongation of the isolation membrane, a double-membrane structure that surrounds the soon-to-be degraded cellular material, eventually forming a vesicle called an autophagosome. The autophagosome then fuses with lysosomes, creating an autolysosome, wherein the lysosomal machinery degrades the isolated cellular material.

Pulses of the steroid hormone 20-hydroxyecdysone (ecdysone) trigger multiple developmental processes in *Drosophila*. These processes include the differentiation and morphogenesis of adult structures and the destruction of larval tissues. In two distinct larval tissues, the larval midgut and larval salivary glands, ecdysone triggers for the induction of autophagy resulting in their subsequent destruction. In the midgut of the larval intestine, autophagy functions in a caspase-independent manner to control cell death (Denton et al., 2009). By contrast, autophagy functions in parallel with caspases to destroy larval salivary gland cells (Berry & Baehrecke, 2007). These two tissues provide excellent experimental systems to investigate the roles autophagy can play either alone or in cooperation with caspases to achieve programmed cell death.

1.2. Genetic approaches to study autophagic cell death in *Drosophila*

Classical genetic and full-genome sequencing approaches resulted in a sophisticated map of the *Drosophila* genome. Furthermore, recent deep sequencing-based transcriptome experiments have led to extremely detailed annotations of genes and gene expression patterns. From these annotations, conserved autophagy (*Atg*) genes have been identified. Furthermore, the well annotated *Drosophila* genome has allowed for the quick identification and study of novel regulatory mechanisms of autophagy during the destruction of the midgut as well as the salivary glands. This knowledge combined with the ability to conduct genetic experiments with single cell resolution in an intact organism make fruit flies an ideal system to study the relationships between autophagy, cell survival, and cell death.

Drosophila is one of the premiere animal model systems that are used to identify new genes that function in processes using forward genetic approaches (Bernards & Hariharan, 2001). In addition, transposable P-element-based genetic engineering has also allowed for the development of multiple important genetic approaches, such as clonal mutant cell analysis, targeted mutagenesis, transgenesis, and RNA interference (RNAi) among many others. Significantly, the development of binary gene expression systems, such as the GAL4/UAS system (Brand & Perrimon, 1993), has granted for the means to study gene function at the cellular, tissue, and organismal level. With this technology, the well annotated *Drosophila* genome allows for robust reverse genetic approaches such as either targeted gene disruption (Rong & Golic, 2000) or RNAi using one of the multiple whole genome collections. Furthermore, recent technologies such as TAL effector nuclease and the clustered, regularly interspersed, short palindromic repeats systems have allowed for the efficient generation of specific mutations (Gratz et al., 2013).

In *Drosophila* and other animals, mutations in essential genes results in lethality making it difficult to study gene function. However, the use of flippase-based mitotic recombination enables the generation of homozygous mutant cell clones so that gene function can be studied at single cell resolution (Lee & Luo, 2001; Xu & Rubin, 1993). These patches of homozygous mutant clone cells are surrounded by homozygous and heterozygous control

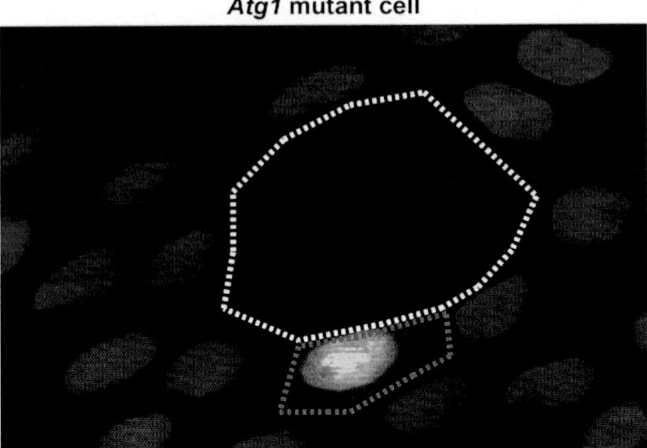

Figure 8.1 *Drosophila* genetic techniques allows for the generation of mutant cells in an otherwise heterozygous animal. In this case, the white-outlined *Atg1* homozygous mutant cell was generated using flippase-based mitotic recombination and is marked by the absence of green fluorescent protein (GFP). The greener GFP-marked cell (red outline) is a control homozygous wild-type cell, while the remaining lighter green GFP cells are control heterozygous wild-type/*Atg1* mutant cells. *Reprinted from Chang et al. (2013) with permission from Nature Publishing Group.* (See the color plate.)

cells (Fig. 8.1), allowing for direct comparison of homozygous mutant cells to neighboring control cells. This powerful genetic tool thus allows the studying of lethal autophagy genes *in vivo*.

Using these genetic tools, programmed autophagic cell death can be studied in great depth. Various transgenic, mutant, and RNAi-based screens have led to the identification of novel genes that function to regulate autophagy during programmed cell death. Transgenic animals that carry fluorescent reporters can be utilized to visualize autophagy in these mutant and RNAi animals. Additionally, epistasis experiments using various mutants, RNAi, and transgenic animals can be used to determine where genes genetically function relative to one another.

2. MATERIALS AND METHODS

2.1. Fly food

The autophagic degradations of the larval midgut and salivary glands occur on a precise developmental schedule (Lee & Baehrecke, 2001; Lee et al., 2002). The timing of this schedule can vary depending on factors, such as temperature and food. Animals raised on nutrient-rich food grow faster, whereas animals raised on nutrient-poor food grow slower. Because of this, the midgut and salivary glands of animals raised on rich food die earlier than they would if animals were raised on less nutritious food. Furthermore, induction of autophagy can occur as a starvation response due to nutrient-poor conditions. Therefore, to minimize the complications associated with the induction of autophagy associated with nutrient restriction and to ensure consistent degradation of the midgut and salivary glands, animals should be grown on nutrient-rich food.

Drosophila nutrient-rich food ingredients: 6.5 g/l agar; 63 g/l yeast; 60 g/l cornmeal; 60 ml/l molasses; 4 ml/l acid mix (41.7% propionic acid and 3.5% phosphoric acid); 0.13% tegosept.

2.2. Staging of animals

The timing of the degradations of the midgut and salivary glands depends on temperature. The optimal temperature for staging animals for the degradation of the midgut and salivary glands is 25 °C. If a different temperature is desired, the new degradation timing of the midgut and salivary glands needs to be determined. Additionally, crowding animals by raising them at high density can cause stress that may influence developmental timing and induce autophagy as a starvation response. To ensure proper degradation of the midgut and salivary glands, food vials should not be overcrowded. At 25 °C in noncrowded conditions, the larval midgut begins to degrade by inducing autophagy at the onset of puparium formation, and the larval salivary glands begin to degrade by inducing autophagy and activating caspases 13 h after puparium formation (APF).

1. Using forceps collect white prepupae (0 h APF) from food vials. *Note: white prepupae and staged pupae are delicate and should be handled with care.*
2. Place the white prepupae on a water-moistened tissue folded inside a Petri dish, cover, label, and place the dish in a 25 °C incubator until desired stage is reached. *Note: pupae should not be placed in standing water as this may alter their development.*

2.3. Histology

Defects in the degradation and clearance of the midgut and salivary glands result in remnants of these tissues remaining at stages during development when they would normally be absent. These remnants can easily be visualized in histological sections.

2.3.1 Preparation of samples

1. Place staged animals in a dissection dish containing phosphate-buffered saline (PBS), hold anterior end of pupa with forceps, and cut posterior end with dissecting scissors exposing the inside of the animal.
2. Fix the pupa by placing it in 1 ml of 80% ethanol, 4% formaldehyde, 5% acetic acid, 1% glutaraldehyde (FAAG) in a 1.5-ml centrifuge tube and store at 4 °C for at least one night. *Note: up to 50 pupae can be placed in one tube, and they can be stored at 4 °C for approximately 1 month.*
3. Dehydrate pupae by removing the FAAG, and incubating them with 1 ml of 80% ethanol for 10 min, 85% ethanol for 15 min, 90% ethanol for 20 min, 100% ethanol for 25 min, 100% ethanol for 85 min, xylenes for 25 min, and xylenes for 85 min. *Note: samples should be rotated while incubating.*
4. Remove xylenes, and using a paint brush place approximately 15 pupae in a scintillation vial containing approximately 20 paraplast x-tra chips (McCormick Scientific 39503002) and 2 ml xylenes.
5. Incubate the vial overnight at 45 °C. While incubating samples, melt paraplast x-tra chips in a beaker at 55 °C.
6. Decant the wax/xylenes solution into an appropriate disposal container while keeping the pupae in the vial. Fill the vial with melted paraplast x-tra and incubate at 55 °C.
7. Repeat paraplast changes two more times over 1 day.
8. Pour wax and samples into a traditional paper boat used for paraffin histology.
9. Before wax solidifies, orient pupae using toothpicks ventral side down and allow the wax to solidify overnight.
10. Remove paper around the wax and cut out a square block around each pupa. *Note: leave approximately 0.5 cm of space around the pupa.*

2.3.2 Sectioning

1. Mount the square block containing the embedded pupa onto a wax chuck by melting the top of the chuck with a heated metal spatula and placing the block into the melted wax. Hold the block on the chuck until the wax has solidified.
2. Mount the chuck into a microtome and section the pupa. Warm water on a microscope slide using a slide warmer set to 45 °C and place ribbons generated from the sectioning paraffin on the water. *Note: to obtain straight paraffin ribbons, trim the paraffin blocks so they are square.*

3. Once the sectioning is complete, remove the water with a paper towel, and incubate the slides overnight on the slide warmer.

2.3.3 *Staining*

1. Fill 11 staining dishes with the following solutions: two dishes with xylenes, two with 100% ethanol, two with 90% ethanol, two with 70% ethanol, one with acid water (0.2% glacial acetic acid in dH_2O), one with Weigert's Hematoxylin (equal parts Weigert's Iron Hematoxylin A and Weigert's Iron Hematoxylin B), and one with Pollak Trichrome (Combine: 150 ml ethanol, 150 ml dH_2O, and 3 ml glacial acetic acid. Split the solution into four beakers with approximately 75 ml each. To beaker 1 add 0.5 g acid fuchsin and 1.0 g panceau 2R. To beaker 2 add 0.45 g light green, SF, yellowish. To beaker 3 add 0.75 g orange G and 1.5 g phophotungstic acid. To beaker 4 add 1.5 g phosphomolybdic acid. When each reagent is maximally dissolved, combine the four beakers and filter). Fill two plastic containers, one with tap water and the other with dH_2O. *Note: the stains should be stored in dark containers and kept in the dark to reduce light exposure.*
2. Place the slides in a slide staining rack and place the rack in the solutions listed above to hydrate, stain, and dehydrate the samples. *Hydration*: first xylenes for 3 min, second xylenes for 3 min, first 100% ethanol for 3 min, second 100% ethanol for 3 min, first 90% ethanol for 3 min, first 70% ethanol for 3 min, and running tap water for 5 min. *Staining*: Weigert's Hematoxylin for 5 min, running tap water for 5 min, Pollak trichrome for 7 min, dH_2O for 10 s, acid water for 10 s. *Dehydration*: second 70% ethanol for 3 min, second 90% ethanol for 3 min, second 100% ethanol for 3 min, first 100% ethanol for 3 min, second xylenes for 3 min, and first xylenes for 3 min,
3. Let the slides dry, add approximately 1 μl Permount for each mm^2 of cover slip area, and place the cover slip on top trying not to allow air bubbles to form.

2.4. Immunochemistry

Immunoblotting and immunofluorescence serve as tools to visualize proteins that are involved in and associated with changes in autophagy activity during the degradation of the midgut and salivary glands. Proteins whose levels are affected during autophagy are visualized by immunoblotting, while proteins whose localization changes are visualized by immunofluorescence microscopy.

2.4.1 Immunoblotting

1. Place appropriately staged animals in a dissection dish containing PBS, dissect the tissue of interest, and using forceps, place the tissue into 1 ml of PBS in a 1.5-ml centrifuge tube. To obtain enough protein for immunoblotting, dissect at least 15 salivary glands or 5 midguts. *Note: if desired, protein concentrations can be measured. However, in the limited time allotted for dissecting properly staged tissues, relatively little protein is extracted, and measuring protein concentration requires a relatively large amount of sample. Tissues dissected from various time points have approximately the same amount of protein.*
2. Lightly centrifuge the samples at 1000 rpm for 1 min, and remove the PBS by pipetting. Add 1.5 μl per salivary gland or 5 μl per midgut of 2× Laemmli buffer (for 1 ml: 100 μl glycerol, 200 μl 10% SDS, 125 μl 1 *M* Tris, pH 6.8, 50 μl B-mercaptoethanol, 0.05% Bromophenol Blue, 525 μl ddH_2O) to the tube, and homogenize the tissue using a pestle.
3. Boil the sample for 5 min.
4. Centrifuge the sample at 14,000 rpm for 5 min and store at −80 °C.
5. Run the sample on a polyacrylamide gel and blot against the protein of interest using standard immunoblotting techniques using appropriate primary and secondary antibodies (Alegria-Schaffer, Lodge, & Vattem, 2009).

2.4.2 Immunofluorescence

1. Place appropriately staged animals in a dissection dish containing PBS, remove the tissue of interest by dissection, and using forceps, place the tissue into 0.5 ml of 4% paraformaldehyde/PBS (PFA) in a 1.5-ml centrifuge tube.
2. Add 500 μl heptane and shake for 20 min at 250 rpm.
3. Remove fixative (lower phase), add 500 μl methanol, and shake vigorously for 1 min by hand.
4. Remove heptane (upper phase) and interphase.
5. Rinse three times with methanol. *Note: tissue samples can be stored in methanol at −20 °C for approximately 2 weeks.*
6. Rinse once with a mixture of 500 μl methanol and 500 μl PBST (PBS and 0.1% Tween-20). *Note: some antibodies do not work well when the sample has been treated with heptane and methanol. Alternatively, the sample in PFA can be incubated at 4 °C overnight and subjected straight to the steps listed below.*

7. Rinse three times with PBST and then four times with PBSBT (PBST and 1% BSA).
8. Block at room temperature in 500 μl PBSBT for 2 h.
9. Add primary antibody at the appropriate dilution in PBSBT, and incubate overnight at 4 °C.
10. Wash the sample four times with PBSBT for 30 min each time.
11. Add secondary antibody at the appropriate dilution in PBSBT, and incubate at room temperature for 2 h. *Note: perform all steps with the secondary antibody in the dark.*
12. Wash the sample three times with PBSBT for 10 min each time.
13. Mount the sample on a slide by addition of Vectashield with DAPI (Vector Labs), addition of a cover slip, and sealing the cover slip with clear nail polish. *Note: slides can be stored at 4 °C in the dark for approximately 1 week.*

2.5. Terminal deoxynucleotidyl transferase dUTP nick end labeling

Terminal deoxynucleotidyl transferase dUTP nick end labeling (TUNEL) is a method used to detect fragmented DNA, which is a hallmark of dying cells in the midgut and salivary glands. In this assay, the nicked ends of fragmented DNA are identified by terminal deoxynucleotidyl transferase (TdT), which adds labeled dUTP to the DNA. TUNEL can be used to detect DNA fragmentation in whole dissected midguts and salivary glands (Denton, Mills, & Kumar, 2008), and here we describe how to perform this procedure on histological paraffin sections:

1. Stage and process pupae for histological sections using the tissue embedding protocol (Section 2.3). Cut sections at 8 μm the day before you wish to do TUNEL staining.
2. Incubate slides containing paraffin sections in two washes of xylenes for 3 min each and two washes of 100% ethanol for 3 min each.
3. Add 1 ml of 10 μl 30% H_2O_2 and 90 μl methanol per slide for 5 min at room temperature and repeat this process a second time.
4. Incubate slides in 75% methanol/25% PBS + 0.1% Triton X-100 (PBSTr) for 5 min, 50% methanol/50% PBSTr for 5 min, 25% methanol/75% PBSTr for 5 min, and 100% PBSTr for 5 min.
5. Add 1 ml/slide of 20 μg/ml Proteinase K in PBSTr and incubate for 12 min at room temperature.
6. Rinse twice for 2 min in PBSTr. *Note: for a positive control, treat sample with DNase followed by rinses in PBSTr.*

7. Add 200 μl 100 m*M* sodium citrate solution in PBSTr, add cover slip, and incubate 65 °C in a humid chamber for 30 min.
8. Rinse in PBSTr for 5 min, add 200 μl ApopTag equilibration buffer (Millipore) and incubate for 5 min at room temperature.
9. Remove equilibration buffer and add 100 μl TdT working buffer (67 μl Reaction Buffer and 33 μl TdT (1 U/μl) from Millipore, 1 μl TdT (20 U/μl) from New England Biolabs, and 0.1% Triton X) per slide, add a cover slip, and incubate at 37 °C in a humid chamber for 2 h.
10. Stop Tdt reaction by incubating in working strength stop/wash buffer (Millipore) at room temperature for 10 min, rinse in PBSTr three times for 1 min each.
11. Block by incubating in PBSBT for 30 min at room temperature, add 150 μl 1:500 dilution of anti-digoxigenin-HRP-POD (Millipore) in PBSBT per slide, add a cover slip, and incubate for 1 h at room temperature in a humid chamber.
12. Wash three times in PBST for 5 min each at room temperature, and add 500 μl DAB mix (10 μl 3% H_2O_2, 10 μl 8% NiCl, 1 ml diaminobenzidine substrate (Millipore)), watch for color reaction, and rinse three times in dH_2O for 1 min each.
13. Counter-stain in Eosin Y (1 μl acetic acid and 300 ml Eosin Y (Sigma)) for 1 min at room temperature and rinse in dH_2O for 5 min and 70% ethanol for 3 min.
14. Dehydrate by dipping slides in 95% ethanol, twice in 100% ethanol, followed by two rinses of xylenes. Add approximately 1 μl Permount for each mm^2 of cover slip area and place the cover slip on top trying not to allow air bubbles to form.

2.6. Transmission electron microscopy

Transmission electron microscopy (TEM) is used to visualize cell ultrastructure and define the morphological forms of programmed cell death. In studies of dying midguts and salivary glands, TEM is important for detecting organelles, membrane structures and integrity, and vesicle structures, including autophagosomes and autolysosomes.

1. Place appropriately staged animals in a dissection dish containing PBS, remove the tissue of interest by dissection, and using forceps, place the tissue into 0.5 ml of 4.0% paraformaldehyde, 2.0% glutaraldehyde, 1% sucrose, and 0.028% $CaCl_2$ in 0.1 *M* sodium cacodylate, pH 7.4, and fix overnight at 4 °C.

2. Thoroughly wash tissues in cacodylate buffer, and postfix in 2.0% osmium tetroxide for 1 h at room temperature.
3. Embed fixed tissues in SPI-pon/Araldite resin (Polysciences) according to manufacturer's recommendations.
4. Cut semithin sections (~2 μm) and stain with 0.1% toluidine blue in dH_2O in order to visualize specimens by light microscopy and select the area of the block to begin cutting ultrathin sections.
5. Cut ultrathin sections (80 nm), collect sections on grids, stain with uranyl acetate and lead citrate, and examine by TEM.

2.7. Atg8 tagged fluorescence

Atg8 (LC3 and GABARAP in mammals) is a protein that decorates the autophagosome membrane via its lipidation by an ubiquitin-like conjugation system (Mizushima & Komatsu, 2011). When tagged with a fluorescent protein such as GFP or mCherry, a reporter is created that can serve to visualize autophagosomes and autolysosomes in the cell. This, therefore, allows for the visualization of autophagy levels in the midgut and salivary glands during their degradation. As degrading tissues can be fragile (particularly the salivary glands), fixing the tissues ensures a stable sample. However, fixing tissues can result in decreased fluorescence, and, if treated carefully, tissues can be imaged without fixation.

1. Place appropriately staged animals in a dissection dish containing PBS and remove the tissue of interest by dissection.
2. Place 20 μl of 4% PFA on a slide and carefully transfer the sample from the PBS to the PFA. *Note: if unfixed tissue is desired, replace PFA with PBS.*
3. Add 20 μl of 2 μ*M* Hoechst stain in PBS to the tissue in PFA, and incubate for 2 min.
4. Remove the Hoechst/PFA solution and wash two times with 20 μl of PBS.
5. Mount in PBS and wait approximately 5 min for DNA to stain before imaging.

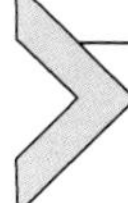

3. DATA ANALYSIS AND INTERPRETATION

3.1. Interpreting histological sections

Histological sections allow for the visualization of the persistence of midgut and salivary gland structures after they would normally be cleared. A rise in the steroid hormone ecdysone at the end of the third larval instar triggers midgut cell death. In control and wild-type animals, the midgut is typically degraded 4 h APF. Therefore, we screen for defects in midgut cell death at 12 h APF. In the midgut, defects in autophagic cell death manifest as a failure of tissue condensation, and persistence of gastric ceca and proventriculus structures at 12 h APF (Fig. 8.2). As autophagy is the only process known to be necessary for midgut cell death, defects in midgut condensation and persistent structures as shown by histology sections generally indicate a failure in the execution of autophagy. Histological sections are generally scored for the presence of midgut structures with a final quantification stating the percentage of animals with remaining midgut structure. Twenty or more animals for control and experimental genotypes are considered acceptable for histological analysis.

A rise in the steroid hormone ecdysone at 10 h APF triggers salivary gland cell death. In control and wild-type animals, salivary glands are mostly cleared by 14 h APF. Therefore, we screen for defects in salivary gland cell death, including the persistence of salivary gland cell and tissue fragments at 24 h APF (Fig. 8.3). Caspase activity and autophagy are both required for salivary gland degradation. It is important to note that during salivary gland destruction, these two processes function independently from each other as well as in an additive manner. Therefore, when either autophagy or caspases are inhibited, partial degradation of the salivary glands occurs, and small, diffused cellular fragments remain. These fragments are diffused as the glands have partially degraded, and the cells have detached from each other. Generally, when caspases are inhibited, the remnants appear as condensed/densely stained cellular fragments (Fig. 8.3B). When autophagy is inhibited,

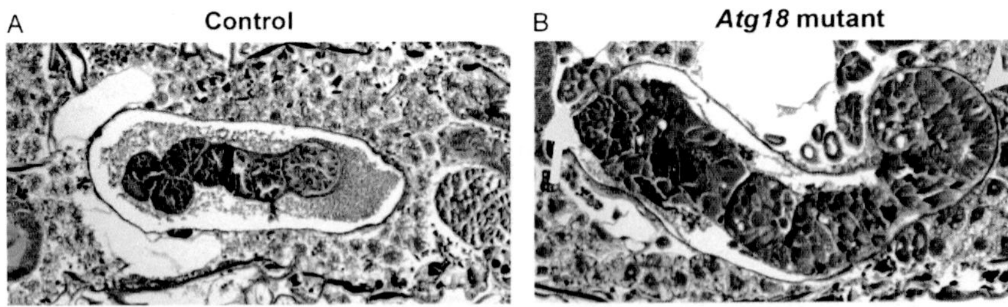

Figure 8.2 Histology of the midgut. (A) A control animal 12 h APF that shows compaction of the midgut and degradation of gastric caeca and proventriculus structures. (B) An autophagy-defective *Atg18* mutant animal that has failed to degrade its midgut properly. The midgut has failed to compact (yellow arrowhead), and the gastric caeca have failed to degrade (yellow arrow). *Reprinted from Denton et al. (2009) with permission from Elsevier.* (See the color plate.)

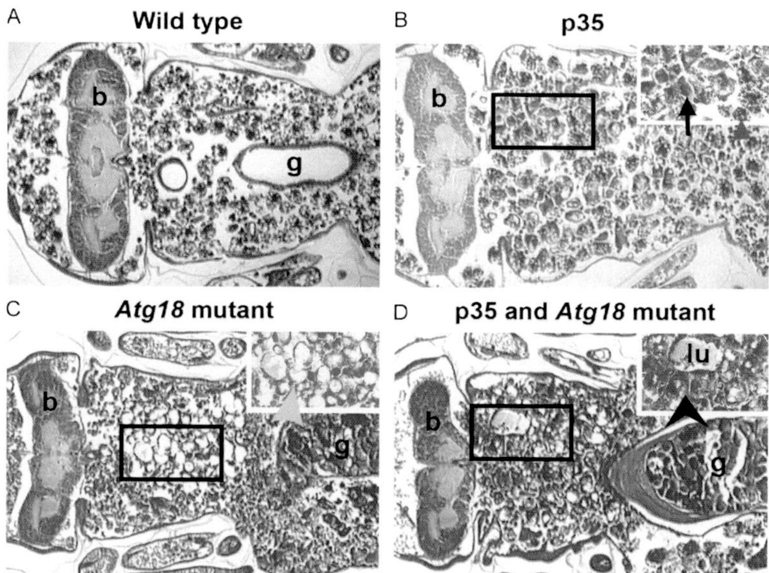

Figure 8.3 Histology of the salivary glands. (A) A wild-type animal in which the salivary glands have been properly degraded and cleared. (B) An animal in which the caspase inhibitor p35 has been expressed specifically in the salivary glands. This animal has condensed cell fragments (black arrow) diffused throughout its thorax. The red arrowhead indicates a fat body cell. (C) An autophagy-defective *Atg18* mutant animal that has failed to degrade its salivary glands. This animal has vacuolated cell fragments (yellow arrowhead) diffused throughout its thorax. (D) An *Atg18* mutant animal in which p35 has been expressed specifically in the salivary glands. This animal has gland fragments (black arrowhead) that also display remnants of salivary gland structure such as a lumen. Symbols are (b) brain, (g) midgut, and (lu) salivary gland lumen. *Reprinted from Berry and Baehrecke, 2007 with permission from Elsevier.* (See the color plate.)

the remnants appear as vacuolated/lightly stained cellular fragments (Fig. 8.3C). However, when both processes are inhibited, a more complete remnant of the salivary glands remains (Fig. 8.3D). These remnants generally retain the shape and structure of the tissue and are called gland fragments. Because of these additive phenotypes, epistasis experiments can be done to determine if a gene functions to regulate autophagy or caspases in the salivary glands. Histological sections are generally scored for the presence of salivary gland material with a final quantification stating the percentage of animals with salivary gland material. Further detailed quantification can be done scoring for the presence of cellular versus glandular fragments. Twenty or more animals for control and experimental genotypes are considered acceptable for histological analysis.

3.2. Quantifying and interpretation of TUNEL

Analyses of TUNEL in midguts and salivary glands require appropriate control samples. In the context of the midgut, we consistently observe TUNEL-positive staining in wild-type and control animals 0 h APF, and this is an appropriate stage for the analysis of mutant animals. In the context of salivary glands, TUNEL-positive staining in wild-type and control animals occurs 12–13.5 h APF, and this is an appropriate stage for the analysis of mutant animals (Fig. 8.4). We typically analyze at least 10 animals per treatment. If results are not consistent with this number of specimens, we increase the sample size.

3.3. Quantifying and interpretation of immunochemistry and fluorescently tagged Atg8

3.3.1 Immunoblotting

Immunoblotting allows for the visualization of protein levels. As autophagy is induced, proteins that are targeted for degradation will show a decrease in levels. When autophagy is defective, the proteins will tend to accumulate. Ref(2)P (p62/SQSTM1 in mammals) is an example of a protein that is directly degraded by autophagy; its levels reflect autophagic activity. Therefore, as the midgut and salivary glands induce autophagy for programmed cell death, the levels of Ref(2)P decrease. When autophagy is defective in these tissues, the levels of Ref(2)P remain high. In addition, detection of the lipidated form of Atg8 (Atg8-II) is a widely used marker of autophagy (Fig. 8.5). These levels can be visualized by immunoblotting, and quantified using standard techniques.

Caspases play an active role in the degradation of the salivary glands, and the detection of cleaved Caspase-3 and cleavage of known caspase substrates

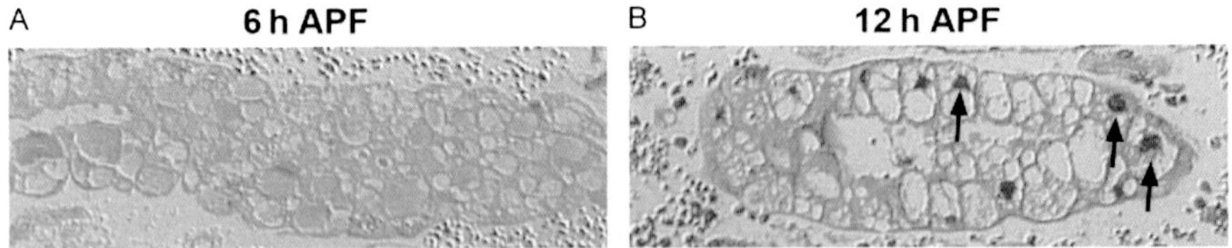

Figure 8.4 TUNEL staining. (A) A wild-type salivary gland 6 h APF showing no TUNEL staining indicating that DNA is not nicked and the salivary gland is not undergoing programmed cell death. (B) A wild-type salivary gland 12 h APF showing TUNEL-positive staining (black arrows) indicating nicked DNA and the salivary gland is undergoing programmed cell death. *Reprinted from Lee and Baehrecke (2001) with permission from Company of Biologists.* (See the color plate.)

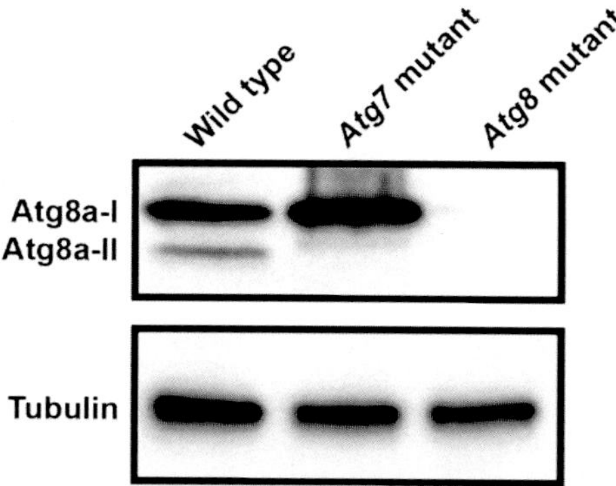

Figure 8.5 Atg8 immunoblot of the midgut. In this blot, autophagy is active in the wild-type sample as indicated by the presence of the Atg8a-II band. Additionally, the lower intensity of the Atg8a-II suggests that autophagic flux is occurring. If flux was inhibited, a buildup of Atg8a-II would occur. This blot also shows that Atg8a cannot be lipidated and, therefore, is defective when the autophagy essential gene *Atg7* is mutated. *Reprinted from Chang et al. (2013) with permission from Nature Publishing Group.*

can be visualized by immunoblotting with antibodies against proteins, such as cleaved-Caspase-3 (Drice in flies), Lamin, and cleaved-Lamin. When caspases are active proteins such as Lamins and Caspase-3 are cleaved. Therefore, immunoblots against Lamins will show a decrease in intact protein levels, while immunoblots against the cleaved forms of Lamins and Caspase-3 will increase. These levels can be visualized by immunoblotting, and quantified using standard immunoblot techniques.

3.3.2 Immunofluorescence and fluorescently tagged Atg8

Immunofluorescence and fluorescently tagged Atg8 can be used to visualize the protein localization and autophagy activity of a cell by microscopy. For example, caspase activity can be visualized through the loss of nuclear Lamins and an increase in processed (cleaved) Caspase-3 in salivary glands (Fig. 8.6A–C). Similarly, using mitotic recombination, homozygous mutant cells can be identified by immunofluorescence, and defects in autophagy in these midgut cells can be visualized through the accumulation of p62 (Fig. 8.6 D). Also, immunofluorescence helps visualize a protein's cellular localization and, potentially, effect as the cell undergoes programmed death. Additionally, with fluorescence, single subcellular level changes of proteins can be observed that would otherwise be missed when using immunoblotting to detect proteins in a mixed population of cells.

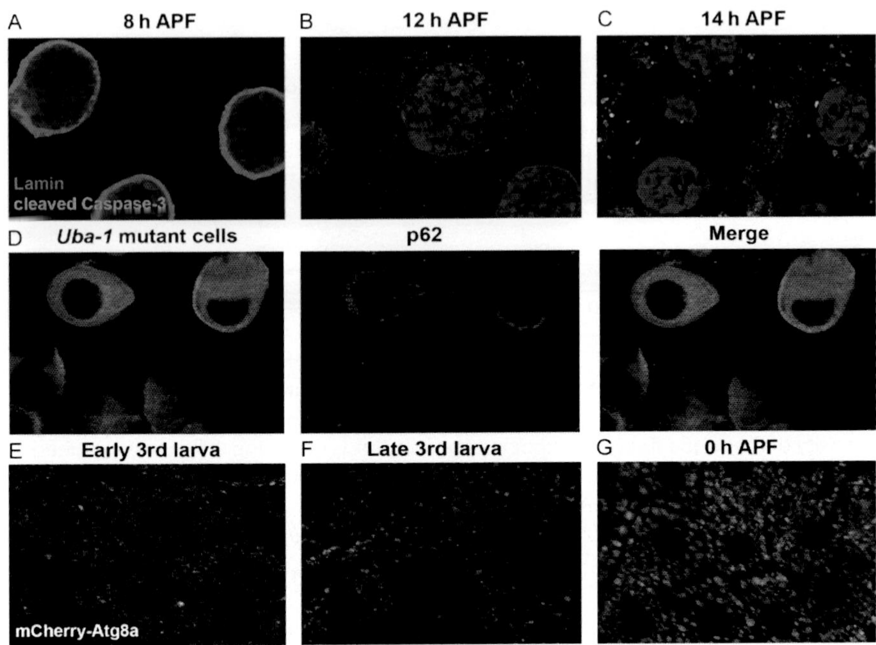

Figure 8.6 Immunofluorescence in the salivary gland and midgut. (A–C) Salivary glands of varying stages stained with anti-Lamin (red) and anti-cleaved-Caspase3 (green) antibodies. (A) A salivary gland 8 h APF when caspases are not active so Lamins remain intact (anti-Lamin staining is highly detectable and surrounds the nucleus) and Caspase-3 remains uncleaved. (B) A salivary gland 12 h APF when caspases are first activated resulting in the cleavage of Lamins and Caspase-3 (anti-cleaved-Caspase3 staining appears as puncta). (C) A salivary gland 14 h APF when Lamins are almost undetected and cleaved Caspase-3 is abundant. (D) A 0 h APF midgut in which mitotic recombination has been induced to create *Uba1* mutant clone cells that are marked with GFP (green). In these cells, autophagy is defective and the autophagy substrate p62 accumulates (red). (E–G) Midguts from early third instar (E), late third instar (F), and 0 h APF (G) larvae that express a mCherry-Atg8a reporter. An increase in reporter puncta occurs as the animals pupate indicating an increase in autophagy levels. *Panels (A)–(C): Reprinted from Martin and Baehrecke (2004) with permission from Company of Biologists. Panels (D)–(G): Reprinted from Chang et al. (2013) with permission from Nature Publishing Group.* (See the color plate.)

During autophagic cell death, quantifying immunofluorescence generally depends on the properties of the protein of interest. For example, if the protein's cellular localization changes. Generally, software such as Zeiss measurement software and ImageJ that measure parameters such as fluorescence intensity can be utilized.

Fluorescently tagged Atg8 is used as a reporter to visualize autophagosomes and autolysosomes. Therefore, when autophagy is active, these fluorescently labeled vesicles show up as puncta structures in the cells; autophagy increases in dying midgut cells at puparium formation (Fig. 8.6E–G). These puncta can easily be quantified using automated software such as Zeiss automated measurement software and ImageJ. To properly quantify autophagic puncta, it is best to consult the software's user manual. It is important to note that for a well controlled experiment, all microscope settings should remain the same when imaging control and experimental samples. Furthermore, to ensure proper quantification, care should be taken to ensure no pixels are saturated in the image.

3.4. Quantifying and interpretation of TEM

TEM images are collected from multiple cells from at least three independent animals per genotype. To quantify the number of structures, such as autophagosomes and autolysosomes (Fig. 8.7), the number of structures is quantified per given area. Data are expressed as the average number of autophagic structures per area based on analyses of multiple cells from each sample. This logic can be applied to other structures in cells, such as mitochondria. Please see Ylä-Anttila, Vihinen, Jokitalo, and Eskelinen (2009) for more details on quantification of autophagy by TEM.

3.5. Caveats to autophagy markers and flux through the pathway

Autophagy is a very dynamic process involving multiple complex steps; defects in autophagy can arise at any of these steps. From the initial formation of the isolation membrane, the protein Atg8 begins to decorate the

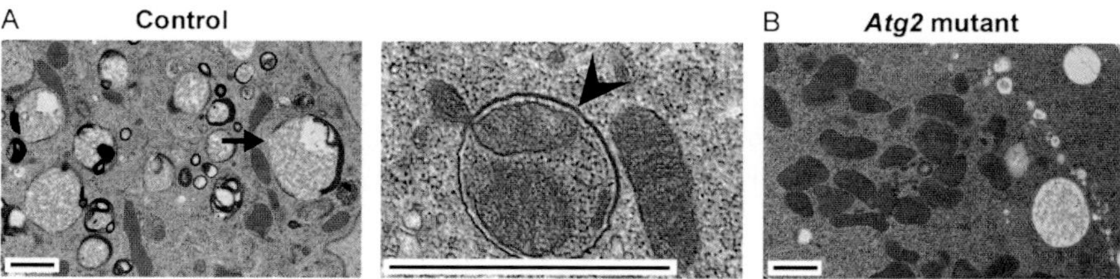

Figure 8.7 Transmission electron microscopy of the midgut. (A) A control midgut sample 2 h APF in which autophagy is occurring as indicated by the presence of autolysosomes (arrow) and double-membrane autophagosomes (arrowhead). (B) An autophagy-defective *Atg2* mutant 2 h APF which has no autophagosome or autolysosome as well as an accumulation of mitochondria, an organelle that is targeted for degradation during midgut cell death. Scale bars represent 20, 1, and 20 μm, respectively. *Reprinted from Chang et al. (2013) with permission from Nature Publishing Group.*

membrane of the autophagosome even before the vesicle has completed forming. Atg8 then remains localized to the membrane after the formation of the autolysosome and stays localized to this autolysosome until it is either degraded or recycled. This entire autophagic process is called autophagic flux. If, for example, the fusion between autophagosomes and lysosomes is inhibited, autophagic flux will be disrupted. In this case, autophagosomes will accumulate, increasing the fluorescently tagged Atg8 signal. This would falsely suggest that there is an increase in autophagy levels, when, in fact, there is actually an inhibition of the autophagic process.

To test autophagic flux, the following can be done. First, use a double-tagged GPF and mCherry Atg8 reporter. This reporter takes advantage of the quenching of GFP fluorescence in the low pH environment of the autolysosome. While mCherry fluorescence remains unaffected at low pH. With this reporter, the autophagosome would appear yellow (green GFP and red mCherry), and autolysosomes would appear red (quenched GFP and red mCherry). If the fusion between autophagosomes and lysosomes was inhibited, red autolysosomes would not appear. Second, perform an immunoblot for Ref(2)P. As Ref(2)P is degraded through autophagy, an accumulation of Ref(2)P would indicate an inhibition of autophagic flux. Third, perform an immunoblot for Atg8. As Atg8 is lipidated to localize to the autophagic membrane, this lipidated form (called Atg8-II) runs at a lower molecular weight (Fig. 8.5). If autophagic flux is inhibited, a buildup of Atg8-II would occur as it cannot be recycled or degraded. Finally, analyze TEM images. TEM images can indicate if there is a buildup of autophagosome failing to fuse with lysosomes.

ACKNOWLEDGMENTS

We thank our colleagues and members of the Baehrecke laboratory for encouraging us to improve our protocols. Research in the Baehrecke laboratory is supported by NIH Grants GM079431 and CA159314 to E. H. B., AI099708 to Neal Silverman and E. H. B., and S10RR027897 to the UMass EM Core. E. H. B. is an Ellison Medical Foundation Scholar.

REFERENCES

Alegria-Schaffer, A., Lodge, A., & Vattem, K. (2009). Performing and optimizing Western blots with an emphasis on chemiluminescent detection. *Methods in Enzymology*, *463*, 573–599.

Bernards, A., & Hariharan, I. K. (2001). Of flies and men—studying human disease in Drosophila. *Current Opinion in Genetics & Development*, *11*, 274–278.

Berry, D. L., & Baehrecke, E. H. (2007). Growth arrest and autophagy are required for salivary gland cell degradation in *Drosophila*. *Cell*, *131*, 1137–1148.

Brand, A. H., & Perrimon, N. (1993). Targeted gene expression as a means of altering cell fates and generating dominant phenotypes. *Development*, *118*, 401–415.

Chang, T. K., Shravage, B. V., Hayes, S. D., Powers, C. M., Simin, R. T., Wade Harper, J., et al. (2013). Uba1 functions in Atg7- and Atg3-independent autophagy. *Nature Cell Biology*, *15*(9), 1067–1078.

Denton, D., Mills, K., & Kumar, S. (2008). Methods and protocols for studying cell death in Drosophila. *Methods in Enzymology*, *446*, 17–37.

Denton, D., Shravage, B., Simin, R., Mills, K., Berry, D. L., Baehrecke, E. H., et al. (2009). Autophagy, not apoptosis, is essential for midgut cell death in Drosophila. *Current Biology*, *19*, 1741–1746.

Gratz, S. J., Cummings, A. M., Nguyen, J. N., Hamm, D. C., Donohue, L. K., Harrison, M. M., et al. (2013). Genome engineering of Drosophila with the CRISPR RNA-guided Cas9 nuclease. *Genetics*, *194*, 17048–17058.

Lee, C.-Y., & Baehrecke, E. H. (2001). Steroid regulation of autophagic programmed cell death during development. *Development*, *128*, 1443–1455.

Lee, C.-Y., Cooksey, B. A. K., & Baehrecke, E. H. (2002). Steroid regulation of midgut cell death during *Drosophila* development. *Developmental Biology*, *250*, 101–111.

Lee, T., & Luo, L. (2001). Mosaic analysis with a repressible cell marker (MARCM) for Drosophila neural development. *Trends in Neurosciences*, *24*, 251–254.

Martin, D. N., & Baehrecke, E. H. (2004). Caspases function in autophagic cell death in *Drosophila*. *Development*, *131*(2), 275–284.

Mizushima, N., & Komatsu, M. (2011). Autophagy: Renovation of cells and tissues. *Cell*, *147*, 728–741.

Rong, Y. S., & Golic, K. G. (2000). Gene targeting by homologous recombination in Drosophila. *Science*, *288*, 2013–2018.

Xu, T., & Rubin, G. M. (1993). Analysis of genetic mosaics in developing and adult *Drosophila* tissues. *Development*, *117*, 1223–1237.

Ylä-Anttila, P., Vihinen, H., Jokitalo, E., & Eskelinen, E. (2009). Monitoring autophagy by electron microscopy in Mammalian cells. *Methods in Enzymology*, *452*, 143–164.

CHAPTER NINE

Structural Studies of Death Receptors

Paul C. Driscoll[1]
Division of Molecular Structure, Medical Research Council, National Institute for Medical Research, London, United Kingdom
[1]Corresponding author: e-mail address: pdrisco@nimr.mrc.ac.uk

Contents

Abstract

This chapter describes reports of the structural characterization of death ligands and death receptors (DRs) from the tumor necrosis factor (TNF) and TNF receptor families. The review discusses the interactions of these proteins with agonist ligands, inhibitors, and downstream signaling molecules. Though historically labeled as being implicated in programmed cell death, the function of these proteins extends to nonapoptotic pathways. The review highlights, from a structural biology perspective, the complexity of DR signaling and the ongoing challenge to discern the precise mechanisms that occur at the point of DR activation, including how the degree to which the receptors are induced

Methods in Enzymology, Volume 545
ISSN 0076-6879
http://dx.doi.org/10.1016/B978-0-12-801430-1.00009-3

to cluster may be related to the nature of the impact upon the cell. The potential for posttranslational modification and receptor internalization to play roles in DR signaling is briefly discussed.

ABBREVIATIONS

CARD caspase recruitment domain
cCRD canonical CRD
CRD cysteine-rich domain
DcR decoy receptor
DD death domain
DED death effector domain
DISC death-inducing signaling complex
DR death receptor
NMR nuclear magnetic resonance
PLAD preligand association domain
PyD pyrin domain
TNF tumor necrosis factor
TNFR TNF receptor
TRAIL TNF apoptosis-inducing ligand
TRAIL-R1/2 TRAIL receptor 1/2

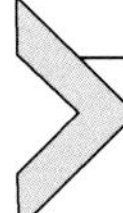

1. INTRODUCTION. SIGNALING BY THE TUMOR NECROSIS RECEPTOR SUPERFAMILY

The tumor necrosis factor receptor (TNFR) superfamily includes a number of type I transmembrane and secreted glycoproteins implicated in signaling in a variety of contexts. The membrane-associated members of the superfamily include TNFR1, TNFR2, CD27, CD40, CD95/Fas, the ectodysplasin receptors, TNF-related apoptosis-inducing ligand (TRAIL) receptors-1 and -2, death receptors-3 and -6 (DR3 and DR6), and several others (Cheng, Kinosaki, Murali, & Greene, 2003; Hehlgans & Pfeffer, 2005; Schrofelbauer & Hoffmann, 2011). In addition, so-called decoy receptors (DcRs) have been identified that are either secreted or lack functional intracellular signaling motifs and yet have a structure similar to that of the extracellular domains of the transmembrane counterparts (Zhan et al., 2011). For example, DcR3 lacks a transmembrane domain and binds to the ligands CD95L, TL1A, and LIGHT. DcR1 is a transmembrane family member that lacks a cytoplasmic domain and would appear to act as a decoy for TRAIL. DcR2 is another apparent DcR for TRAIL with transmembrane and cytoplasmic domains, but lacks a full death domain (DD) sequence that is associated with the downstream intracellular signaling of the majority of the TNFR superfamily (*vide infra*).

A subset of these receptors have long been associated with signaling programmed cell death (apoptosis) though it has subsequently become apparent that the precise outcome of activation is more varied, and dependent upon both the specific receptor and the nature of its interactions with ligand(s) (Ashkenazi & Dixit, 1998; Wilson, Dixit, & Ashkenazi, 2009). Thus, it is now well established that so-called death receptors can also signal for survival or proliferation, depending upon the context (Hueber, Zornig, Bernard, Chautan, & Evan, 2000; Newton, Harris, Bath, Smith, & Strasser, 1998; Newton, Kurts, Harris, & Strasser, 2001; Peter et al., 2007; Zornig, Hueber, & Evan, 1998). The content of this review is biased toward those members of the TNFR superfamily (and their associated ligands) that have been implicated in cell death. A defining characteristic of these receptors is that they each include a cytoplasmic region predicted to contain a so-called DD (see Section 10ff.), a six-helix bundle with propensity to take part in homophilic and heterophilic interactions with other DD-containing proteins (Park, Lo, et al., 2007). Despite the "death" label some of what follows is more relevant to the functions of these proteins in nonapoptotic signaling; it is sometimes difficult to divorce the two aspects.

A canonical example of DR signaling is provided by CD95 (also commonly referred to as Fas, less commonly as Apo1) (Krammer, 2000; Krueger, Fas, Baumann, & Krammer, 2003; Nagata, 1997; Strasser, Jost, & Nagata, 2009). Binding of the ligand CD95L (FasL and Apo1L) leads to serial recruitment of the immediate adaptor protein Fas-associated with a DD (FADD) and procaspase-8 and -10, two closely related cysteine proteases, and regulatory homologues of the caspase-8/-10 from the cellular FLICE inhibitory protein (cFLIP) family (FLICE: FADD-like interleukin 1β-converting enzyme). The receptor-accreted assembly of proteins is called the death-inducing signaling complex (DISC) (Algeciras-Schimnich et al., 2002; Kischkel et al., 1995; Walczak & Sprick, 2001). Induced proximity of the procaspases within the DISC leads to autoproteolysis and release into the cytosol of active caspase heterotetramers that initiate signaling cascades that can proceed to programmed cell death (Medema et al., 1997). In circumstances approximately described as a subthreshold ligand-dependent clustering of DRs, the signaling outcome is qualitatively distinct, leading to cell survival and proliferation rather than death, usually through downstream activation of the nuclear factor κ-light-chain-enhancer of activated

B cells (NF-κB) transcription machinery. Despite the Janus-like role played by these receptors, signaling for either cellular survival or demise, these proteins have acquired the moniker of "death receptors," a descriptor that will be adopted for convenience in the following account.

Experimental structural analysis of DRs has contributed significantly to our understanding of the ligand specificity and mechanisms by which ligation is associated with downstream signaling. However, it is arguable that the precise mechanism through which signaling of any one of the DRs takes place remains obscure. As well as providing an overview of the output of attempts to structurally characterize aspects of DR signaling, this review attempts to highlight significant deficiencies in the integration of these results into a coherent picture and to identify some of the challenges that lie in store.

2. OUTLINE DEATH LIGAND AND DR DOMAIN STRUCTURE

The outline structure of a DR comprises a cysteine-rich, glycosylated N-terminal extracellular domain linked to single hydrophobic transmembrane helix and a C-terminal cytoplasmic domain that includes a 80–100 amino acid region predicted to be rich in α-helical secondary structure. By dint of amino acid sequence analysis and a variety of structural analyses of both death receptor and nonreceptor examples, this latter region is identified as a death domain, often abbreviated to "DD". Where the 3D structures of DDs are known these almost invariably (*vide infra*) adopt a globular six-helix "Greek-key" topology, common to other homologous protein sequence families such as death effector domains (DEDs), caspase recruitment domains (CARDs), and pyrin domains (PyDs) (Fairbrother et al., 2001; Park, Lo, et al., 2007). Apart from DR6 (655 residues), the DRs range in size from 335 to 468 residues, a remarkably compact size compared to other transmembrane receptors such as receptor tyrosine kinases.

DR ligands comprise a homologous protein family in their own right. These ~230–280 residue type II proteins have in common a short N-terminal cytoplasmic domain, a transmembrane domain, a short-stalk segment, and a globular C-terminal ectodomain. A feature of the DR ligands is that they can be subject to release from the membrane, due to the action of metalloproteases on the membrane-proximal stalk region (Mariani, Matiba, Baumler, & Krammer, 1995; Reilly et al., 2009;

Schneider et al., 1998). The resulting soluble forms of the ligands can be isolated as homotrimers.

The 3D structure of the canonical member of the death ligand protein family, TNFα, was obtained by X-ray crystallography in 1989 (PDB1TNF) (Eck & Sprang, 1989; Jones, Stuart, & Walker, 1989) (Fig. 9.1A and B). The structure reveals a rubber bung-shaped homotrimer, with the 144 residue ectodomain protomer dominated by a globular "jelly

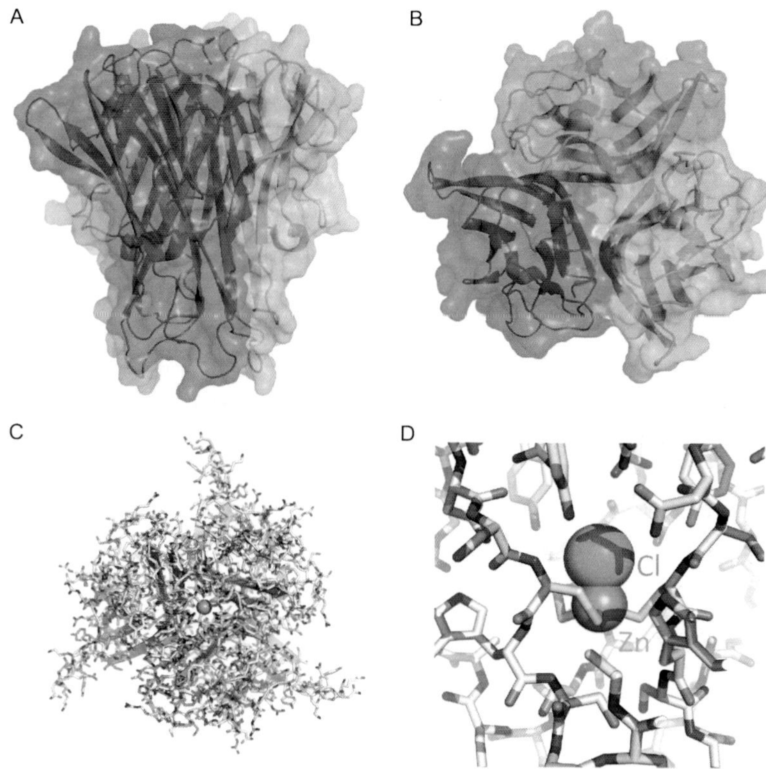

Figure 9.1 The 3D structure of death ligands. (A and B) The homotrimer structure of TNFα (PDB 1TNF) (Eck & Sprang, 1989), showing the three chains in transparent van der Waals surface representation, and secondary structure in ribbon format. In the orientation shown in (A), the N-termini are at the top of the structure; the stalk regions connect the N-terminus to the transmembrane regions that are not part of the crystal structure. The structure has been rotated by 90° in (B) to highlight the N-terminal face and the threefold symmetry. (C and D) The structure of TRAIL (PDB 1DG6) (Hymowitz et al., 2000) shown in stick format, with (C) the N-terminal face to the front (C), and the zinc atom highlighted in orange. (D) The coordination of the zinc and the associated chlorine atom in side-on view; the threefold symmetry axis is collinear with the Zn—Cl bond. (See the color plate.)

roll" β-barrel comprising almost entirely antiparallel β-strands. The core is made up of two four-stranded β-sheets: a flat inner sheet with strands A, H, C, and F, and a curved outer sheet with strands B, G, D, and E. The inner sheets constitute the inter-protomer contacts, the outer sheets the exposed surface of the trimeric assembly. The first and last strands A and H come together in the inner sheet. A 20-residue insertion between strands A and B folds into an additional pair of β-strands denoted A″ and B′ flanking the inner and outer sheet structures, respectively. Contacts between the TNFα protomers are formed by acute butt joints in which the C-terminal edge of one chain packs against the inner face of the neighboring subunit. Thus, strands E and F of one molecule abut strands A, H, C, and F of the next chain at a 30° angle, making a contact that involves 40 mostly hydrophobic side chains. Overall, the homotrimer structure buries approximately 2200 $Å^2$ of protomer solvent accessible area, consistent with the experimentally determined high level of stability with respect to chemical denaturation.

Subsequent 3D structures of the ligand for the homologous cell death agonist TRAIL showed an essentially identical protomer topology and homotrimer arrangement, despite low sequence identity (<30%). A striking difference between TRAIL and other TNF-family members is the insertion of ~12–16 residues in the A–A″ loop that traverses the width of the entire subunit structure twice, and covers a large proportion of the upper part of the B–G–D–E outer surface (PDB 1D2Q) (Cha et al., 1999). In addition, the intersubunit contacts are dominated by hydrophobic contacts rather than polar or electrostatic interactions witnessed for TNFα and -β. The cleaved, soluble TRAIL homotrimer was initially reported to be relatively unstable over time. However, the structure of TRAIL obtained by Hymowitz et al. revealed a zinc ion located on the threefold symmetry axis coordinated by the three unique cysteine side chains (one per TRAIL protomer) and a chloride anion (PDB 1DG6) (Fig. 9.1C and D), and associated experiments demonstrated that zinc coordination is critical for the biological activity (Hymowitz et al., 2000).

3. DR ECTODOMAIN STRUCTURE

Typical text-book representations of the interaction between a DR and its ligand depict the homotrimeric ligand sitting inside three chains of the receptor, as illustrated in Fig. 9.2. This concept is based upon the 2.85 Å crystal structure of lymphotoxin-α (a.k.a. TNFβ) bound to the

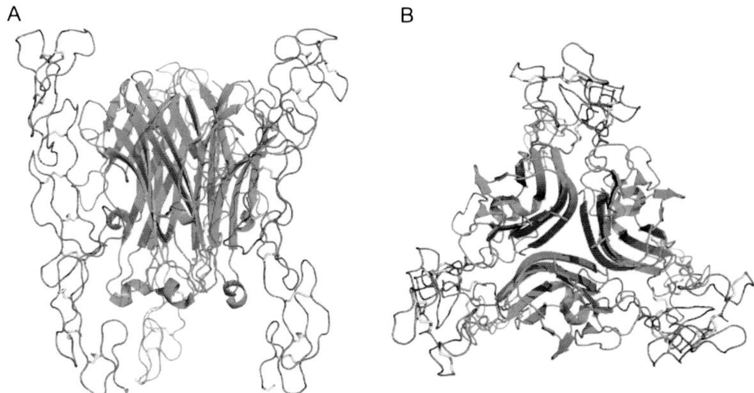

Figure 9.2 The structure of the complex formed between lymphotoxin-α (TNFβ) and the ectodomain of TNFR1 (PDB 1TNR) (Banner et al., 1993). The lymphotoxin-α homotrimer is shown in green, and the three chains of TNFR1 in mauve. Disulfide connections are highlighted in yellow. In the orientation shown in (A), the ligand would be attached to a membrane at the top of the picture, and the receptor chains to a membrane at the bottom. The structure has been rotated through 90° in (B) so that the top (N-terminal) face of the complex is to the front, and to highlight that the each receptor chain to two chains of the ligand, but does not contact any other receptor chain. (See the color plate.)

ectodomain of TNFR1 that was reported in 1993 (PDB 1TNR) (Banner et al., 1993). The structure revealed for the first time the cysteine-rich, disulfide ladder nature of a DR ectodomain. Further, crystallographic investigations of both the TNFR1 ectodomain alone and other DRs, notably the TRAIL-R2 receptor bound to TRAIL (*vide infra*), have provided a more complete picture of the nature and variance of the ectodomain organization. It is convenient to discuss these latter receptor-only results first before contemplating the nature of the interaction with DR ligands.

The combined approaches of amino acid sequence and the structural biology analysis have revealed that DR extracellular regions are comprised of tandem cysteine-rich domains (CRDs) of ~40 residues (Fig. 9.3A and B). The CRDs link together to constitute long, narrow rod-like structures, sometimes bent. Each of the CRDs lacks regular secondary structure, has little by way of a hydrophobic core, but buries a high proportion of the surface area of the cysteine sulfur atoms. The canonical CRD (cCRD) contains six cysteine residues paired into three disulfide bonds with the following connectivity: Cys1–Cys2, Cys3–Cys5, and Cys4–Cys6. TNFR1 has three such cCRDs (CRD1–CRD3), but this is followed by a fourth (CRD4) that possesses an alternative disulfide connectivity: Cys1–Cys2, Cys3–Cys6, and Cys4–Cys5 (Fig. 9.3A and B). The common structural

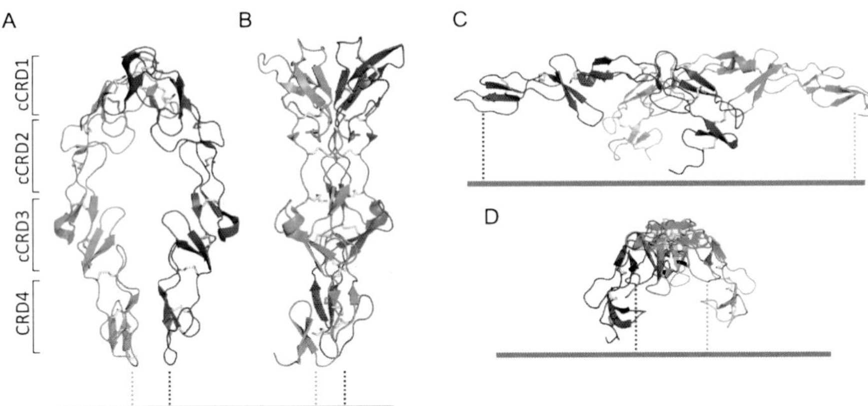

Figure 9.3 The 3D structure of the ectodomain of TNFR1 crystallized in the absence of a ligand (PDB 1NCF) (Naismith et al., 1995). The structure is composed of three canonical cysteine-rich domains (cCRD1–3) and a fourth cysteine-rich domain (CRD4) that has different disulfide bond connectivities (A). The disulfide bond linkages are depicted in yellow. Within the crystal lattice two types of inter-chain contacts that suggest the potential for homodimerization can be discerned. (A and B) Two orientations, rotated by 90° about the vertical symmetry axis, of the parallel homodimer, and (C and D) the antiparallel homodimer. The approximate locations of the linking segments to the transmembrane regions are indicated by dashed lines. The transmembrane domains are likely to be much further apart (by up to 100 Å) in the case of the antiparallel homodimer. (See the color plate.)

aspects of the N-terminal portions of the canonical and alternative CRDs have been christened "A1 modules," whereas the C-terminal portion of a cCRD is a B2 module and that of the alternative CRD in TNFR1 is a C2 module (Naismith, Devine, Kohno, & Sprang, 1996). The ectodomain of TRAIL-R2 has a similar structure to that of TNFR1, but lacks CRD4 altogether, and possesses only the last of the three disulfides present in TNFR1 CRD1; TRAIL-R1 is expected to have a similar organization (*vide infra*).

The crystal structure of CD95 ectodomain bound to an antibody F_{ab} (PDB 3TJE and also 3THM) reveals three CRDs, corresponding to CRD1–CRD3 of TNFR1 (Chodorge et al., 2012). CRD1 lacks the first canonical disulfide and has an "extra" disulfide bond (Cys1A–Cys1B) in the N-terminal part of CRD3. The electron density for the more distal C-terminal part of CRD3 is less well defined and lacks direct evidence for the Cys3–Cys5 disulfide, though the Cys4–Cys6 linkage is observed. Though the 3D structure of DR3 has yet to be experimentally elucidated, the amino acid sequence of DR3 has sufficient cysteine residues and sequence similarity to TNFR1 to have a conserved structure with four CRDs, each with three disulfides.

The DR6 disulfide ladder contains nine disulfide bonds in four CRDs: CRD1 lacks the A1 motif and possesses two disulfide bonds in its "B1-motif"; CRD3 and CRD4 both lack the Cys3–Cys5 disulfide in their B-motifs. The overall structure of DR6 CRD1–4 (residues 51–214), which in its unliganded state is a monomer, has the overall shape of a bent rod with an obtuse angle subtended by the long axes of the tandem CRD1–CRD2 and CRD3–CRD4 domains (PDB 3QO4) (Kuester, Kemmerzehl, Dahms, Roeser, & Than, 2011). When examined by SAXS, a glycosylated construct corresponding to the entire DR6 ectodomain (residues 1–348) appears as a dimer, suggesting the potential for interaction between the membrane-proximal region following CRD4 (PDB 3U3P, 3U3Q, 3U3S, 3U3T, and 3U3V) (Ru et al., 2012).

3.1. The TNFR1 ectodomain

To date, only two of the DR family ectodomains have been crystallized as single entity without a binding partner. Crystals of the TNFR1 ectodomain have been obtained under different buffer conditions and in different crystal forms. Intriguingly, given that there is little evidence for TNFR ectodomain self-association *in vitro*, in each of these lattices the protein is observed to make inter-chain crystal contacts of sufficient extent that could be relevant to interaction between full-length TNFR1 chains in the native context of cellular membranes. Naismith et al. originally described the 2.25 Å structure of bacterially expressed, nonglycosylated TNFR1(11–172) in tetragonal crystals obtained at pH 8.0 (PDB 1NCF) (Naismith, Devine, Brandhuber, & Sprang, 1995). Two distinct homodimers are apparent in the lattice, described as parallel and antiparallel (Fig. 9.3). The latter arrangement has the two TNFR1 chains in a head-to-tail overlap in which the CRD2 domains make pairwise reciprocal hydrogen bonds and conceal a total of 1475 $Å^2$ of solvent accessible surface area (Fig. 9.3C and D). This structure places the C-terminus of each chain at a separation of greater than 100 Å. On the other hand, the parallel contact is a head-to-head arrangement in which the CRD1 domains come together to bury a total of 2143 $Å^2$ of solvent accessible surface area, and the C-termini are relatively close together (Fig. 9.3A and B). In each case, surface shape complementarity is assessed as similar to that observed in the complex of TNFR1 with TNFβ and in antibody–antigen complexes; there are comparable numbers of ion pairs (2 and 3, respectively, for parallel and antiparallel dimers), hydrogen bonds (10 and 14), and van der Waals contacts (149 and 101), though a divergent population of bridging water molecules (10 and 5). The presence of N-glycans would be unlikely to interfere with either interaction surface. Separately the TNFR1 ectodomain was crystallized in hexagonal and orthorhombic (PDB 1EXT) crystals from a high-salt buffer with pH 3.7 that both diffracted to better than 1.9 Å (Naismith et al., 1996). Although the details of the packing arrangement vary between the two crystal forms, the dominant aspect is the presence of a common TNFR1 homodimer, different from either of those obtained at higher pH. Again the arrangement is antiparallel, but here the two protomers overlap lengthwise to yield a structure with long axis ~90 Å, similar to that of a single chain. Intermolecular contacts are made along the entire length of each chain, with total solvent accessible surface burial of 2880 Å^2. Again the presence of N-glycosylation would likely not block the contact that, similar to the high-pH antiparallel dimer, appears to obscure the TNF ligand binding surface (*vide infra*). Given that the low-pH homodimer arrangement is conserved within the context of different crystal lattices, it is possible that this structure might be of biological relevance for TNFR1 molecules when being trafficked through endosomal compartments of the cell.

3.2. The TRAIL-R2 ectodomain

Monoclonal antibody complexes of the TRAIL-R2 receptor have also been crystallized. Fellouse et al. derived a TRAIL-R2 binding F_{ab} ($K_D = 34$ nM) by phage display (Fellouse et al., 2005). This group randomized the complementarity-determining regions (CDRs) in a humanized F_{ab} framework with a binary degenerate codon that yielded equal proportions of Ser and Tyr residues. The structure of the F_{ab}, christened YSd1, bound to the TRAIL-R2 ectodomain was solved at 3.35 Å resolution (PDB 1ZA3). The binding site on TRAIL-R2 overlaps partly with that of TRAIL on the N-terminal CRD1 and CRD2 domains. One of the heavy chain CDR tyrosines is in a position that imperfectly mimics the role of TRAIL Tyr216, known to contribute significantly to binding. The three independent TRAIL-R2 chains in the lattice suggest the potential for the molecule to flex at the CRD2–CRD3 linkage. The same group extended the phage display activities to derive both small cyclized peptides, single-chain variable antibody fragments (scF_Vs), and monovalent F_{ab}s (Li et al., 2006). Although a series of apparently tight binding

(IC_{50}=6–80 n*M*) cyclized peptides was recovered in the screen, these were not active as an agonist of TRAIL-R2-mediated cell death in leucine zipper-appended dimer or trimer form; a tetramic peptide was significantly cytotoxic (EC_{50}=170 n*M*). The crystal structure of the phage display derived F_{ab} BDF1 was co-crystallized with TRAIL-R2 (PDB 2H9G). The structure shows that CDR regions home to a surface on CRD1 and CRD2 similar to that targeted by YSd1, though the details of the interaction are divergent. The authors suggest that a Leu-Ala-Leu tripeptide motif embedded in the CDR-H3 loop, which binds to TRAIL-R2 in an extended conformation, could provide a basis to rationalize the binding affinity of the cyclized peptides, some of which contain a Leu-X-Leu motif.

Blockade of DR function can be exploited by viruses. For example, human cytomegalovirus (HMCV) codes for a glycoprotein UL141 that blocks cell surface expression of TRAIL receptors by effecting retention in the endoplasmic reticulum, and thereby contributes to the attenuation of host immune effector pathways. Nemčovičová et al. characterized the binding of HMCV UL141 to TRAIL-R1 and -R2, and the poliovirus receptor CD155 that is structurally distinct from DRs (Nemcovicova, Benedict, & Zajonc, 2013). The crystal structure of UL141 bound to the TRAIL-R2 ectodomain (PDB 4I9X) is a 2:2 heterotetramer in which the protomers of a UL141 dimer each contact a TRAIL-R2 chain in a mode that mimics, in part, the interaction with TRAIL (Fig. 9.4). Thus, while TRAIL binding to TRAIL-R2 leads to head-to-head (180°) trimerization of the receptor, UL141 binds two TRAIL-R2 protomers in a diagonal fashion across the whole UL141 ectodomain, yielding an approximate 90° relative orientation of the rod-like TRAIL-R2 ectodomains. The structure of each UL141 chain includes an N-terminal immunoglobulin-like domain

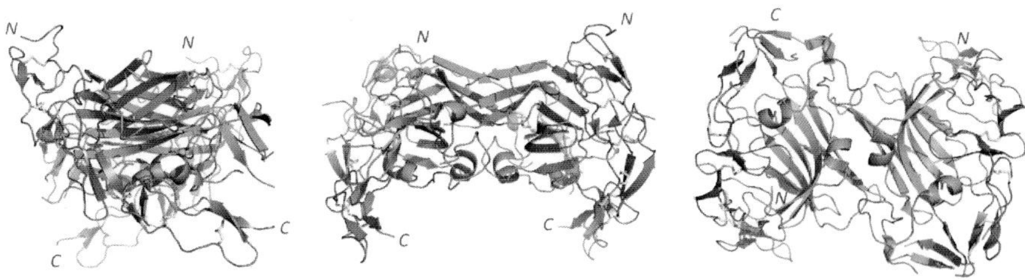

Figure 9.4 The 3D structure of the complex formed between the human cytomegalovirus glycoprotein UL141 homodimer (teal) and the ectodomain of TRAIL-R2 (gray-brown) that blocks cell surface expression of the receptor, viewed from three orthogonal directions (PDB 4I9X) (Nemcovicova et al., 2013). (See the color plate.)

followed by a C-terminal β-sheet domain. Each of the two intermolecular contacts amounts to burial of around 1350 $Å^2$ surface area, similar to that of a single TRAIL/TRAIL-R2 interface. The regions of TRAIL-R2 involved in UL141 and TRAIL binding share about half of the respective surfaces. The overall structure of TRAIL-R2 is conserved, with only a small deviation in the structure of CRD3 β1β2 loop, residues 143–157 in the center of the contact zone. The authors argue on the basis of extensive mutational analysis that UL141 mimics many aspects of the mode in which TRAIL binds to TRAIL-R2. Further structural studies are required to fully comprehend the basis for the 400-fold weaker binding of UL141 to TRAIL-R1 and CD155. In the latter case, evidence from surface plasmon resonance experiments suggests that the CD155 binding site on UL141 is distinct from that of TRAIL-R2.

3.3. CD95 ectodomain

Chodorge et al. reported the derivation of a series of agonist monoclonal antibodies against the CD95 ectodomain using phage display methods in scF_v format (Chodorge et al., 2012). One scF_V, E09, was identified as possessing antiproliferative activity and was converted to human IgG1 form. This antibody was potent at inducing CD95-dependent caspase-3 and -7 activation, and DNA fragmentation ($EC_{50}=0.7$ n*M*). The 1.9 Å crystal structure of the E09 F_{ab} bound to CD95 (PDB 3TJE) shows that the CD95 epitope extends over one face of both CRD1 and CRD2, partially overlapping that predicted for CD95L which covers CRD2 and CRD3. Though both E09 and CD95L are both CD95 agonists, the predicted binding pose of the latter minimally overlaps that of the antibody—a single loop—leading to the question how agonism is achieved by E09. Ribosome display methods were used to derive a variant of E09 with 49-fold improved K_D and seven amino acid substitutions relative to the parent F_{ab}. The complex of the F_{ab} of this antibody, EP6b_B01, with CD95 revealed a conserved binding site, but double the number of intermolecular hydrogen bonds (PDB 3THM). Surprisingly, EP6b_B01 was not active in a cell-killing assay and overall the researchers noted a negative correlation between CD95 affinity and apoptotic efficiency for four IgG variants. Based on kinetic measurements, the authors proposed that potency of agonism by CD95 mAbs requires a reasonably fast off-rate and partial dissociation of the receptor. High-affinity potent mAbs might lead to bivalent receptor-mAb "dead-ends". Consistent with their concept, the negative effect of high affinity could be counteracted by increasing the avidity with protein A-dependent F_c-cross-linking.

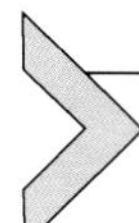

4. PHYSIOLOGICAL COMPLEXES OF DEATH LIGANDS WITH DRs

To date, the inventory of crystal structures of complexes between DRs and their physiological ligands is limited to TNFβ/TNFR1 (PDB 1TNR) (Banner et al., 1993) and TRAIL/TRAIL-R2, the latter solved independently by three groups (PDB codes 1D0G, 1DV4, and 1DU3) (Cha et al., 2000; Hymowitz et al., 1999; Mongkolsapaya et al., 1999). The two assemblies have many characteristics in common that are also shared with the structure of TNFα bound to the ectodomain of TNFR2 (PDB 3ALQ) (Mukai et al., 2010); TNFR2 lacks a cytoplasmic DD and strictly is not classified as a DR. Each of these structures shows a structure with effectively an axial threefold symmetry axis with the homotrimeric ligand clasped between three receptor chain "rods" (Fig. 9.2). The truncated three sided pyramid of the ligand trimer sits with its wider "base" between the membrane-proximal receptor CRDs. The receptor ectodomains are aligned along the interstices between neighboring ligand subunits, and do not directly contact one another. It appears that combination with the ligand might fix the distance between receptor chains (a separation of ~50 Å in the case of TRAIL-R2), and thereby organize or restrict the spatial arrangement of the appended transmembrane and cytoplasmic regions.

Detailed analysis shows that the contacts between the different ligands and the receptor chains share topological characteristics but are significantly different in terms of the physicochemical characteristics, which is perhaps unsurprising given that the amino acid sequence identity between TNFβ and TRAIL is only 19% and between TNFR1 and TRAIL-R2 23%. The main contacts occur between the ligand and the receptor CRD2 and CRD3. TRAIL-R2 CRD1 lacks the A1 motif and is essentially a small, single disulfide-linked cap for the rest of the chain. In the receptors, CRD2 and CRD3 are linked by a Cys-X-Cys tripeptide that appears to act as an articulation point: the relative orientation of CRD2 and CRD3 in the two DR structures differs by a tilt and rotation of 36°. Each contact between a receptor and the cleft between ligand protomers buries a substantial surface area: ~2170 $Å^2$ in TNFβ/TNFR1 and ~2960 $Å^2$ for TRAIL/TRAIL-R2 (PDB 1D4V). In the first instance, the contact is shared equally between the two neighboring subunits on the ligand side, but is strongly biased (~70:30) toward the CRD2 domain on the receptor side. In the latter case the contact zones are equally divided between the two ligand domains, and between the CRD2 and CRD3 receptor repeats (dominated by the "50s loop" and "90s loop," respectively; numbering refers to the receptor chain lacking the 55-residue signal peptide). The difference in tilt angle between CRD2 and CRD3 in TRAIL-R2 allows for closer approach of CRD3 to the TRAIL surface. In terms of both topological and chemical characteristics the CRD3–ligand contact zones are highly divergent between the TNFβ/TNFR1 and TRAIL/TRAIL-R2 complexes. The CRD2-ligand contact zones are topologically more highly similar, but the physicochemical conservation is limited to a hydrophobic contact between Tyr142 in TNFβ (Tyr216 in TRAIL) and Leu79/100 in TNFR1 (Leu59/114 in TRAIL-R2). The importance of this interaction is reflected in the fact that mutagenesis of the TNFβ or TRAIL Tyr residue (and similarly of the corresponding residue in TNFα and CD95L) abolishes receptor binding (Goh, Loh, & Porter, 1991; Schneider et al., 1997; Yamagishi et al., 1990). However, the remainder of this contact region is comprised of polar or charged moieties, the details of which differ significantly. Of note, a CRD3 residue (Glu147) that is conserved among TRAIL-binding proteins (TRAIL-R1, and DcR1 and DcR2) makes a salt–bridge interaction with Arg149 in the A–A″ loop extension that is present in TRAIL, but not in TNFβ. Replacement of the A–A″ loop in TRAIL with the shorter region from TNFβ leads to almost complete loss of binding to TRAIL-R2 (Cha et al., 1999; Mongkolsapaya et al., 1999). Cha et al. report that in their model for the structure of TRAIL bound to TRAIL-R2 (PDB 1DU3) an N-terminal portion of the TRAIL A–A″ loop that is missing or poorly defined in the other two structures essentially fills a void between the C-terminal half of CRD2 and the ligand (Cha et al., 2000). They highlight that comparison of the structures of free and bound TRAIL suggests significant ligand-dependent remodeling of this first part of the A–A″ loop and the likely flexible nature of this region of the ligand in free solution.

5. A DECOY RECEPTOR–LIGAND COMPLEX

The DcR3 is able to neutralize three different TNF-family ligands, CD95L, LIGHT, and TL1A. Zhan et al. determined the crystal structure of a major portion of DcR3 including the four CRD regions bound to TL1A, an endothelial cell-derived agonist of DR3 implicated in inflammation of the gut (Zhan et al., 2011). The structure of isolated DcR3 (PDB

3MHD) is a familiar bent Cys-rich rod. The overall architecture of the complex (PDB 3MI8 and 3K51) is similar to that of TNF/TNFR and TRAIL/TRAIL-R2. The contact between TL1A and DcR3 is concentrated on CRD1 and CRD2 and involves recognition of the only invariant side chain in the ligand, and interactions with the backbone atoms of the TL1A DE loop. The connection between the ligand and the receptor CRD3 is much less intimate than is the case of the other ligand–receptor pairs: such as they are, the contacts are not present in either of the two crystal forms of the complex that were analyzed. The authors of this study argue that the lack of a contact with CRD3, thought to be important in the other systems, coupled with the DcR3 contact with invariant and backbone atoms in the DE loop, enable TL1A to bind three distinct ligands. It is perhaps worth noting that these structures lack the C-terminal region of DcR3, reported to reverse signal to dendritic cells by cross-linking cell surface proteoglycans.

6. THE DR PRELIGAND ASSOCIATION DOMAIN

Despite the fact that in general DR ectodomains have been observed to form apparently self-associated homodimers within crystal lattices, direct measurement of the hydrodynamic properties yields results that indicate that such interactions are weak in solution. Nevertheless, a variety of observations argue that in a transmembrane context, the ectodomains do interact, and in such a way as to regulate signaling. For example, Lenardo and coworkers identified a domain in the N-terminus of TNFR1, TNFR2, and TRAIL-R1 that could mediate specific assembly of receptor homotrimers in the absence of a heterologous binding partner (Chan et al., 2000). The region of the protein that can effect these interactions is distinct from that involved in direct contacts with the ligand (CRD2 and CRD3). This region of the protein is described as both necessary and sufficient to permit TNFα binding and has been called the preligand association domain (PLAD). In a related manner, the Lenardo group was able to rationalize in terms of an N-terminal PLAD the effects of heterozygous mutations in the CD95 receptor that lead to dominant interference of CD95L-dependent signaling (Siegel et al., 2000); the mutant and wild-type receptors bind to each other in a ligand-independent fashion. Similar findings were reported by Papoff et al. (Cascino, Papoff, DeMaria, Testi, & Ruberti, 1996; Papoff et al., 1996, 1999). Fluorescence resonance energy transfer measurements in cells demonstrated the presence of preassociated CD95 chains. Clancy et al. also reported that DcR2 acts less as a DcR by mopping up TRAIL, but as a regulator of the function of the ligand by binding to TRAIL-R2 via an N-terminal PLAD (Clancy et al., 2005). Deng et al. reported that soluble versions of TNFR PLADs able to inhibit TNFα action *in vitro* were potent blockers of arthritis in animal models (Deng, Liu, & Tsokos, 2010; Deng, Zheng, Chan, & Lenardo, 2005). In each case, the PLAD function is associated with the CRD1 region of the receptor, and in the case of CD95, the major site of self-interaction has been claimed to reside between residues 43 and 66 (Edmond et al., 2012). The implication of these findings is that DRs are present as ligand-free preassociated oligomers on the cell surface and suggest that the action of an activating ligand is to impose a different organization of the receptor chains in order that a transmembrane signal can be promulgated. At present, the structural basis of PLAD interactions is speculative. The data are consistent with the early suggestion (Naismith et al., 1995) that in the absence of ligands, the receptor ectodomains interact to inhibit downstream signaling by the associated DDs. In this respect, the parallel dimer structure of the TNFR1 ectodomain (PDB 1NCF), in which the protein chains make reciprocal interactions in the CRD1 regions, is a potential model for the model for the nonproductive self-association. However, this model is at odds with the results of cross-linking experiments which suggest the presence of TNFR1 homotrimers on the cell surface (Chan et al., 2000). Importantly, it is claimed that the PLAD is required for ligand binding, despite lack of evidence from structural data that there are any specific interactions between the CRD1 region and the ligand. The evidence is that deletion mutants lacking the PLAD can still bind to antireceptor antibodies. It would be helpful to be able to characterize in more detail the structure and conformational stability of ligand binding domains of such truncated receptors. Intriguingly, Richter et al. have suggested that the membrane-proximal stalk region of TNFR2, and potentially other related receptors, can influence the self-association and ligand responsiveness (Richter et al., 2012).

Other modes of regulation of DR signaling have been reported for which we have little understanding at the structural level. For example, it has been demonstrated that TRAIL-receptor signaling can be affected by O-linked glycosylation. Wagner et al. found that expression of the peptidyl O-glycosyltransferase GALNT14 correlated with TRAIL sensitivity in a number of cancer cell lines (Wagner et al., 2007). O-Glycosylation of the TRAIL-R1 and -R2 ectodomains promotes receptor clustering and ligand responsiveness. On a different level, it has been suggested that DR ectodomains are influenced by the presence of specific plasma membrane components. In many instances, DR migration to lipid rafts and agonist-initiated endocytosis is invoked as part of the signaling mechanism. Relevant to this scenario, Chakrabandhu et al. have presented evidence that CD95 and, potentially other DRs, possess a glycolipid-binding motif within a loop of the CRD3 domain that is not involved in CD95L interactions (Chakrabandhu et al., 2008). Mutation of this region has a significantly deleterious effect on CD95-dependent cell death.

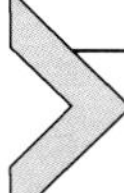

7. DEATH LIGAND STRUCTURE–ACTIVITY RELATIONSHIPS

In a variety of contexts, it would prove valuable to block or modify the action of death ligands at DRs. Often this in the context where the ligand–receptor combination acts in a manner that is not pro-apoptotic, and where the "death" moniker is, therefore, not strictly appropriate. For example, the action of excess amounts of TNFα at TNFR1 elicits an inflammatory response implicated as a causative or contributory mechanism in various auto-immune diseases such as Crohn's disease, rheumatoid arthritis, and ulcerative colitis. TNFα action at TNFR2, which occurs on a more limited set of cell types, is required to maintain T-cells for antimicrobial responses. Recognition of this aspect of TNF-signaling has led to a need to understand the basis of TNFα binding at both TNFR1 and TNFR2, and the development of antiTNF therapies.

Reed et al. determined the structures of both wild-type TNFα and a mutant including the substitution Arg31Asp (PDB 1A8M) that displayed reduced binding to both TNFR1 and TNFR2 but a sevenfold preferential binding to the TNFR1 receptor over TNFR2 (Reed et al., 1997). The structures of the two proteins are highly similar, and the lower affinity is rationalized based on a model that invokes electrostatic repulsion between the TNFα Asp31 and the conserved receptor glutamic acid residue (Glu56 in TNFR1). The differential impact with respect to TNFR1 and TNFR2 binding is postulated to be a result of the presence of potential H-bond interactions between Asp31 with TNFR1 Ser59 and Cys70, whereas interactions with the corresponding TNFR2 Thr60 and Cys71 side chains are weaker.

Phage display methodology was adopted by Shibata et al. to select a TNFR1-selective antagonist of TNFα binding (Shibata et al., 2008). The TNFα variant, denoted R1antTNF, contained substitutions at six positions within the receptor binding site: Ala84Ser, Val85Thr, Ser86Thr, Tyr87His, Gln88Asn, and Thr89Qln. Interestingly, R1antTNF binds TNFR1 as avidly as the wild-type TNFα yet does not elicit TNFR1-mediated cellular responses. Moreover, the mutant can block the activity of wild-type agonist at TNFR1 but does not affect TNFR2-binding. R1antTNF was reported to block hepatic injury in an animal model of hepatitis. The crystal structure of R1antTNF (PDB 2E7A) showed few differences compared to wild-type TNFα. Previously, it had been reported that mutation of the Tyr87 residue was deleterious to receptor binding. The authors of this study proposed that the substitution of Tyr with His at this position has lesser impact, at least in the context of the other five mutations, and that the difference in biological activity of R1antTNF is likely due to the elevated on- and off-rates for TNFR1 binding (as assessed by surface plasmon resonance). Potentially a faster recycling of receptor occupancy limits the ability of TNFR1 receptors to establish downstream signaling complexes. Similar conclusions were drawn for a related mutant denoted R1antTNF-T8 (TNFα Ala84Thr Val85Pro Ser86Ala Tyr87Ile Gln88Asn Thr89Arg) (PDB 2ZPX) (Mukai, Nakamura, et al., 2009).

Mukai et al. also used phage display technology to develop TNFα-based reagents with differential TNFR1 and TNFR2 binding properties (Mukai, Shibata, et al., 2009). They targeted six residues in noncontiguous regions of the receptor binding site near TNFα residues 30, 80, and 140. Ultimately, 16 TNFR1- and 4 TNFR2-selective variants were derived. The former group was highly mutated around residue 30, the latter near residue 140. The TNFR1-selective mutant R1-6 (TNFα Leu29Lys Arg31Ala Arg32Gly Glu146Ser Ser147Thr) was characterized in detail including by structural analysis (PDB 2ZJC). A small variation of the loop structure between residues 30 and 34 was identified. Using models for R1-6/TNFR1 and R1-6/TNFR2 based on the known structure of TNFβ bound to TNFR1 and a model of TNFα bound to TNFR2, the authors speculated that the Arg31Ala and Arg32Gly substitutions impact electrostatic interactions with TNFR2 (that possesses the negatively charged Asp54, Glu57, and Glu70 triad in its binding site—these residues are Arg53, Glu56, His69 in TNFR1) and that the Leu29Lys mutation contributes to binding to TNFR1, potentially by the Lys29 providing compensation for the loss of Arg32.

Cha et al. developed a variant TNFα, MS3, which displays various advantageous properties (Cha et al., 1998). The MS3 construct omits the seven N-terminal residues that are disordered in the structure of the wild type and includes substitutions Leu29Ser, Ser52Ile, and Tyr56Phe. MS3 is highly stable with respect to proteolysis and exhibits 11- and 71-fold lower binding affinity for the TNFR1 and TNFR2 receptors, respectively. MS3 was able to suppress solid tumor growth in mice more efficiently than the wild-type protein, and both the *in vitro* cytotoxic and *in vivo* systemic toxicity of M3S are substantially reduced compared to TNFα. Structural characterization of MS3 in two crystal forms (PDB 4TSW and 5TSW) suggests reconfiguration of the loop containing residues 29–36 into a rigid structure, supported by intra- and intersubunit contacts. An "MS3 mutant" lacking the Leu29Ser substitution binds TNFR1 more tightly than MS3. Together, the data suggest that the lower toxicity of MS3 is attributable to its attenuated receptor binding capability, and that its enhanced antitumor activity is due to its extended half-life *in vivo*.

8. STRUCTURAL ANALYSIS OF AntiTNF AGENTS

Cunningham and coworkers were able to develop a small-molecule inhibitor of TNFα function that operates by displacement of a subunit from assembled TNFα homotrimers (He et al., 2005). The compound, comprised of a trifluoromethylphenyl indole and dimethyl chromone moieties joined by a dimethylamine spacer, was active in the low-micromolar range in assays of TNFα-receptor binding and downstream inhibitor of NF-kappaB degradation in cells. A variety of kinetic measurements demonstrate that formation of an intermediate complex between the compound and the intact homotrimer results in a 600-fold accelerated subunit dissociation rate. The co-crystal structure of the compound with TNFα (PDB 2AZ5) showed that a single compound molecule displaces a subunit of the TNFα trimer to form a complex with a subunit homodimer. The compound is essentially folded in half and contacts 16 residues in the 2 protein chains that would be buried in the intact homotrimer. The interaction involves 330 $Å^2$ of protein surface area and appears to be essentially hydrophobic in nature.

Therapeutic blockade of TNF action has been brought to the clinic in the form protein-based biologics such as a soluble TNFR2-Fc chimera (etanercept) and antiTNFα monoclonal antibodies (infliximab and adalimumab). Structural analysis of both of the antibody complexes has been pursued. The structure of the mouse–human chimera infliximab F_{ab} bound to TNFα has been solved at 2.6 Å resolution (PDB 4G3Y), enabling identification of the binding epitope that is dominated by the C–D and E–F loops (Liang et al., 2013). The E–F region of TNFα is apparently disordered in the complex with TNFR2 and is the only segment of the mAb-bound structure with significant deviation of its coordinates compared to the isolated protein. Each infliximab F_{ab} binds only one chain in the TNFα homotrimer, though each homotrimer can bind three F_{ab}s without steric hindrance. The total buried surface area in each contact between F_{ab} and TNFα is 1977 $Å^2$. Although the E–F loop is not implicated in interactions with TNFRs, it is presumed that the presence of the F_{ab} works to sterically hinder access of TNFα to the TNFR1 and TNFR2 binding sites.

Similarly, the structure of the F_{ab} region of the fully humanized mAb adalimumab complexed to TNFα has been determined at 3.10 Å resolution (PDB 3WD5) (Hu et al., 2013). In this case, each F_{ab} binds to TNFα through a larger interface with a total buried surface area of 2540 $Å^2$ distributed across a number of discontinuous segments of two adjacent TNFα protomers and directly coinciding with the TNFR-binding site. This characteristic may contribute to the reported enhanced efficacy of adalimumab over other antiTNFα biologics.

Poxviruses encode a variety of proteins that interfere or modulate host immune responses. A common target of these molecules is TNFα, and poxviruses are known to express viral TNF receptor homologues. Tanapox virus protein 2L and its relatives constitute a different category of antiTNFα protein, distantly related to the ectodomain of class I major histocompatibility (MHC) proteins (13% sequence identity, with predicted MHC heavy chain-type α_1–α_3 domains but failing to bind β2-microglobulin). The Tanapox 2L protein binds TNFα with 43 p*M* affinity. Bjorkman and coworkers determined the crystal structure of the highly related Yaba-like disease virus 2L protein bound to the TNFα homotrimer (PDB 3IT8) (Yang, West, & Bjorkman, 2009). The structure shows that 2L molecules bind across each of the TNFα interprotomer contacts, similarly to TNF receptors, and occluding 2060 $Å^2$ of protein surface. The interaction surface of 2L involves the canonical helical regions of the MHC-like α1 and α2 domains, though there is no MHC-like groove between these two structural elements. The mechanism of action of the 2L proteins would appear to involve straightforward competition for the TNF receptor binding site on TNFα.

9. STRUCTURAL ANALYSIS OF THE BLOCKADE OF DR FUNCTION

In principle, blockade of death-ligand signaling can be effected at the level of the receptor, rather than of the ligand. Aspects of this approach have been covered in the descriptions of the structures of DR ectodomains in Section 3 (*vide supra*). One intriguing additional example is the discovery of a small-molecule inhibitor of TNFR1, based on the screening of chemical libraries for blockers of TNFα signaling. Carter et al. reported that IW927, a *N*-alkyl 5-arylidene-2-thioxo-1,3-thiazolidin-4-one, could selectively block TNFα binding to TNFR1 and signaling at mid-to-high nanomolar levels (Carter et al., 2001). The authors of this study revealed that this class of compound is photochemically labile: initially the compounds bind weakly to the protein target with mid-micromolar IC_{50} and in the presence of light become covalently (though reversibly) linked to the binding partner. The crystal structure of IV703, a close analog of IW927, bound to TNFR1 (PDB 1FT4) shows the compound joined to the backbone amide nitrogen atom of Ala62 via its phenyl ring. The position of the thiazolidinone adduct on the surface of TNFR1 is well placed to interfere with TNFα binding.

10. DR CYTOPLASMIC DOMAINS

The differentiation of the DR subset of TNFR superfamily proteins from the other members is rooted in the observation that they each contain a homologous region of ca. 80–100 residues within the cytoplasmic region named a death domain (DD). The DD sequence signature was recognized to be present in a number of other nonmembrane-associated adaptor and signaling proteins including in organisms that lack evidence for DD-containing transmembrane receptors (e.g., the Pelle and Tube proteins in *Drosophila*) and it was quickly established that DDs have a tendency for homotypic interaction, either through a propensity for self-association, heterodimer formation, or higher order oligomerization. A combination of sequence analysis and structural studies (*vide infra*) has revealed that DDs are a part of a larger family of topologically related protein modules that are present in proteins engaged in cell death signaling pathways: DEDs, CARDs, and PyDs (Fairbrother et al., 2001; Park, Lo, et al., 2007; Weber & Vincenz, 2001a). The critical importance of the DD for DR signaling was established at an early stage. For example, *lpr*^{cg}-mice that exhibit a lupus-like lymphoproliferative autoimmune disorder were found to have a point mutation in the DD region of CD95 (Watanabefukunaga, Brannan, Copeland, Jenkins, & Nagata, 1992). Furthermore, deletion mutagenesis indicated that the cell-killing capability of both CD95 and TNFR1 is dependent upon the presence of an intact DD (Itoh & Nagata, 1993; Tartaglia, Ayres, Wong, & Goeddel, 1993). For each receptor, the DD is separated from the transmembrane region by a more or less extended peptide that is presumed to lack ordered structure, but contains membrane-proximal Cys and basic residues that may play significant roles in interactions with the plasma membrane (see below). DR6 represents an exception: the cytoplasmic domain appears to contain a membrane-proximal DD separated from a C-terminal globular domain that has some similarity to CARDs (Pan et al., 1998). A solution structure of this CARD-like domain has been deposited in the PDB (2DBH), but no further details have been published to date. Other members of the DR family have C-terminal tail regions that follow the DD and are presumed to be unstructured but may play a role in interactions with other proteins. An example is human CD95, which has been reported to interact with postsynaptic density protein, *Drosophila* disc large tumor suppressor, and zonula occludens-1 protein domain-containing proteins via its extreme C-terminal Ser-Leu-Val motif (Saras, Engstrom, Gonez, & Heldin, 1997; Yanagisawa et al., 1997). However, the mouse CD95 ortholog lacks this motif and the role of such interactions in CD95 signaling has been disputed (Cuppen, Nagata, Wieringa, & Hendriks, 1997).

Yeast two-hybrid screening was used to demonstrate that the CD95-DD binds to the 208 residue protein FADD that comprises a DD coupled to an N-terminal DED (Chinnaiyan, Orourke, Tewari, & Dixit, 1995). The FADD-DD, expressed on its own, binds to the CD95-DD, and can act in a dominant-negative (DN) fashion to block agonist-induced CD95-signaling; hence, FADD-DD is often referred to as FADD-DN (Chinnaiyan et al., 1996; Hueber et al., 2000; Newton et al., 1998, 2001; Zornig et al., 1998).

11. DD STRUCTURE

Fesik and coworkers first revealed the structure of a DD using nuclear magnetic resonance (NMR) spectroscopy applied to human CD95-DD at low pH (4.0), conditions under which the protein has a reduced tendency to aggregate at high concentrations (Huang, Eberstadt, Olejniczak, Meadows, & Fesik, 1996). The structure includes six amphipathic α-helices arranged in an antiparallel, Greek-key topology, with highly flexible and N- and C-terminal tail regions (PDB 1DDF) (Fig. 9.5A). The authors reported that a mutant form of the construct (Asp244Ala) was nonaggregating at pH 6.0 and possesses essentially the same structure. The study also presented data on a number of other variants with surface residue alanine substitutions that demonstrated that the ability to bind to FADD *in vitro* is often, but not always, coupled with the tendency for CD95 self-association. Two variants (Glu245Ala and Lys247Ala) were nonaggregating and still able to bind FADD. Based on these measurements, the apparent focus of the

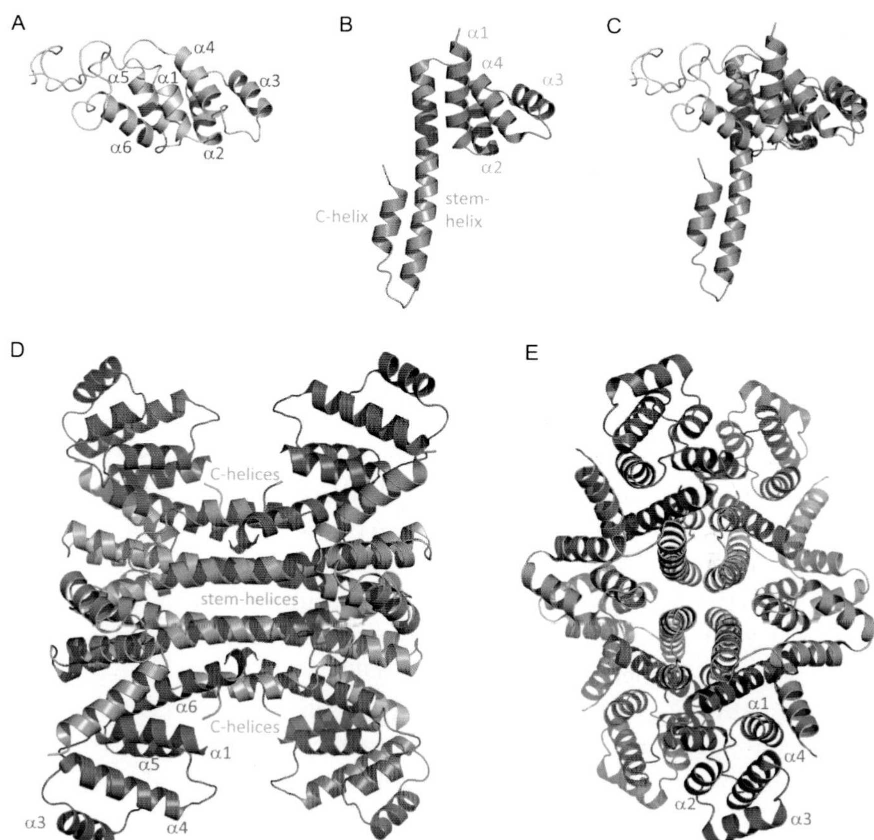

Figure 9.5 Solution structure of the DD of CD95 and the crystallographic complex with the DD of the adaptor protein FADD. (A) The NMR-derived solution structure of the CD95-DD (PDB 1DDF) (Huang et al., 1996) illustrating the characteristic six-helix Greek-key topology. (B) The structure of CD95-DD extracted from the heterotetramic complex with FADD-DD (D and E) obtained by Scott et al. (2009) (PDB 3EZQ) showing the restructuring of helices α5 and α6 into the stem- and C-helix αα hairpin. In (B–E), CD95-DD from 3EZQ is shown in orange; in (D and E) FADD-DD is shown in blue. (C) A superposition of the solution structure of CD95-DD with that observed in the complex with FADD-DD. The orientation shown in (E) is rotated by 90° relative to that in (D). (See the color plate.)

FADD-binding site is situated on α-helices α2 and α3 and the connecting loop and includes the site that corresponds to the *lpr*^{cg} mutation (Ile246Asn) in mice. When a similar substitution is introduced into the human CD95 protein, NMR evidence points to a local unfolding of the α3 helix and an overall lowering of the thermodynamic stability of the DD (Eberstadt, Huang, Olejniczak, & Fesik, 1997). Martin et al. used a series of recombinant CD95-DD constructs to screen by NMR and other methods the impact of type Ia ALPS mutations (Martin et al., 1999). For three missense mutations (Arg234Gln, Ala241Asp, and Asp244Val), the NMR data indicated that the protein was folded normally. For the missense mutants Thr225Pro, Arg234Pro, and Ile294Ser, and the nonsense mutants Gln257Stop and Gln260STOP, the NMR spectra indicated global destabilization of the protein fold. All of these variants, and another not examined by NMR, Leu278STOP, failed to bind FADD *in vitro* or *in vivo*.

Several other DDs have been structurally characterized by NMR, including those of TNFR1 (PDB 1ICH) (Sukits et al., 2001; Telliez et al., 2000) and the immediate DR adaptor protein FADD (PDBs 1E3Y and 1E41 (human); 1FAD (mouse)) (Berglund et al., 2000; Jeong et al., 1999). These and other DDs from a variety of proteins each adopt the same all-α secondary structure and topological fold (Park, Lo, et al., 2007). Similar to the case with CD95, Sukits et al. had to work with a mutant of TNFR1-DD that was more tractable by NMR than the wild type (Sukits et al., 2001). Mutagenesis studies showed that residues on the surface of α2 as well as parts of α3 and α4 are involved in self-association of TNFR1-DD and its interactions with the immediate adaptor protein TRADD. Jeong et al. reported that mouse FADD-DD has a small tendency to self-associate, a characteristic not found for the human orthologs which is stably monomeric, and that any of a number of surface residue substitutions was able to reduce that behavior (Jeong et al., 1999). Only a subset of those substitutions, corresponding to residues on the surface of helices α2 and α3, had any impact on binding to human FADD, as assessed by surface plasmon resonance. Together, these structural and biophysical studies suggested that homotypic DD–DD interactions likely take place by antiparallel interaction of the domain surfaces involving the α2–α3 motif and that the specificity of the interactions would be based upon the relatively poorly conserved composition of the surface exposed side chains (Sukits et al., 2001). Slightly counter to this model, Werner and coworkers reported that, based on both *in vitro* and *in vivo* binding assays, a larger surface of the FADD-DD was involved in interactions with CD95, ranging over two sites: a major one encompassing residues in α-helices 1, 2, 3, 5, and 6, and a minor one involving residues in α3 and α4 (Hill et al., 2004). Noting that in the crystal structure of the DD–DD complex formed between the unrelated *Drosophila* proteins, Pelle and Tube helices α2 and α6 and the α5–α6 loop of Tube-DD interact with α3 and α4 of Pelle-DD (PDB 1D2Z) (Xiao, Towb, Wasserman, & Sprang, 1999), and the same FADD-DD α2–α3 surface was implicated in TNFR-mediated apoptotic signaling (Bang, Jeong, Kim, Jung, & Kim, 2000), these authors argued that the DD might represent a promiscuous protein–protein interaction module. Consistent with this view, Thomas et al. found that the interaction of the cytoplasmic domain of TRAIL-R2 with FADD depends upon residues extending over helices α2–α6, though valine at position 108 in FADD is required for binding to CD95, but not TRAIL-R2 (Thomas, Bender, Morgan, & Thorburn, 2006; Thomas, Henson, Reed, Salsbury, & Thorburn, 2004).

12. THE DD SUPERFAMILY

Contemporaneous with advances in understanding DD structure and function, similar progress was being made to obtain structures and functional characterization of related domain types. The Fesik team obtained the 3D solution structure of nonaggregating variants of the FADD–DED (PDB 1A1W and 1A1Z) (Eberstadt et al., 1998), Wagner and coworkers obtained the solution structure of the CARD of receptor-interacting protein (RIP)-associated protein with a DD (RAIDD) (PDB 3CRD) (Chou, Matsuo, Duan, & Wagner, 1998), and the binary complex of procaspase-9 and apoptotic protease-activating factor 1 (APAF-1) CARDs was solved by X-ray crystallography by Qin et al. (1999) (PDB 3YGS). Somewhat later, Hiller et al. (2003) and Liepinsh et al. (2003) reported the NMR-based solution structures of the PyDs of the apoptosis- and inflammation-related NACHT, LRR, and PYD domains-containing protein 1 (NALP1) (PDB 1PN5), and apoptosis-associated speck-like protein containing a CARD (ASC) (PDB 1UCP), respectively. These proteins share with DDs a common antiparallel all-α topology, though the α1 helix in the CARDs is strongly bent and best described as α1 and α1b. Importantly, the homotypic CARD–CARD interaction is not antiparallel: the positively charged surface of the procaspase-9 helices α1/α1b and α4 is recognized by the negatively charged surface of helices α2 and α3 of APAF-1. These studies were supported by many others; several focused on the homotypic interaction of FADD–DED with the DED-containing pro-domain of procaspase-8, cFLIP, and viral homologues (Carrington et al., 2006; Eberstadt et al., 1998; Li, Jeffrey, Yu, & Shi, 2006; Muppidi et al., 2006; Thome et al., 1997; Yang et al., 2005). Since in this "death domain superfamily" self-association is commonly observed and a given member of one subtype tends to bind only the same or another member of the same subtype (e.g., DD–DD, DED–DED, and CARD–CARD), a reasonable projection is that DDs, DEDs, CARDs, and PyDs likely evolved from a common ancestor, and that within each subgroup both amino acid sequences and binding modes have diverged in history (Park, Lo, et al., 2007).

13. DD ASSEMBLY REVEALED BY THE STRUCTURE OF THE PIDDosome CORE

Despite the progress with various members of the DD protein domain superfamily, it has proved challenging to understand the structural and mechanistic basis of the interaction between DR cytoplasmic domains (and their DDs) and immediate adaptor and downstream signaling proteins. In large part, this is because it has been difficult to obtain functional forms of the receptor DDs at concentrations required for structural and biophysical analysis. In 2001, Weber and Vincenz proposed a model for a hexameric complex of CD95-DD with FADD-DD, inspired by the homotrimer model for CD95L–CD95 ectodomain based on results for TNF/TNFRs and TRAIL/TRAIL-R2, and that integrated known sequence and structure–activity data for a variety of DD superfamily proteins, including the interaction modes in the heterodimer Pelle-DD:Tube-DD and APAF-1: procaspase-9 crystal structures (Weber & Vincenz, 2001b). These two authors defined the Pelle–Tube type interface as type I, and the APAF-1: procaspase-9 interface as type II, and invoked the formation of a third interface, type III, that forms as the result of the juxtaposition of a type I and a type II interface. In 2007, Park et al. reported the 3.2 Å resolution crystal of a binary DD assembly comprising seven DDs from the protein RAIDD and five domains from p53-induced protein with a DD (PIDD), a 157 kDa particle hypothecated to represent the core of a high-molecular-weight complex involved in caspase-2 activation and dubbed the PIDDosome (PDB 2OF5) (Park, Logette, et al., 2007) (Fig. 9.6). The structure lacks formal symmetry, and the description of the structure in terms of three layers—five PIDD-DDs (bottom), five RAIDD-DDs (middle), two RAIDD-DDs (top)—is only loosely accurate. Looking down from the top, the two stacked five-membered rings of RAIDD-DDs and the PIDD-DDs around are related by a rotational offset around a common central axis. Intriguingly, despite the low symmetry, each component DD has a quasi-equivalent environment, with between three and six neighboring domains. The eight kinds of interdomain interfaces can be classified according to the scheme proposed by Weber and Vincenz (2001b) (Fig. 9.6C). Namely, in the type I interfaces, residues from α1 and α4 of the first DD (type Ia surface) interact with residues from α2 and α3 of the second DD (type Ib surface); in the type II

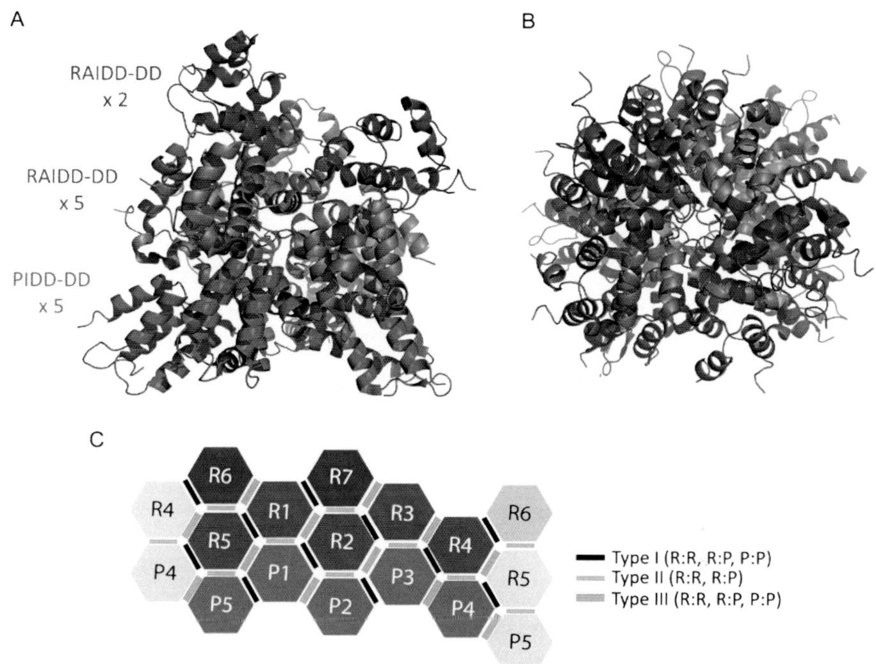

Figure 9.6 The 3D structure of the PIDDosome core particle (PDB 2OF5) (Park, Logette, et al., 2007). The structure is comprised of seven RAIDD-DD domains and five PIDD-DD domains arranged approximately as three layers (A) around a screw axis evident in the center of the structure shown in the orientation (B) rotated by 90° relative to (A). The structure lacks formal symmetry but contacts between domains can be characterized in three classes types I–III as summarized schematically in (C) with is adapted from Park, Logette, et al. (2007). In (C), light colors indicate domains that appear twice in the diagram, due to the pseudo-cylindrical nature of the structure. (See the color plate.)

interfaces, residues in the α4 helix and the α4–α5 loop of the first DD (type IIa surface) interact with residues at the α5–α6 loop and α6 helix of the second DD (type IIb surface); and in the type III interactions, residues from α3 of the first DD (type IIIa surface) contact residues near the α1–α2 and α3–α4 loops of the second DD (type IIIb surface).

The PIDDosome core structure provides an appealing model for other DD–DD binding pairs, particularly in the manner in which it might explain the observations that multiple surfaces of, for example, FADD-DD are implicated in binding CD95 or TRAIL-R2 DDs. Evidence for high-molecular-weight complex formation between a solubilized form of CD95-DD and FADD-DD was reported by Ferguson et al. (2007), who exploited the observation that appending a target protein to the B1 domain of streptococcal protein G (GB1) often improves its solubility (Zhou, Lugovskoy, & Wagner, 2001; Zhou & Wagner, 2010). A combination of heteronuclear NMR, analytical size exclusion chromatography (SEC), and electrospray-ionization mass spectrometry (ESI-MS) measurements indicated that the complex formed between GB1-CD95-DD and FADD-DD was ~160 kDa. Extension of this work in which the solubility-enhancing influence of the GB1 domain was swapped for the CD95 Lys247Ala mutation (Huang et al., 1996) showed that the complex formed between the two proteins likely comprises 10 protein chains with a 5:5 composition, as assessed by multi-angle light scattering (MALS) coupled to SEC, and ESI-MS (Esposito et al., 2010). Ile/Leu/Val ^{13}C-methyl transverse relaxation optimized spectroscopy (ILV-^{13}CH3-TROSY) (Kay, 2011; Ruschak & Kay, 2010; Tugarinov, Hwang, Ollerenshaw, & Kay, 2003) of the complex demonstrated that the environments of individual methyl groups within the globular parts of the component domains are non-uniform, leading the authors of this study to argue that structure of the CD95-DD:FADD-DD complex could possess low-symmetry characteristics similar to those of the PIDDosome core.

14. STRUCTURAL CHARACTERIZATION OF CD95:FADD-DD COMPLEXES

In 2008, Scott et al. reported the 2.7 Å resolution structure of a complex formed between wild-type CD95 and FADD-DDs obtained from crystals prepared from material purified from the combined lysates of separate bacterial cultures in high-salt and low-pH buffer (0.95 *M* citric acid, 1.9 *M* ammonium sulfate, pH 4.0) (Scott et al., 2009). The structure contains a tetrameric arrangement of four CD95-DDs and four FADD-DDs, with a dimer of two CD95-DD:FADD-DD complex dimers (Fig. 9.5D and E). Each FADD-DD contacts only a single CD95-DD, and the contacts between the heterodimers are made solely by CD95-DD molecules. Critically, the structure of the CD95-DD is significantly different from the canonical six-helix Greek-key topology (Fig. 9.5B and C). The C-terminal part of the domain is substantially restructured such that the α6 helix has shifted and fused with α5 to form a long helix, called the stem helix, and the extreme C-terminus of CD95, regarded as being outside of the DD homology region, is ordered into an α-helix (the C-helix) that folds back to contact the stem helix in an antiparallel fashion. A pair of stem/

C-helix hairpins combine in an antiparallel arrangement to fashion bridges between CD95 molecules leading to the 2:2 CD95-DD:FADD-DD dimer of dimers that the authors surmise is the relevant structure in the physiological context of a juxtamembrane position. The restructuring of the CD95-DD exposes hydrophobic residues that contact a hydrophobic patch on helices α1 and α6 of FADD-DD. Compared to the isolated protein, FADD-DD helix α6 is shifted in this complex, apparently to avoid steric overlap with the CD95-DD C-helix. This feature is attractive in the sense that it could herald a binding-induced rearrangement of the structure of full-length FADD that exposes a caspase-8 binding site on the FADD–DED. The authors sought to demonstrate the functional relevance of CD95-opening by mutating the normally buried α6 residue Ile313 to aspartic acid in intact CD95 expressed in Huh7 cells. Compared to the wild type, cells expressing the mutant CD95 were apparently more susceptible to cell death both in the presence and in the absence of CD95 agonists. These results led to a model that invokes CD95 clustering-induced restructuring of the receptor DD, followed by association with FADD and propagated conformational change leading to the recruitment of caspase-8.

In striking contrast to this work, Wu and coworkers subsequently demonstrated using a combination of negative-stain electron microscopy (EM), ESI-MS, and MALS that combinations of both human CD95-DD and human FADD-DD, and mouse CD95-DD and human FADD-DD predominantly form particles with 5:5 composition at pH 8.0 and 100 m*M* NaCl (Wang et al., 2010). Crystals of the chimeric complex were obtained that diffracted to 6.8 Å. Since the EM images of the particles were reminiscent of those obtained for the PIDDosome core, the structure of the complex was solved by molecular replacement using a model based on a part of that structure (PDB 3OQ9) (Fig. 9.7). Simulated EM projections based on the crystal structure agreed well with the experimental results. Extensive mutagenesis experiments were performed to validate the interpretation of the diffraction data. Substitutions made at the type I and type III interfaces had a dramatic effect on particle formation as did CD95-DD mutations at the type IIa surface and FADD-DD mutations on the type IIb surface; CD95-DD type IIb and FADD type IIa mutations had little or no effect. The authors noted a semiquantitative correlation between the number of occurrences of an interaction type in the CD95-DD:FADD-DD complex and the effect of mutations in the respective interfaces, both in *in vitro* and *in vivo* measurements.

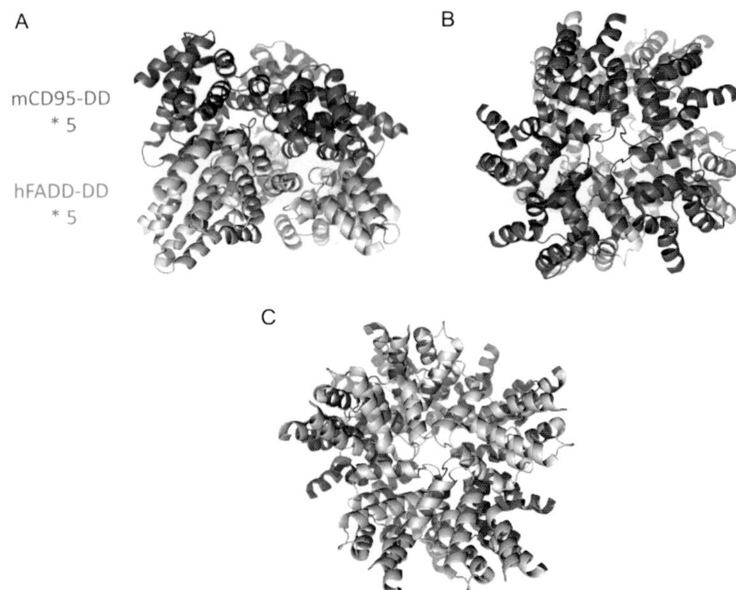

Figure 9.7 The low-resolution (6.9 Å) crystal structure of the chimeric complex between mouse CD95-DD and human FADD-DD (PDB 3OQ9) (Wang et al., 2010). (A) The two-layer arrangement of five CD95-DD chains (blue) atop five FADD-DD chains (light brown). The model is based upon the arrangement of the bottom two layers of the PIDDosome core (Fig. 9.6). (B and C) Two different 90° rotations of the structure, relative to that in (A) to highlight the pseudo-helical symmetry axis with either the CD95-DD or FADD-DD domains to the fore, respectively. (See the color plate.)

15. RELEVANCE OF CD95:FADD-DD ASSEMBLIES TO PHYSIOLOGICAL CD95 SIGNALING

The heterotetramer and heterodecamer structures of the CD95-DD: FADD-DD complex present very different bases for the rationalization of events critical for DISC formation. Since in the cell FADD is not constitutively associated with DRs at the cell membrane, it is necessary to invoke a signaling trigger that leads to such recruitment. Agonist binding to the receptor must lead to either a reorganization or restructuring (or both) of the receptors such that an effective FADD-binding surface is created. It is now widely acknowledged that the qualitative nature of the signaling output from CD95 depends upon the degree of cross-linking of the agonist (Feig, Tchikov, Schutze, & Peter, 2007; Holler et al., 2003; Reilly et al., 2009), such that nonapoptotic signals pertain in the presence of soluble CD95L (presumed to be homotrimeric) and apoptotic signaling depends upon higher level clustering and potentially internalization of the receptor. Yet all pathways appear to invoke initial recruitment of FADD and other DISC components to the ligand-bound receptor.

In this context, Scott et al. proposed that ligand-dependent clustering of CD95 leads to conformational change in the CD95-DD that enhances intermolecular CD95 oligomerization through mutual stem/C-helix interactions and FADD association via an exposed hydrophobic patch (Scott et al., 2009). Wang et al. suggested that simple ligand-driven aggregation of the receptor may be sufficient to drive both cooperative mutual CD95-DD and heterophilic CD95-DD:FADD-DD interactions (Wang et al., 2010). It is appropriate to examine the manner in which the sites of disruptive and nondisruptive CD95 and FADD substitutions map on the respective crystal structures. It is well known that type Ia ALPS mutants of CD95 act in a DN fashion to disrupt CD95 signaling, and Wang et al. demonstrated explicitly that their structure correctly predicted the effects of both structure-based and ALPS CD95 mutants on CD95 signaling in jurkat cells (Wang et al., 2010). Scott et al. presented rationalization of the known effects of mutations on CD95 signaling that relied in part on the presumed higher order clustering of CD95-DD:FADD-DD heterotetramers; many of the FADD-DD mutations known to affect signaling are on a surface of the domain that is not engaged with CD95-DD in their structure (Esposito et al., 2010; Scott et al., 2009). A number of other aspects of these studies are worthy of consideration. Both Wang et al. and Esposito et al. obtained 5:5 CD95-DD:FADD-DD complexes under pseudo-physiological conditions. Wang et al. used CD95-DD constructs lacking the extreme C-terminal tail region, downstream of the α6 helix, and demonstrated to be highly disordered in solution by NMR (Eberstadt et al., 1997; Esposito et al., 2010; Huang et al., 1996). The data obtained by Esposito et al. showed the formation of an asymmetric high-molecular-weight particle that was independent of the presence of the C-terminal 12 residue of CD95-DD (Esposito et al., 2010). In the heterotetramer structure, the nature of the intermolecular CD95-DD stem/C-helix bridge would appear to depend absolutely on the presence of this region. At the same time, it has been reported that the C-terminal tail region of CD95 is inhibitory to interactions with FADD, for example by Chinnaiyan et al. (1995). It has also been pointed out that not all species possess an extensive C-terminal region outside of the CD95-DD homology region, arguing that stem/C-helix formation could not be a common mechanism underpinning CD95–FADD interactions and DISC formation (Esposito et al., 2010). In addition, it is curious that the heterotetramer structure was obtained under crystallization conditions with high salt and low pH, whereas others have reported that independently high salt or low pH is disruptive to interactions between CD95 and FADD, for example, Esposito et al. (2010).

At present, it is extremely difficult to reconcile the findings of structural analysis of DR structure and interactions into a satisfactorily coherent picture. While it has proved possible, after considerable effort, to demonstrate and characterize *in vitro* interactions between, for example, CD95-DD and FADD-DD, the results are not always concordant (*vide supra*) and difficult to integrate into a bigger picture of DISC formation. At present, one cannot be sure that either of the CD95-DD:FADD-DD structures is entirely relevant to the situation that pertains in the cell. Moreover, extension of the structural analysis of this binary DD combination to larger parts of the proteins involved or to other members of the DR family appears to be challenging. For example, we could show that CD95-DD binds FADD-DD by NMR and other methods (Esposito et al., 2010; Ferguson et al., 2007), yet in our hands the same CD95-DD construct fails to bind full-length FADD despite the presence in the latter protein of the intact FADD-DD (Esposito, D., Richter, C., Driscoll, P. C., et al., unpublished). Our attempts to extend the NMR-based approach to interaction of FADD and FADD-DD with the cytoplasmic domains of TRAIL-R1 and TRAIL-R2 have been met with failure to obtain soluble, folded forms of the corresponding DDs, despite extensive screening of domain boundaries, mutations, species orthologs, and co-expression with FADD-DD (Ragan, T. J., Garza-Garcia, A., Driscoll, P.C., et al., unpublished). Our lack of success in this respect, coupled with the anecdotal reports of similar difficulties from other researchers, observation that the CD95-DD must possess some degree of conformational plasticity (as detected in the heterotetramer complex with FADD) (Scott et al., 2009) and that other purported DDs can demonstrably lack stable 3D structure, for example, the DDs of death-associated protein kinase (Dioletis, Dingley, & Driscoll, 2013) and RIP kinase-1-DD (Nemotollahi, L., Garza-Garcia, A., Driscoll, P. C., et al., unpublished), has led us to consider whether all predicted DDs can adopt a stable structure on their own. It is conceivable that it is the context of the whole cytoplasmic domain, in a juxtamembrane environment, that provides a stabilizing influence on DR structure. Moreover, it is worth noting multiple reports that palmitoylation of the membrane-proximal cysteine residue in the cytoplasmic domain of CD95 is required for functional signaling (Chakrabandhu et al., 2007; Feig et al., 2007; Lee et al., 2006). In addition, a number of reports implicate CD95-DD in interactions with proteins that link to the actin cytoskeleton, for example, via ezrin (Fais, De Milito, & Lozupone, 2005; Hebert et al., 2008; Lozupone et al., 2004; Luciani et al., 2004; Mielgo et al., 2007; Parlato et al., 2000; Rebillard et al., 2010), and that phosphorylation of tyrosine residue (Tyr291) in CD95-DD has a role in CD95 signaling (Daigle, Yousefi, Colonna, Green, & Simon, 2002; Eberle, Reinehr, Becker, Keitel, & Haussinger, 2007; Kennedy & Budd, 1998; Kleber et al., 2008; Reinehr, Becker, Hongen, & Haussinger, 2004; Reinehr, Gorg, Hongen, & Haussinger, 2004; Reinehr, Schliess, & Haussinger, 2003; Sancho-Martinez & Martin-Villalba, 2009). Intriguingly, the Tyr291 side-chain hydroxyl is exposed in the 3D structure of CD95-DD and could be available to a protein tyrosine kinase. However, the residues immediately to the C-terminus of this residue, including Leu294, which is invoked as part of a classical pY-X-X-L Src homology 2 (SH2) domain-binding motif, are completely buried; for SH2 binding to occur by canonical means the structure of the CD95-DD would have to be very significantly disrupted.

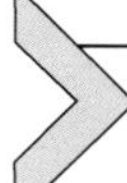

16. UNANSWERED QUESTIONS AND FUTURE PROSPECTS

In summary, structural analysis of the apical components of DR signaling has revealed the ground-state conformations of many of the fundamental building blocks involved in receptor activation and DISC assembly. More work is required to better understand how these blocks become engaged in the act of signaling, not least in how these structures relate to: different signaling outcomes arising from the action of soluble versus membrane-associated or otherwise cross-linked agonists (Feig et al., 2007; Holler et al., 2003; Reilly et al., 2009); the observed formation of denaturant-resistant high-molecular weight aggregates (Feig et al., 2007; Kamitani, Nguyen, & Yeh, 1997; Kischkel et al., 1995); the roles played by lipid microdomain localization and receptor internalization (Lee et al., 2006; Legembre et al., 2005; Muppidi & Siegel, 2004; Muppidi, Tschopp, & Siegel, 2004; Rossin et al., 2010; Schneider-Brachert et al., 2006; Schutze, Tchikov, & Schneider-Brachert, 2008); and the recently reported ligand-dependent biases in the compositional profiles of DISC preparations in which the numbers of caspase-8 molecules significantly outweigh those of FADD (Dickens et al., 2012; Schleich, Krammer, & Lavrik, 2013; Schleich et al., 2012). It seems likely that the further elucidation of relationship between structure and mechanism in this area will have to appeal to techniques beyond the traditional structural biology paradigms of X-ray crystallography and NMR spectroscopy, to new, frontier techniques such as single-particle analysis and high-resolution optical measurements.

ACKNOWLEDGMENTS

I thank the members of my research group over many years for valuable discussions concerning the structure and mechanism of DR signaling. This work is supported by the Medical Research Council (File Reference U117574559).

REFERENCES

Algeciras-Schimnich, A., Shen, L., Barnhart, B. C., Murmann, A. E., Burkhardt, J. K., & Peter, M. E. (2002). Molecular ordering of the initial signaling events of CD95. *Molecular and Cellular Biology*, *22*(1), 207–220.

Ashkenazi, A., & Dixit, V. M. (1998). Death receptors: Signaling and modulation. *Science*, *281*(5381), 1305–1308. http://dx.doi.org/10.1126/science.281.5381.1305.

Bang, S., Jeong, E. J., Kim, I. K., Jung, Y. K., & Kim, K. S. (2000). Fas- and tumor necrosis factor-mediated apoptosis uses the same binding surface of FADD to trigger signal transduction—A typical model for convergent signal transduction. *Journal of Biological Chemistry*, *275*(46), 36217–36222.

Banner, D. W., Darcy, A., Janes, W., Gentz, R., Schoenfeld, H. J., Broger, C., et al. (1993). Crystal-structure of the soluble human 55 Kd TNF receptor-human TNF-beta complex—Implications for TNF receptor activation. *Cell*, *73*(3), 431–445.

Berglund, H., Olerenshaw, D., Sankar, A., Federwisch, M., McDonald, N. Q., & Driscoll, P. C. (2000). The three-dimensional solution structure and dynamic properties of the human FADD death domain. *Journal of Molecular Biology*, *302*(1), 171–188.

Carrington, P. E., Sandu, C., Wei, Y. F., Hill, J. M., Morisawa, G., Huang, T., et al. (2006). The structure of FADD and its mode of interaction with procaspase-8. *Molecular Cell*, *22*(5), 599–610.

Carter, P. H., Scherle, P. A., Muckelbauer, J. A., Voss, M. E., Liu, R. Q., Thompson, L. A., et al. (2001). Photochemically enhanced binding of small molecules to the tumor necrosis factor receptor-1 inhibits the binding of TNF-alpha. *Proceedings of the National Academy of Sciences of the United States of America*, *98*(21), 11879–11884.

Cascino, I., Papoff, G., DeMaria, R., Testi, R., & Ruberti, G. (1996). Fas/Apo-1 (CD95) receptor lacking the intracytoplasmic signaling domain protects tumor cells from Fas-mediated apoptosis. *Journal of Immunology*, *156*(1), 13–17.

Cha, S. S., Kim, J. S., Cho, H. S., Shin, N. K., Jeong, W., Shin, H. C., et al. (1998). High resolution crystal structure of a human tumor necrosis factor-alpha mutant with low systemic toxicity. *Journal of Biological Chemistry*, *273*(4), 2153–2160. http://dx.doi.org/10.1074/jbc.273.4.2153.

Cha, S. S., Kim, M. S., Choi, Y. H., Sung, B. J., Shin, N. K., Shin, H. C., et al. (1999). 2.8 angstrom resolution crystal structure of human TRAIL, a cytokine with selective antitumor activity. *Immunity*, *11*(2), 253–261.

Cha, S. S., Sung, B. J., Kim, Y. A., Song, Y. L., Kim, H. J., Kim, S., et al. (2000). Crystal structure of TRAIL-DR5 complex identifies a critical role of the unique frame insertion in conferring recognition specificity. *Journal of Biological Chemistry*, *275*(40), 31171–31177.

Chakrabandhu, K., Herincs, Z., Huault, S., Dost, B., Peng, L., Conchonaud, F., et al. (2007). Palmitoylation is required for efficient Fas cell death signaling. *The EMBO Journal*, *26*(1), 209–220.

Chakrabandhu, K., Huault, S., Garmy, N., Fantini, J., Stebe, E., Mailfert, S., et al. (2008). The extracellular glycosphingolipid-binding motif of Fas defines its internalization route, mode and outcome of signals upon activation by ligand. *Cell Death and Differentiation*, *15*(12), 1824–1837.

Chan, F. K. M., Chun, H. J., Zheng, L. X., Siegel, R. M., Bui, K. L., & Lenardo, M. J. (2000). A domain in TNF receptors that mediates ligand-independent receptor assembly and signaling. *Science*, *288*(5475), 2351–2354.

Cheng, X., Kinosaki, M., Murali, R., & Greene, M. I. (2003). The TNF receptor superfamily: Role in immune inflammation and bone formation. *Immunologic Research*, *27*(2–3), 287–294. http://dx.doi.org/10.1385/IR:27:2-3:287.

Chinnaiyan, A. M., Orourke, K., Tewari, M., & Dixit, V. M. (1995). Fadd, a novel death domain-containing protein, interacts with the death domain of Fas and initiates apoptosis. *Cell*, *81*(4), 505–512.

Chinnaiyan, A. M., Tepper, C. G., Seldin, M. F., ORourke, K., Kischkel, F. C., Hellbardt, S., et al. (1996). FADD/MORT1 is a common mediator of CD95 (Fas/APO-1) and tumor necrosis factor receptor-induced apoptosis. *Journal of Biological Chemistry*, *271*(9), 4961–4965.

Chodorge, M., Zuger, S., Stirnimann, C., Briand, C., Jermutus, L., Grutter, M. G., et al. (2012). A series of Fas receptor agonist antibodies that demonstrate an inverse correlation between affinity and potency. *Cell Death and Differentiation*, *19*(7), 1187–1195.

Chou, J. J., Matsuo, H., Duan, H., & Wagner, G. (1998). Solution structure of the RAIDD CARD and model for CARD/CARD interaction in caspase-2 and caspase-9 recruitment. *Cell*, *94*(2), 171–180. http://dx.doi.org/10.1016/S0092-8674(00)81417-8.

Clancy, L., Mruk, K., Archer, K., Woelfel, M., Mongkolsapaya, J., Screaton, G., et al. (2005). Preligand assembly domain-mediated ligand-independent association between TRAIL receptor 4 (TR4) and TR2 regulates TRAIL-induced apoptosis. *Proceedings of the National Academy of Sciences of the United States of America*, *102*(50), 18099–18104.

Cuppen, E., Nagata, S., Wieringa, B., & Hendriks, W. (1997). No evidence for involvement of mouse protein-tyrosine phosphatase-BAS-like Fas-associated phosphatase-1 in Fas-mediated apoptosis. *Journal of Biological Chemistry*, *272*(48), 30215–30220. http://dx.doi.org/10.1074/jbc.272.48.30215.

Daigle, I., Yousefi, S., Colonna, M., Green, D. R., & Simon, H. U. (2002). Death receptors bind SHP-1 and block cytokine-induced anti-apoptotic signaling in neutrophils. *Nature Medicine*, *8*(1), 61–67. http://dx.doi.org/10.1038/Nm0102-61.

Deng, G. M., Liu, L., & Tsokos, G. C. (2010). Targeted tumor necrosis factor receptor I preligand assembly domain improves skin lesions in MRL/lpr mice. *Arthritis and Rheumatism*, *62*(8), 2424–2431. http://dx.doi.org/10.1002/art.27534.

Deng, G. M., Zheng, L., Chan, F. K., & Lenardo, M. (2005). Amelioration of inflammatory arthritis by targeting the pre-ligand assembly domain of tumor necrosis factor receptors. *Nature Medicine*, *11*(10), 1066–1072. http://dx.doi.org/10.1038/nm1304.

Dickens, L. S., Boyd, R. S., Jukes-Jones, R., Hughes, M. A., Robinson, G. L., Fairall, L., et al. (2012). A death effector domain chain DISC model reveals a crucial role for caspase-8 chain assembly in mediating apoptotic cell death. *Molecular Cell*, *47*(2), 291–305.

Dioletis, E., Dingley, A. J., & Driscoll, P. C. (2013). Structural and functional characterization of the recombinant death domain from death-associated protein kinase. *PLoS One*, *8*(7), e70095. http://dx.doi.org/10.1371/journal.pone.0070095, ARTN e70095.

Eberle, A., Reinehr, R., Becker, S., Keitel, V., & Haussinger, D. (2007). CD95 tyrosine phosphorylation is required for CD95 oligomerization. *Apoptosis*, *12*(4), 719–729.

Eberstadt, M., Huang, B. H., Chen, Z. H., Meadows, R. P., Ng, S. C., Zheng, L. X., et al. (1998). NMR structure and mutagenesis of the FADD (Mort1) death-effector domain. *Nature, 392*(6679), 941–945.

Eberstadt, M., Huang, B. H., Olejniczak, E. T., & Fesik, S. W. (1997). The lymphoproliferation mutation in Fas locally unfolds the Fas death domain. *Nature Structural Biology, 4*(12), 983–985.

Eck, M. J., & Sprang, S. R. (1989). The structure of tumor necrosis factor-alpha at 2.6-a resolution—Implications for receptor-binding. *Journal of Biological Chemistry, 264*(29), 17595–17605.

Edmond, V., Ghali, B., Penna, A., Taupin, J. L., Daburon, S., Moreau, J. F., et al. (2012). Precise mapping of the CD95 pre-ligand assembly domain. *PLoS One*, 7(9), e46236.

Esposito, D., Sankar, A., Morgner, N., Robinson, C. V., Rittinger, K., & Driscoll, P. C. (2010). Solution NMR investigation of the CD95/FADD homotypic death domain complex suggests lack of engagement of the CD95 C terminus. *Structure, 18*(10), 1378–1390.

Fairbrother, W. J., Gordon, N. C., Humke, E. W., O'Rourke, K. M., Starovasnik, M. A., Yin, J. P., et al. (2001). The PYRIN domain: A member of the death domain-fold superfamily. *Protein Science, 10*(9), 1911–1918. http://dx.doi.org/10.1110/Ps.13801.

Fais, S., De Milito, A., & Lozupone, F. (2005). The role of FAS to ezrin association in FAS-mediated apoptosis. *Apoptosis, 10*(5), 941–947.

Feig, C., Tchikov, V., Schutze, S., & Peter, M. E. (2007). Palmitoylation of CD95 facilitates formation of SDS-stable receptor aggregates that initiate apoptosis signaling. *The EMBO Journal, 26*(1), 221–231.

Fellouse, F. A., Li, B., Compaan, D. M., Peden, A. A., Hymowitz, S. G., & Sidhu, S. S. (2005). Molecular recognition by a binary code. *Journal of Molecular Biology, 348*(5), 1153–1162.

Ferguson, B. J., Esposito, D., Jovanovic, J., Sankar, A., Driscoll, P. C., & Mehmet, H. (2007). Biophysical and cell-based evidence for differential interactions between the death domains of CD95/Fas and FADD. *Cell Death and Differentiation, 14*(9), 1717–1719.

Goh, C. R., Loh, C. S., & Porter, A. G. (1991). Aspartic acid 50 and tyrosine 108 are essential for receptor binding and cytotoxic activity of tumour necrosis factor beta (lymphotoxin). *Protein Engineering, 4*(7), 785–791.

He, M. M., Smith, A. S., Oslob, J. D., Flanagan, W. M., Braisted, A. C., Whitty, A., et al. (2005). Small-molecule inhibition of TNF-alpha. *Science, 310*(5750), 1022–1025. http://dx.doi.org/10.1126/science.1116304.

Hebert, M., Potin, S., Sebbagh, M., Bertoglio, J., Breard, J., & Hamelin, J. (2008). Rho-ROCK-dependent ezrin–radixin–moesin phosphorylation regulates Fas-mediated apoptosis in jurkat cells. *Journal of Immunology, 181*(9), 5963–5973.

Hehlgans, T., & Pfeffer, K. (2005). The intriguing biology of the tumour necrosis factor/tumour necrosis factor receptor superfamily: Players, rules and the games. *Immunology, 115*(1), 1–20. http://dx.doi.org/10.1111/j.1365-2567.2005.02143.x.

Hill, J. M., Morisawa, G., Kim, T., Huang, T., Wei, Y., Wei, Y. F., et al. (2004). Identification of an expanded binding surface on the FADD death domain responsible for interaction with CD95/Fas. *Journal of Biological Chemistry, 279*(2), 1474–1481.

Hiller, S., Kohl, A., Fiorito, F., Herrmann, T., Wider, G., Tschopp, J., et al. (2003). NMR structure of the apoptosis- and inflammation-related NALP1 pyrin domain. *Structure, 11*(10), 1199–1205. http://dx.doi.org/10.1016/j.str.2003.08.009.

Holler, N., Tardivel, A., Kovacsovics-Bankowski, M., Hertig, S., Gaide, O., Martinon, F., et al. (2003). Two adjacent trimeric Fas ligands are required for Fas signaling and formation of a death-inducing signaling complex. *Molecular and Cellular Biology, 23*(4), 1428–1440.

Hu, S., Liang, S., Guo, H., Zhang, D., Li, H., Wang, X., et al. (2013). Comparison of the inhibition mechanisms of adalimumab and infliximab in treating tumor necrosis factor alpha-associated diseases from a molecular view. *Journal of Biological Chemistry*, *288*(38), 27059–27067. http://dx.doi.org/10.1074/jbc.M113.491530.

Huang, B. H., Eberstadt, M., Olejniczak, E. T., Meadows, R. P., & Fesik, S. W. (1996). NMR structure and mutagenesis of the Fas (APO-1/CD95) death domain. *Nature*, *384*(6610), 638–641.

Hueber, A. O., Zornig, M., Bernard, A. M., Chautan, M., & Evan, G. (2000). A dominant negative Fas-associated death domain protein mutant inhibits proliferation and leads to impaired calcium mobilization in both T-cells and fibroblasts. *Journal of Biological Chemistry*, *275*(14), 10453–10462.

Hymowitz, S. G., Christinger, H. W., Fuh, G., Ultsch, M., O'Connell, M., Kelley, R. F., et al. (1999). Triggering cell death: The crystal structure of Apo2L/TRAIL in a complex with death receptor 5. *Molecular Cell*, *4*(4), 563–571.

Hymowitz, S. G., O'Connell, M. P., Ultsch, M. H., Hurst, A., Totpal, K., Ashkenazi, A., et al. (2000). A unique zinc-binding site revealed by a high-resolution X-ray structure of homotrimeric Apo2L/TRAIL. *Biochemistry*, *39*(4), 633–640.

Itoh, N., & Nagata, S. (1993). A novel protein domain required for apoptosis—Mutational analysis of human Fas antigen. *Journal of Biological Chemistry*, *268*(15), 10932–10937.

Jeong, E. J., Bang, S., Lee, T. H., Park, Y. I., Sim, W. S., & Kim, K. S. (1999). The solution structure of FADD death domain—Structural basis of death domain interactions of Fas and FADD. *Journal of Biological Chemistry*, *274*(23), 16337–16342.

Jones, E. Y., Stuart, D. I., & Walker, N. P. (1989). Structure of tumour necrosis factor. *Nature*, *338*(6212), 225–228. http://dx.doi.org/10.1038/338225a0.

Kamitani, T., Nguyen, H. P., & Yeh, E. T. (1997). Activation-induced aggregation and processing of the human Fas antigen. Detection with cytoplasmic domain-specific antibodies. *Journal of Biological Chemistry*, *272*(35), 22307–22314.

Kay, L. E. (2011). Solution NMR spectroscopy of supra-molecular systems, why bother? A methyl-TROSY view. *Journal of Magnetic Resonance*, *210*(2), 159–170. http://dx.doi.org/10.1016/j.jmr.2011.03.008.

Kennedy, N. J., & Budd, R. C. (1998). Phosphorylation of FADD/MORT1 and Fas by kinases that associate with the membrane-proximal cytoplasmic domain of Fas. *Journal of Immunology*, *160*(10), 4881–4888.

Kischkel, F. C., Hellbardt, S., Behrmann, I., Germer, M., Pawlita, M., Krammer, P. H., et al. (1995). Cytotoxicity-dependent Apo-1 (Fas/Cd95)-associated proteins form a death-inducing signaling complex (Disc) with the receptor. *The EMBO Journal*, *14*(22), 5579–5588.

Kleber, S., Sancho-Martinez, I., Wiestler, B., Beisel, A., Gieffers, C., Hill, O., et al. (2008). Yes and PI3K bind CD95 to signal invasion of glioblastoma. *Cancer Cell*, *13*(3), 235–248.

Krammer, P. H. (2000). CD95's deadly mission in the immune system. *Nature*, *407*(6805), 789–795. http://dx.doi.org/10.1038/35037728.

Krueger, A., Fas, S. C., Baumann, S., & Krammer, P. H. (2003). The role of CD95 in the regulation of peripheral T-cell apoptosis. *Immunological Reviews*, *193*(1), 58–69. http://dx.doi.org/10.1034/j.1600-065X.2003.00047.x.

Kuester, M., Kemmerzehl, S., Dahms, S. O., Roeser, D., & Than, M. E. (2011). The crystal structure of death receptor 6 (DR6): A potential receptor of the amyloid precursor protein (APP). *Journal of Molecular Biology*, *409*(2), 189–201. http://dx.doi.org/10.1016/j.jmb.2011.03.048.

Lee, K. H., Feig, C., Tchikov, V., Schickel, R., Hallas, C., Schutze, S., et al. (2006). The role of receptor internalization in CD95 signaling. *The EMBO Journal*, *25*(5), 1009–1023.

Legembre, P., Daburon, S., Moreau, P., Ichas, F., de Giorgi, F., Moreau, J. F., et al. (2005). Amplification of Fas-mediated apoptosis in type II cells via microdomain recruitment. *Molecular and Cellular Biology*, *25*(15), 6811–6820.

Li, F. Y., Jeffrey, P. D., Yu, J. W., & Shi, Y. G. (2006). Crystal structure of a viral FLIP—Insights into FLIP-mediated inhibition of death receptor signaling. *Journal of Biological Chemistry*, *281*(5), 2960–2968.

Li, B., Russell, S. J., Compaan, D. M., Totpal, K., Marsters, S. A., Ashkenazi, A., et al. (2006). Activation of the proapoptotic death receptor DR5 by oligomeric peptide and antibody agonists. *Journal of Molecular Biology*, *361*(3), 522–536.

Liang, S. Y., Dai, J. X., Hou, S., Su, L. S., Zhang, D. P., Guo, H. Z., et al. (2013). Structural basis for treating tumor necrosis factor alpha (TNF alpha)-associated diseases with the therapeutic antibody infliximab. *Journal of Biological Chemistry*, *288*(19), 13799–13807. http://dx.doi.org/10.1074/jbc.M112.433961.

Liepinsh, E., Barbals, R., Dahl, E., Sharipo, A., Staub, E., & Otting, G. (2003). The death-domain fold of the ASC PYRIN domain, presenting a basis for PYRIN/PYRIN recognition. *Journal of Molecular Biology*, *332*(5), 1155–1163. http://dx.doi.org/10.1016/j.jmb.2003.07.007.

Lozupone, F., Lugini, L., Matarrese, P., Luciani, F., Federici, C., Iessi, E., et al. (2004). Identification and relevance of the CD95-binding domain in the N-terminal region of ezrin. *Journal of Biological Chemistry*, *279*(10), 9199–9207.

Luciani, F., Matarrese, P., Giammarioli, A. M., Lugini, L., Lozupone, F., Federici, C., et al. (2004). CD95/phosphorylated ezrin association underlies HIV-1 GP120/IL-2-induced susceptibility to CD95(APO-1/Fas)-mediated apoptosis of human resting CD4(+) T lymphocytes. *Cell Death and Differentiation*, *11*(5), 574–582.

Mariani, S. M., Matiba, B., Baumler, C., & Krammer, P. H. (1995). Regulation of cell-surface Apo-1/Fas (Cd95) ligand expression by metalloproteases. *European Journal of Immunology*, *25*(8), 2303–2307. http://dx.doi.org/10.1002/eji.1830250828.

Martin, D. A., Zheng, L. X., Siegel, R. M., Huang, B. H., Fisher, G. H., Wang, J., et al. (1999). Defective CD95/APO-1/Fas signal complex formation in the human autoimmune lymphoproliferative syndrome, type Ia. *Proceedings of the National Academy of Sciences of the United States of America*, *96*(8), 4552–4557.

Medema, J. P., Scaffidi, C., Kischkel, F. C., Shevchenko, A., Mann, M., Krammer, P. H., et al. (1997). FLICE is activated by association with the CD95 death-inducing signaling complex (DISC). *The EMBO Journal*, *16*(10), 2794–2804. http://dx.doi.org/10.1093/emboj/16.10.2794.

Mielgo, A., Brondani, V., Landmann, L., Glaser-Ruhm, A., Erb, P., Stupack, D., et al. (2007). The CD44 standard/ezrin complex regulates Fas-mediated apoptosis in jurkat cells. *Apoptosis*, *12*(11), 2051–2061.

Mongkolsapaya, J., Grimes, J. M., Chen, N., Xu, X. N., Stuart, D. I., Jones, E. Y., et al. (1999). Structure of the TRAIL-DR5 complex reveals mechanisms conferring specificity in apoptotic initiation. *Nature Structural Biology*, *6*(11), 1048–1053.

Mukai, Y., Nakamura, T., Yoshikawa, M., Yoshioka, Y., Tsunoda, S., Nakagawa, S., et al. (2010). Solution of the structure of the TNF-TNFR2 complex. *Science Signaling*, *3*(148), ra83.

Mukai, Y., Nakamura, T., Yoshioka, Y., Shibata, H., Abe, Y., Nomura, T., et al. (2009). Fast binding kinetics and conserved 3D structure underlie the antagonistic activity of mutant TNF: Useful information for designing artificial proteo-antagonists. *Journal of Biochemistry*, *146*(2), 167–172.

Mukai, Y., Shibata, H., Nakamura, T., Yoshioka, Y., Abe, Y., Nomura, T., et al. (2009). Structure–function relationship of tumor necrosis factor (TNF) and its receptor interaction based on 3D structural analysis of a fully active TNFR1-selective TNF mutant. *Journal of Molecular Biology*, *385*(4), 1221–1229.

Muppidi, J. R., Lobito, A. A., Ramaswamy, M., Yang, J. K., Wang, L., Wu, H., et al. (2006). Homotypic FADD interactions through a conserved RXDLL motif are required for death receptor-induced apoptosis. *Cell Death and Differentiation*, *13*(10), 1641–1650.

Muppidi, J. R., & Siegel, R. M. (2004). Ligand-independent redistribution of Fas (CD95) into lipid rafts mediates clonotypic T cell death. *Nature Immunology*, *5*(2), 182–189. http://dx.doi.org/10.1038/Ni1024.

Muppidi, J. R., Tschopp, J., & Siegel, R. M. (2004). Life and death decisions: Secondary complexes and lipid rafts in TNF receptor family signal transduction. *Immunity*, *21*(4), 461–465.

Nagata, S. (1997). Apoptosis by death factor. *Cell*, *88*(3), 355–365.

Naismith, J. H., Devine, T. Q., Brandhuber, B. J., & Sprang, S. R. (1995). Crystallographic evidence for dimerization of unliganded tumor-necrosis-factor receptor. *Journal of Biological Chemistry*, *270*(22), 13303–13307.

Naismith, J. H., Devine, T. Q., Kohno, T., & Sprang, S. R. (1996). Structures of the extracellular domain of the type I tumor necrosis factor receptor. *Structure*, *4*(11), 1251–1262.

Nemcovicova, I., Benedict, C. A., & Zajonc, D. M. (2013). Structure of human cytomegalovirus UL141 binding to TRAIL-R2 reveals novel, non-canonical death receptor interactions. *PLoS Pathogens*, *9*(3), e1003224.

Newton, K., Harris, A. W., Bath, M. L., Smith, K. G., & Strasser, A. (1998). A dominant interfering mutant of FADD/MORT1 enhances deletion of autoreactive thymocytes and inhibits proliferation of mature T lymphocytes. *The EMBO Journal*, *17*(3), 706–718. http://dx.doi.org/10.1093/emboj/17.3.706

Newton, K., Kurts, C., Harris, A. W., & Strasser, A. (2001). Effects of a dominant interfering mutant of FADD on signal transduction in activated T cells. *Current Biology*, *11*(4), 273–276.

Pan, G. H., Bauer, J. H., Haridas, V., Wang, S. X., Liu, D., Yu, G. L., et al. (1998). Identification and functional characterization of DR6, a novel death domain-containing TNF receptor. *FEBS Letters*, *431*(3), 351–356. http://dx.doi.org/10.1016/S0014-5793(98)00791-1.

Papoff, G., Cascino, I., Eramo, A., Starace, G., Lynch, D. H., & Ruberti, G. (1996). An N-terminal domain shared by Fas/Apo-1 (CD95) soluble variants prevents cell death in vitro. *Journal of Immunology*, *156*(12), 4622–4630.

Papoff, G., Hausler, P., Eramo, A., Pagano, M. G., Di Leve, G., Signore, A., et al. (1999). Identification and characterization of a ligand-independent oligomerization domain in the extracellular region of the CD95 death receptor. *Journal of Biological Chemistry*, *274*(53), 38241–38250. http://dx.doi.org/10.1074/jbc.274.53.38241.

Park, H. H., Lo, Y. C., Lin, S. C., Wang, L., Yang, J. K., & Wu, H. (2007). The death domain superfamily in intracellular signaling of apoptosis and inflammation. *Annual Review of Immunology*, *25*, 561–586.

Park, H. H., Logette, E., Raunser, S., Cuenin, S., Walz, T., Tschopp, J., et al. (2007). Death domain assembly mechanism revealed by crystal structure of the oligomeric PIDDosome core complex. *Cell*, *128*(3), 533–546.

Parlato, S., Giammarioli, A. M., Logozzi, M., Lozupone, F., Matarrese, P., Luciani, F., et al. (2000). CD95 (APO-1/Fas) linkage to the actin cytoskeleton through ezrin in human T lymphocytes: A novel regulatory mechanism of the CD95 apoptotic pathway. *The EMBO Journal*, *19*(19), 5123–5134.

Peter, M. E., Budd, R. C., Desbarats, J., Hedrick, S. M., Hueber, A. O., Newell, M. K., et al. (2007). The CD95 receptor: Apoptosis revisited. *Cell*, *129*(3), 447–450.

Qin, H. X., Srinivasula, S. M., Wu, G., Fernandes-Alnemri, T., Alnemri, E. S., & Shi, Y. G. (1999). Structural basis of procaspase-9 recruitment by the apoptotic protease-activating factor 1. *Nature*, *399*(6736), 549–557.

Rebillard, A., Jouan-Lanhouet, S., Jouan, E., Legembre, P., Pizon, M., Sergent, O., et al. (2010). Cisplatin-induced apoptosis involves a Fas-ROCK-ezrin-dependent actin remodelling in human colon cancer cells. *European Journal of Cancer, 46*(8), 1445–1455.

Reed, C., Fu, Z. Q., Wu, J., Xue, Y. N., Harrison, R. W., Chen, M. J., et al. (1997). Crystal structure of TNF-alpha mutant R31D with greater affinity for receptor R1 compared with R2. *Protein Engineering, 10*(10), 1101–1107. http://dx.doi.org/10.1093/protein/10.10.1101.

Reilly, L. A. O., Tai, L., Lee, L., Kruse, E. A., Grabow, S., Fairlie, W. D., et al. (2009). Membrane-bound Fas ligand only is essential for Fas-induced apoptosis. *Nature, 461*(7264), 659–663.

Reinehr, R., Becker, S., Hongen, A., & Haussinger, D. (2004). The Src family kinase yes triggers hyperosmotic activation of the epidermal growth factor receptor and CD95. *Journal of Biological Chemistry, 279*(23), 23977–23987.

Reinehr, R., Gorg, B., Hongen, A., & Haussinger, D. (2004). CD95-tyrosine nitration inhibits hyperosmotic and CD95 ligand-induced CD95 activation in rat hepatocytes. *Journal of Biological Chemistry, 279*(11), 10364–10373.

Reinehr, R., Schliess, F., & Haussinger, D. (2003). Hyperosmolarity and CD95L trigger CD95/EGF receptor association and tyrosine phosphorylation of CD95 as prerequisites for CD95 membrane trafficking and disc formation. *FASEB Journal, 17*(6), 731–733.

Richter, C., Messerschmidt, S., Holeiter, G., Tepperink, J., Osswald, S., Zappe, A., et al. (2012). The tumor necrosis factor receptor stalk regions define responsiveness to soluble versus membrane-bound ligand. *Molecular and Cellular Biology, 32*(13), 2515–2529.

Rossin, A., Kral, R., Lounnas, N., Chakrabandhu, K., Mailfert, S., Marguet, D., et al. (2010). Identification of a lysine-rich region of Fas as a raft nanodomain targeting signal necessary for Fas-mediated cell death. *Experimental Cell Research, 316*(9), 1513–1522.

Ru, H., Zhao, L. X., Ding, W., Jiao, L. Y., Shaw, N., Liang, W. G., et al. (2012). S-SAD phasing study of death receptor 6 and its solution conformation revealed by SAXS. *Acta Crystallographica Section D, Biological Crystallography, 68*, 521–530.

Ruschak, A. M., & Kay, L. E. (2010). Methyl groups as probes of supra-molecular structure, dynamics and function. *Journal of Biomolecular NMR, 46*(1), 75–87. http://dx.doi.org/10.1007/s10858-009-9376-1.

Sancho-Martinez, I., & Martin-Villalba, A. (2009). Tyrosine phosphorylation and CD95: A FAScinating switch. *Cell Cycle, 8*(6), 838–842.

Saras, J., Engstrom, U., Gonez, L. J., & Heldin, C. H. (1997). Characterization of the interactions between PDZ domains of the protein-tyrosine phosphatase PTPL1 and the carboxyl-terminal tail of Fas. *Journal of Biological Chemistry, 272*(34), 20979–20981. http://dx.doi.org/10.1074/jbc.272.34.20979.

Schleich, K., Krammer, P. H., & Lavrik, I. N. (2013). The chains of death: A new view on caspase-8 activation at the DISC. *Cell Cycle, 12*(2), 193–194. http://dx.doi.org/10.4161/Cc.23464.

Schleich, K., Warnken, U., Fricker, N., Ozturk, S., Richter, P., Kammerer, K., et al. (2012). Stoichiometry of the CD95 death-inducing signaling complex: Experimental and modeling evidence for a death effector domain chain model. *Molecular Cell, 47*(2), 306–319. http://dx.doi.org/10.1016/j.molcel.2012.05.006.

Schneider, P., Bodmer, J. L., Holler, N., Mattmann, C., Scuderi, P., Terskikh, A., et al. (1997). Characterization of Fas (Apo-1, CD95)-Fas ligand interaction. *Journal of Biological Chemistry, 272*(30), 18827–18833.

Schneider, P., Holler, N., Bodmer, J. L., Hahne, M., Frei, K., Fontana, A., et al. (1998). Conversion of membrane-bound Fas(CD95) ligand to its soluble form is associated with downregulation of its proapoptotic activity and loss of liver toxicity. *Journal of Experimental Medicine, 187*(8), 1205–1213.

Schneider-Brachert, W., Tchikov, V., Merkel, O., Jakob, M., Hallas, C., Kruse, M. L., et al. (2006). Inhibition of TNF receptor 1 internalization by adenovirus 14.7 K as a novel immune escape mechanism. *Journal of Clinical Investigation*, *116*(11), 2901–2913. http://dx.doi.org/10.1172/JCI23771.

Schrofelbauer, B., & Hoffmann, A. (2011). How do pleiotropic kinase hubs mediate specific signaling by TNFR superfamily members? *Immunological Reviews*, *244*(1), 29–43. http://dx.doi.org/10.1111/j.1600-065X.2011.01060.x.

Schutze, S., Tchikov, V., & Schneider-Brachert, W. (2008). Regulation of TNFR1 and CD95 signalling by receptor compartmentalization. *Nature Reviews. Molecular Cell Biology*, *9*(8), 655–662.

Scott, F. L., Stec, B., Pop, C., Dobaczewska, M. K., Lee, J. J., Monosov, E., et al. (2009). The Fas-FADD death domain complex structure unravels signalling by receptor clustering. *Nature*, *457*(7232), 1019–1022.

Shibata, H., Yoshioka, Y., Ohkawa, A., Minowa, K., Mukai, Y., Abe, Y., et al. (2008). Creation and X-ray structure analysis of the tumor necrosis factor receptor-1-selective mutant of a tumor necrosis factor-alpha antagonist. *Journal of Biological Chemistry*, *283*(2), 998–1007.

Siegel, R. M., Frederiksen, J. K., Zacharias, D. A., Chan, F. K. M., Johnson, M., Lynch, D., et al. (2000). Fas preassociation required for apoptosis signaling and dominant inhibition by pathogenic mutations. *Science*, *288*(5475), 2354–2357.

Strasser, A., Jost, P. J., & Nagata, S. (2009). The many roles of FAS receptor signaling in the immune system. *Immunity*, *30*(2), 180–192. http://dx.doi.org/10.1016/j.immuni.2009.01.001.

Sukits, S. F., Lin, L. L., Hsu, S., Malakian, K., Powers, R., & Xu, G. Y. (2001). Solution structure of the tumor necrosis factor receptor-1 death domain. *Journal of Molecular Biology*, *310*(4), 895–906. http://dx.doi.org/10.1006/jmbi.2001.4790.

Tartaglia, L. A., Ayres, T. M., Wong, G. H. W., & Goeddel, D. V. (1993). A novel domain within the 55 kd TNF receptor signals cell-death. *Cell*, *74*(5), 845–853.

Telliez, J. B., Xu, G. Y., Woronicz, J. D., Hsu, S., Wu, J. L., Lin, L., et al. (2000). Mutational analysis and NMR studies of the death domain of the tumor necrosis factor receptor-1. *Journal of Molecular Biology*, *300*(5), 1323–1333. http://dx.doi.org/10.1006/jmbi.2000.3899.

Thomas, L. R., Bender, L. M., Morgan, M. J., & Thorburn, A. (2006). Extensive regions of the FADD death domain are required for binding to the TRAIL receptor DR5. *Cell Death and Differentiation*, *13*(1), 160–162. http://dx.doi.org/10.1038/sj.cdd.4401714.

Thomas, L. R., Henson, A., Reed, J. C., Salsbury, F. R., & Thorburn, A. (2004). Direct binding of Fas-associated death domain (FADD) to the tumor necrosis factor-related apoptosis-inducing ligand receptor DR5 is regulated by the death effector domain of FADD. *Journal of Biological Chemistry*, *279*(31), 32780–32785. http://dx.doi.org/10.1074/jbc.M401680200.

Thome, M., Schneider, P., Hofmann, K., Fickenscher, H., Meinl, E., Neipel, F., et al. (1997). Viral FLICE-inhibitory proteins (FLIPs) prevent apoptosis induced by death receptors. *Nature*, *386*(6624), 517–521. http://dx.doi.org/10.1038/386517a0.

Tugarinov, V., Hwang, P. M., Ollerenshaw, J. E., & Kay, L. E. (2003). Cross-correlated relaxation enhanced 1H[bond]13C NMR spectroscopy of methyl groups in very high molecular weight proteins and protein complexes. *Journal of the American Chemical Society*, *125*(34), 10420–10428. http://dx.doi.org/10.1021/ja030153x.

Wagner, K. W., Punnoose, E. A., Januario, T., Lawrence, D. A., Pitti, R. M., Lancaster, K., et al. (2007). Death-receptor O-glycosylation controls tumor-cell sensitivity to the proapoptotic ligand Apo2L/TRAIL. *Nature Medicine*, *13*(9), 1070–1077.

Walczak, H., & Sprick, M. R. (2001). Biochemistry and function of the DISC. *Trends in Biochemical Sciences*, *26*(7), 452–453. http://dx.doi.org/10.1016/S0968-0004(01)01895-3.

Wang, L. W., Yang, J. K., Kabaleeswaran, V., Rice, A. J., Cruz, A. C., Park, A. Y., et al. (2010). The Fas-FADD death domain complex structure reveals the basis of DISC assembly and disease mutations. *Nature Structural & Molecular Biology*, *17*(11), 1324–1329.

Watanabefukunaga, R., Brannan, C. I., Copeland, N. G., Jenkins, N. A., & Nagata, S. (1992). Lymphoproliferation disorder in mice explained by defects in Fas antigen that mediates apoptosis. *Nature*, *356*(6367), 314–317. http://dx.doi.org/10.1038/356314a0.

Weber, C. H., & Vincenz, C. (2001a). The death domain superfamily: A tale of two interfaces? *Trends in Biochemical Sciences*, *26*(8), 475–481. http://dx.doi.org/10.1016/S0968-0004(01)01905-3.

Weber, C. H., & Vincenz, C. (2001b). A docking model of key components of the DISC complex: Death domain superfamily interactions redefined. *FEBS Letters*, *492*(3), 171–176. http://dx.doi.org/10.1016/S0014-5793(01)02162-7.

Wilson, N. S., Dixit, V., & Ashkenazi, A. (2009). Death receptor signal transducers: Nodes of coordination in immune signaling networks. *Nature Immunology*, *10*(4), 348–355. http://dx.doi.org/10.1038/ni.1714.

Xiao, T., Towb, P., Wasserman, S. A., & Sprang, S. R. (1999). Three-dimensional structure of a complex between the death domains of Pelle and Tube. *Cell*, *99*(5), 545–555.

Yamagishi, J., Kawashima, H., Matsuo, N., Ohue, M., Yamayoshi, M., Fukui, T., et al. (1990). Mutational analysis of structure–activity relationships in human tumor necrosis factor-alpha. *Protein Engineering*, *3*(8), 713–719.

Yanagisawa, J., Takahashi, M., Kanki, H., YanoYanagisawa, H., Tazunoki, T., Sawa, E., et al. (1997). The molecular interaction of Fas and FAP-1—A tripeptide blocker of human Fas interaction with FAP-1 promotes Fas-induced apoptosis. *Journal of Biological Chemistry*, *272*(13), 8539–8545.

Yang, J. K., Wang, L. W., Zheng, L. X., Wan, F. Y., Ahmed, M., Lenardo, M. J., et al. (2005). Crystal structure of MC159 reveals molecular mechanism of DISC assembly and FLIP inhibition. *Molecular Cell*, *20*(6), 939–949.

Yang, Z. R., West, A. P., & Bjorkman, P. J. (2009). Crystal structure of TNF alpha complexed with a poxvirus MHC-related TNF binding protein. *Nature Structural & Molecular Biology*, *16*(11), 1189–1191. http://dx.doi.org/10.1038/Nsmb.1683.

Zhan, C. Y., Patskovsky, Y., Yan, Q. R., Li, Z. H., Ramagopal, U., Cheng, H. Y., et al. (2011). Decoy strategies: The structure of TL1A:DcR3 complex. *Structure*, *19*(2), 162–171.

Zhou, P., Lugovskoy, A. A., & Wagner, G. (2001). A solubility-enhancement tag (SET) for NMR studies of poorly behaving proteins. *Journal of Biomolecular NMR*, *20*(1), 11–14. http://dx.doi.org/10.1023/A:1011258906244.

Zhou, P., & Wagner, G. (2010). Overcoming the solubility limit with solubility-enhancement tags: Successful applications in biomolecular NMR studies. *Journal of Biomolecular NMR*, *46*(1), 23–31. http://dx.doi.org/10.1007/s10858-009-9371-6.

Zornig, M., Hueber, A. O., & Evan, G. (1998). p53-dependent impairment of T-cell proliferation in FADD dominant-negative transgenic mice. *Current Biology*, *8*(8), 467–470.

CHAPTER TEN

Use of E2~Ubiquitin Conjugates for the Characterization of Ubiquitin Transfer by RING E3 Ligases Such as the Inhibitor of Apoptosis Proteins

Adam J. Middleton, Rhesa Budhidarmo, Catherine L. Day[1]

Department of Biochemistry, Otago School of Medical Sciences, University of Otago, Dunedin, New Zealand
[1]Corresponding author: e-mail address: catherine.day@otago.ac.nz

Contents

Abstract

Ubiquitylation of proteins is a versatile posttranslational modification that can serve to promote protein degradation, or it can have nondegradative roles, such as mediating protein–protein interactions. The Inhibitor of APoptosis (IAP) proteins are important regulators of pathways that control cell death, proliferation, and differentiation. A number of IAP family members are RING E3 ubiquitin–protein ligases, which promote direct transfer of ubiquitin from charged E2 enzymes, or E2 ~ ubiquitin (E2 ~ Ub) conjugates, to substrate proteins. This results in the attachment of nondegradative ubiquitin signals to other proteins, or the autoubiquitylation and degradation of IAPs. Modulating ubiquitin transfer by IAPs is the focus of a number of drug development initiatives and these studies require a detailed understanding of ubiquitylation. Here, we describe

Methods in Enzymology, Volume 545
ISSN 0076-6879
http://dx.doi.org/10.1016/B978-0-12-801430-1.00010-X

preparation of stable E2 ~ Ub conjugates that can be used in biochemical and biophysical experiments to examine RING domain function. In the last 2 years, the availability of these conjugates has helped unveil a molecular understanding of the process of ubiquitin transfer by IAPs. The approaches described here will be suitable for studying other RING E3 ligases.

1. INTRODUCTION

Ubiquitylation is a posttranslational modification that regulates many events in eukaryotic cells. Most famously, ubiquitylation is involved in directing proteins for degradation by the 26S proteasome (Varshavsky, 2012). Attachment of the 8.5 kDa protein ubiquitin to target proteins involves three classes of enzymes: (i) ubiquitin activating E1 enzymes that activate ubiquitin in an ATP-dependent process; (ii) ubiquitin conjugating (UBC) E2 enzymes, which are charged with ubiquitin by the E1; and (iii) the ubiquitin E3 ligases, which provide specificity for the final transfer of ubiquitin to a substrate Lys or N-terminal Met residue (Pickart, 2001). Of the >600 E3 ligases encoded in the human genome most can be classified into one of two families based on the presence of either, a homologous to the E6AP carboxyl terminus (HECT) domain, or a really interesting new gene (RING) domain. The HECT family of ligases is involved in catalysis as they receive ubiquitin from the E2 and transfer it to the substrate (Kee & Huibregtse, 2007). In contrast, the RING family bring the ubiquitin conjugated E2 enzyme (E2 ~ Ub) and the substrate together, allowing transfer of ubiquitin directly from the E2 to the substrate (Deshaies & Joazeiro, 2009). However, recently the RING-in-between-RING (RBR) E3 proteins have been shown to utilise a hybrid RING/HECT mechanism (Wenzel, Lissounov, Brzovic, & Klevit, 2011). These proteins depend on a RING domain for E2 ~ Ub recruitment, but ubiquitin is transferred to a cysteine in the E3 before attachment to the substrate.

The Inhibitor of APoptosis (IAP) proteins were first identified as gene products of baculovirus that inhibited apoptosis of infected insect cells (Clem, Fechheimer, & Miller, 1991; Crook, Clem, & Miller, 1993). Subsequently, the mammalian IAP family was identified (Rothe, Pan, Henzel, Ayres, & Goeddel, 1995; Uren, Pakusch, Hawkins, Puls, & Vaux, 1996). All members of the IAP family possess at least one baculovirus IAP repeat (BIR) domain, a protein–protein interaction module. Many IAPs also contain a RING domain that confers on them the ability to promote their own ubiquitylation (autoubiquitylation), as well as the ubiquitylation of substrate proteins (Mace, Shirley, & Day, 2010; Vaux & Silke, 2005). Ubiquitin transfer by IAPs is central to their ability to modulate signaling pathways. For example, cIAP1 promotes the addition of nondegradative ubiquitin signals on RIPK1, and these are required for assembly of the prosurvival signaling complex (Bertrand et al., 2008). In contrast, small molecule compounds that bind to the BIR domains promote cell death by enhancing cIAP autoubiquitylation and subsequent degradation (Varfolomeev et al., 2007; Vince et al., 2007). As well as modulating cell survival, important roles for IAPs in innate immune signaling have also been established (Vandenabeele & Bertrand, 2012), and the RING domain-mediated E3 ligase activity of XIAP is required for nucleotide-binding and oligomerization domain signaling (Damgaard et al., 2012, 2013).

The ligase domain in IAPs resembles other RING domains and, as expected, it has a conserved flat surface that interacts with E2 enzymes (Mace et al., 2008). Ubiquitin transfer by IAPs depends on both the integrity of the E2-binding site and RING dimerization (Feltham et al., 2011). Disruption of the dimer abrogates activity and although the isolated RING domain from all IAP proteins forms a stable dimer, the longer forms of some IAPs, such as cIAP1, are largely monomeric and ubiquitin transfer is impeded. The structure of the autoinhibited monomeric form of cIAP1 showed that the RING dimer interface is occluded due to interactions with the third BIR domain (Dueber et al., 2011). Remarkably, the interaction interface on the BIR domain includes the pocket to which a number of proteins and small molecule compounds bind, and this structure also explained why addition of small molecule BIR-binding compounds promotes RING dimerization and autoubiquitylation of cIAP1 (Dueber et al., 2011).

These studies established IAP proteins as dimeric RING E3 ligases, but did not account for the essential role of dimerization. In IAPs and related E3s, such as RNF4 and MDM2, dimerization not only depends on contacts from the core RING domain but also residues N- and C-terminal to the RING domain (Budhidarmo, Nakatani, & Day, 2012). A number of mutagenesis studies suggested that the C-terminal residues were required for ubiquitin transfer, and for MDM2 and RNF4 mutations were identified that disrupted ubiquitin transfer but not RING dimerization (Plechanovova et al., 2011; Uldrijan, Pannekoek, & Vousden, 2007). Together, these studies suggested that a conserved solvent-exposed C-terminal aromatic residue played an essential role in ubiquitylation. Until recently, the purpose of this aromatic side chain remained uncertain.

In the last few years, a molecular understanding of why RING dimerization and the C-terminal residues are critical for ubiquitin transfer by IAPs has become clearer. This appreciation of RING domain function has depended on the availability of stable E2∼Ub conjugates that are suitable for *in vitro* and biophysical studies. Here, we describe the preparation of several E2∼Ub conjugates, and experimental approaches that can be used to uncover RING domain function.

2. SYNTHESIS OF E2∼UB CONJUGATES

All E2 enzymes share a central UBC domain that includes the catalytic Cys that is charged with ubiquitin by the E1. Once charged, a thioester bond between the side chain of the Cys and the C-terminal carboxylic group of Gly76 in ubiquitin links the two proteins (Fig. 10.1). The thioester bond is relatively unstable making this conjugate difficult to purify in significant quantities. Therefore, to undertake many biochemical and biophysical experiments, it is necessary to prepare long-lived conjugates.

Here, we describe preparation of three stable conjugates that are linked by either oxyester, disulfide, or isopeptide bonds (Fig. 10.1). Each of these conjugates depends upon the prior purification of E2 and ubiquitin proteins that have been engineered to favor specific linkages. For the oxyester- and isopeptide-linked conjugates wild-type ubiquitin is used, and it is only necessary to mutate the active site Cys of the E2 to either Ser or Lys, respectively (Fig. 10.1).

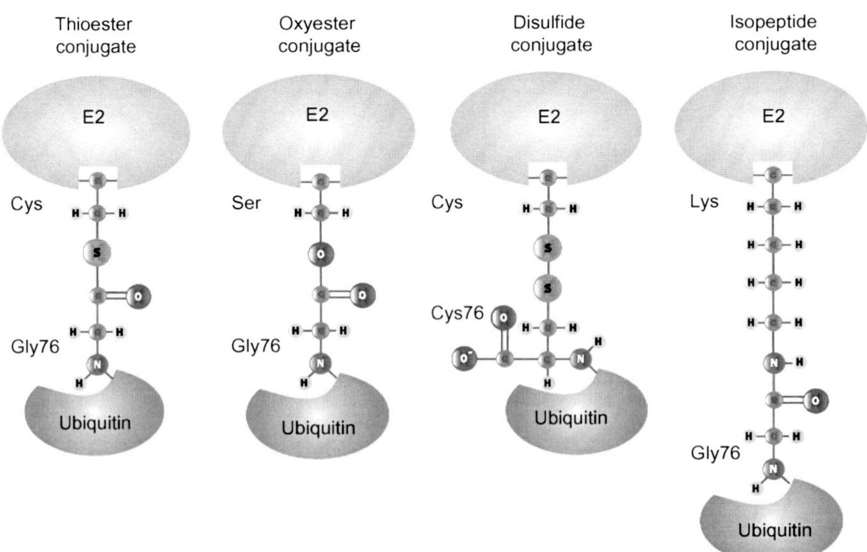

Figure 10.1 Schematic showing the different E2∼Ub conjugates discussed. (See the color plate.)

However, formation of the disulfide-linked conjugate requires mutation of the C-terminal Gly in ubiquitin to Cys. Some E2s, such as UBE2B, only possess one Cys at the catalytic site, and the wild-type protein can be used for conjugation. Whereas others, such as UbcH5b, have several additional Cys that must be mutated to Ser to avoid formation of cross-linked E2s, rather than the desired E2~ubiquitin disulfide. Formation of the disulfide-linked conjugate is not dependent on E1. However, active E1 is required for the preparation of both the oxyester- and isopeptide-linked conjugates.

In a number of E2s, the UBC domain contains an additional noncatalytic "backside" ubiquitin-binding site (Brzovic, Lissounov, Christensen, Hoyt, & Klevit, 2006; Sakata et al., 2010). Interaction of ubiquitin with this site can result in the noncovalent association of conjugate molecules, which may promote chain formation and complicate some analyses. Therefore, mutations are often introduced to disrupt this interaction, and in the case of the widely studied E2, UbcH5b, replacement of Ser 22 with Arg (S22R) is sufficient to disrupt ubiquitin binding to the backside site.

2.1. Purification of the E1

In our laboratory, we use either the *Saccharomyces cerevisiae* E1 enzyme, Uba1, or the human enzyme, Ube1. Both E1 proteins are expressed with an N-terminal His_6-tag and can be purified using immobilized metal ion affinity chromatography (IMAC) followed by size-exclusion chromatography (SEC) as described previously, except that removal of the His tag by thrombin is omitted (Lee & Schindelin, 2008). In our hands, the *S. cerevisiae* enzyme is purer, but the yield lower than that obtained for human Ube1. Because production of the conjugates requires a significant amount of E1, but the purity is less important, we routinely use the human E1 for large-scale conjugation reactions.

2.2. Purification of the E2

A number of approaches can be used to obtain purified E2, including purchasing it from a commercial supplier. However, we routinely express and purify E2s as glutathione-*S*-transferase (GST) fusion proteins (Lorick, Jensen, & Weissman, 2005). In brief, the fusion protein is purified using glutathione sepharose affinity chromatography and then after extensive washing, the resin-bound fusion protein is digested with a protease that cleaves between GST and the E2. The soluble E2 protein is then recovered and is often >90% pure at this stage, although we typically use SEC as a final purification step. Following purification, the E2 is concentrated to 1–2 mg/mL, supplemented with ~25% glycerol, flash frozen in liquid nitrogen and stored at −80 °C (Feltham et al., 2011).

It is also possible to produce the E2 with a C-terminal His_6-tag (Lorick et al., 2005). If this approach is taken the first step would involve IMAC, and then following elution from the resin and cleavage of the tag, or just elution of the protein, the final purification method would be the same as for the GST-tagged variant.

2.3. Purification of ubiquitin

Several different, yet similar, protocols for purifying ubiquitin have been reported (e.g., Raasi and Pickart, 2005). However, we routinely purify untagged human ubiquitin according to the method of Sato et al. (2008). Briefly, following expression in *Escherichia coli* BL21 (DE3) the cells are harvested, resuspended, and lysed in 50 m*M* ammonium acetate pH 4.5, 1 m*M* EDTA. Following centrifugation, the supernatant, which contains ubiquitin, is recovered and then incubated at 55–60 °C for 30 min, resulting in formation of white precipitant that can be separated by centrifugation. The resulting supernatant is then loaded onto a 5-mL HiTrap SP column equilibrated in the lysis buffer and ubiquitin is eluted using a 30-mL linear 0–600 m*M* NaCl gradient. Finally, ubiquitin is loaded onto a Superdex 75 16/60 HiLoad column that has been equilibrated in 20 m*M* Tris–HCl pH 7.5, 50 m*M* NaCl. Purified ubiquitin is concentrated to 4 mg/mL, snap-frozen and stored at −80 °C.

2.4. Formation of disulfide-linked E2~ubiquitin conjugate

Following purification of ubiquitin that has a C-terminal Cys (G76C), and an E2 in which all the noncatalytic Cys residues have been replaced by Ser, the disulfide conjugate can be formed according to the method reported by the Shaw laboratory (Merkley, 2005). This conjugate is relatively easy to form. However, reducing agents (e.g., DTT, TCEP) must not be used and nonreducing SDS PAGE must be used to analyze the conjugate.

Experimental procedure:

1. The disulfide-linked conjugation is relatively inefficient and typically ~40% of the starting material is recovered. For the oxidation reaction, a 2 × molar excess of ubiquitin over E2 is required, and we typically use 0.1 m*M* UbcH5b and 0.2 m*M* G76C ubiquitin.
2. Fully reduce purified G76C ubiquitin and E2 with 2 m*M* TCEP or 2 m*M* DTT then mix together in a reaction volume of ~5 mL.

3. Using a dialysis membrane with a molecular weight cut off of <5000 Da dialyze the sample against 100 m*M* Na phosphate pH 7.5, 100 m*M* NaCl, 10 μ*M* $CuCl_2$.
4. Take a 5 μL sample after 0, 1, 3, 8, 16 h, or until the conjugation is complete (often 1–2 days, but can take 4 days) (Fig. 10.2A). Be sure to change the dialysis buffer three to four times.
5. Equilibrate a HighLoad 26/60 Superdex S75 column (GE) with 50 m*M* Na phosphate pH 7.2, 100 m*M* NaCl. Recover the conjugate from the dialysis tubing and centrifuge for 10 min to remove precipitate prior to loading on the column. The size-exclusion column should separate the

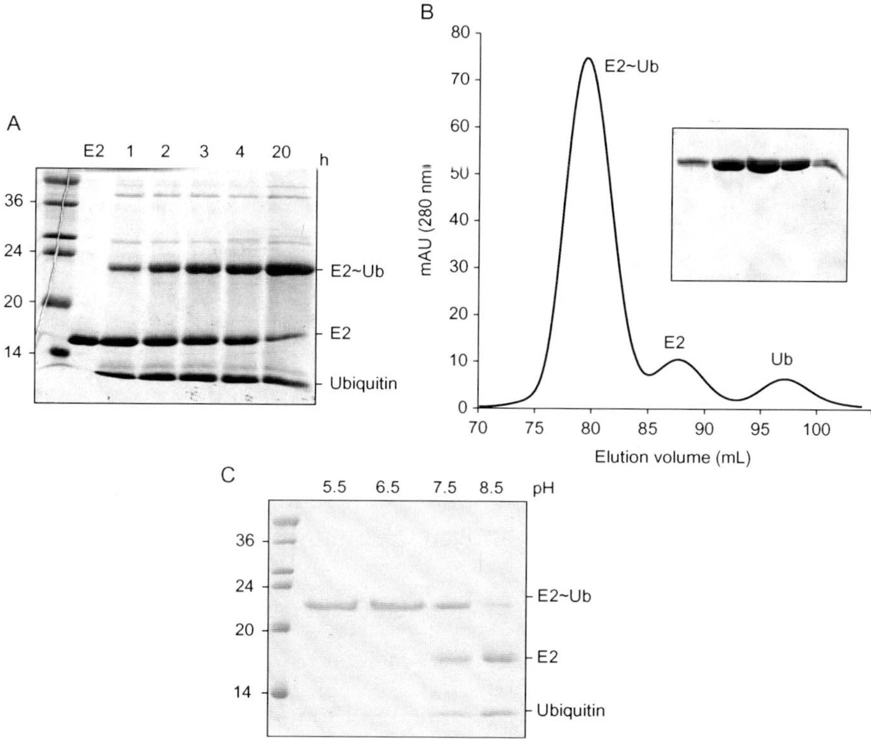

Figure 10.2 Purification and stability of E2 ~ Ub conjugates. (A) A time course showing charging of UbcH5b with ubiquitin. After 20 h, the charging was complete and the reaction was terminated by addition of MES buffer at pH 6.5 prior to separation by SDS PAGE. (B) Size-exclusion chromatography elution profile showing separation of UbcH5b ~ Ub, UbcH5b, and ubiquitin following completion of a conjugation reaction as in (A). Inset is a gel of the conjugate peak, which shows that the protein is pure and contains no contaminating E2 or ubiquitin. (C) The oxyester conjugate is more stable at lower pH. Samples were incubated at the indicated pHs.

conjugate, E2 and ubiquitin (Fig. 10.2B). Pool pure samples, concentrate and snap freeze before storage at −80 °C.

2.5. Formation of oxyester- and isopeptide-linked conjugate

Like the native thioester-linked conjugate, both the oxyester and isopeptide-linked conjugates depend upon activation of ubiquitin by the E1, and significant quantities of active E1 are required. The oxyester conjugate closely mimics the native thioester-linked conjugate (Fig. 10.1), and addition of highly active RING family E3 ligases can promote hydrolysis of the oxyester bond, allowing the relative E3 ligase activity of RING proteins to be assessed (Section 3.2). When a highly stable conjugate is required, such as for crystallization with an E3, the lability of the oxyester conjugate can be a disadvantage. However, mutation of a conserved Asn residue, which promotes catalysis by stabilizing an active site loop (Berndsen, Wiener, Yu, Ringel, & Wolberger, 2013), increases the stability of the oxyester conjugate. For example, in UbcH5b mutation of Asn 77 to Ala (N77A) results in formation of a relatively stable conjugate that has been used for structural studies with ML-IAP, also known as BIRC7 (see Section 3.3) (Dou, Buetow, Sibbet, Cameron, & Huang, 2012).

The isopeptide-linked conjugate differs in that although its formation is E1 dependent, the ubiquitin is linked by an isopeptide bond between a Lys residue at the active site of the E2 and the C-terminal carboxyl of ubiquitin. As a consequence, this conjugate can be more difficult to form because the E1 does not normally catalyze formation of an isopeptide bond. However, once the E2 is charged it is stable and well suited for binding and structural studies (Plechanovova, Jaffray, Tatham, Naismith, & Hay, 2012).

Experimental procedure:

1. This conjugation reaction requires significant amounts of ubiquitin, E2 and E1; these should be purified first.
2. Although the oxyester conjugate can be prepared in assay buffer (3 m*M* ATP, 5 m*M* $MgCl_2$, 50 m*M* Tris pH 7.5, 150 m*M* NaCl), we found that both conjugates were more efficiently formed when creatine and phosphocreatine, which can regenerate ATP from ADP, were included (Raasi & Pickart, 2005; Sakata et al., 2010). Recipe for 5× "cycling buffer": 250 m*M* Tris–HCl pH 7.5, 25 m*M* $MgCl_2$, 50 m*M* phosphocreatine (Sigma), and 3 U/mL creatine phosphokinase (Sigma).
3. Formation of the oxyester conjugate is typically successful at neutral pH; however, the isopeptide conjugate requires a higher pH and the cycling buffer should be adjusted to pH 9.0.

4. To increase the yield of purified conjugate, use a 1.5–2 × molar excess of ubiquitin to E2 as this will ensure complete charging, which improves purification of the conjugate on the S75 size-exclusion column.
5. To prepare a 10 mL conjugation reaction, mix 2 mL of 5 × cycling buffer, 1 mL of E1 at ~10 μ*M* (final concentration 1 μ*M*), ATP for a final concentration of 1 m*M*, 0.2 m*M* E2, and 0.4 m*M* ubiquitin. Make up the volume to 10 mL and incubate at 37 °C for 16–24 h. Take a 5 μL aliquot after 0, 2, 4, 8, 16 h until the reaction is complete (Fig. 10.2A). The oxyester conjugate can discharge (return to E2 and ubiquitin), but this is reduced at lower pH (Fig. 10.2C), and when charging is complete the pH of the reaction should be lowered by addition of MES pH 6.5.
6. Purify conjugate by SEC as outlined for the disulfide-linked conjugate except use a buffer at pH 6.5 for the oxyester conjugate.

3. CHARACTERIZATION OF RING-E2~UB COMPLEXES

The structures of several conjugates have been determined. These show that, in the absence of any binding partners, ubiquitin can occupy distinct positions relative to the E2; and depending on the orientation of ubiquitin, the conjugate is referred to as adopting either an "open" or a "closed" conformation (Fig. 10.3; Wenzel, Stoll, & Klevit, 2010). The first structural model of an isolated E2~Ub conjugate was obtained of a thioester-linked conjugate (Hamilton et al., 2001). For these heroic studies, the NMR data were acquired using a sample that contained E1 and ATP, thereby allowing the unstable E2~ubiquitin thioester to be maintained, while the NMR data were collected. This showed that yeast Ubc1 (the yeast homologue of UBE2K) could adopt a closed conformation, whereby the Ile 44 (I44) centered face of ubiquitin interacts with the E2 adjacent to the active site Cys (Fig. 10.3).

In contrast, crystal structures of the oxyester-linked Ubc13 and UbcH5b conjugates (Eddins, Carlile, Gomez, Pickart, & Wolberger, 2006; Sakata et al., 2010), and the NMR structure of the disulfide-linked UbcH8 conjugate (Serniwka & Shaw, 2009), revealed distinct open conformations (Fig. 10.3). In part, differences in the position of ubiquitin may be due to crystal packing. In the UbcH5b structure, the ubiquitin moiety of one conjugate is bound to the backside surface of another E2 molecule (Sakata et al., 2010). This interaction precludes formation of the closed conformation because ubiquitin contacts the backside of UbcH5b with the same surface that contacts E2 in the closed Ubc1~Ub structure. It is also possible that the favored orientation of ubiquitin differs for each E2. Comparison of

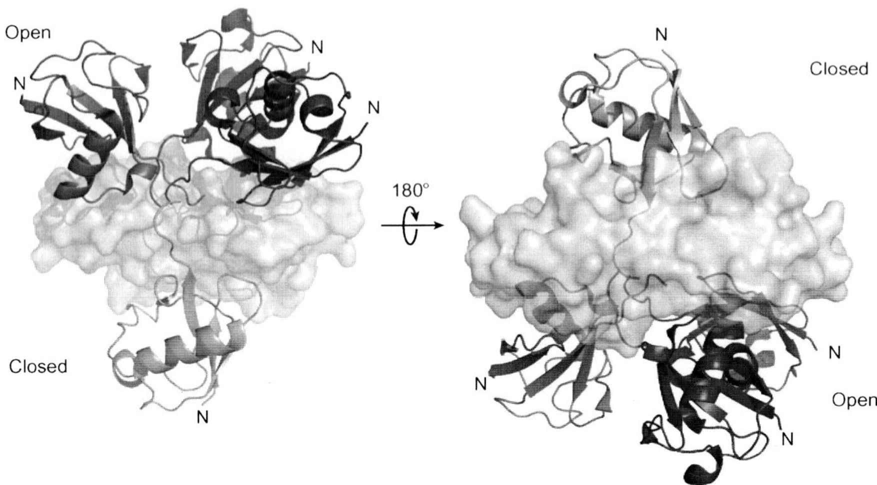

Figure 10.3 E2 ~ Ub conjugate can adopt open and closed conformations. The figure was prepared by overlaying the E2 structures from the UbcH5b, Ubc1, UbcH8, and Ubc13 conjugate structures. The surface of UbcH5b is shown in pale blue. The ubiquitin moieties for each conjugate are shown as ribbons (dark blue is from PDB ID: 2KJH, red: 3A33, gray: 2JMI, and pink: 1FXT). The left panel is oriented with the active site at front, and the right panel is a 180° rotation. (See the color plate.)

the dynamic properties of Ubc13 and UbcH5b conjugates supports this hypothesis, as Ubc13 adopts the closed conformation more frequently than UbcH5c (Pruneda, Stoll, Bolton, Brzovic, & Klevit, 2011).

With various forms of the E2 ~ ubiquitin conjugates in hand, as well as an initial understanding of the dynamic properties of these molecules, it is possible to characterize ubiquitin transfer by RING E3 ligases. Current studies focus on understanding how RING domains promote ubiquitin transfer, and formation of polyubiquitin chains. Here, we describe techniques routinely used in our laboratory to study RING domains from IAP proteins, although the approaches described could be readily applied to other E3s.

3.1. Binding studies

Given the variable nature of the intracellular molar ratio between uncharged E2 and E2 ~ Ub conjugates (Wiener et al., 2013), a complete understanding of RING-mediated ubiquitin transfer requires characterization of both RING-E2 interactions, as well as RING-E2 ~ Ub interactions. Most approaches that are widely used to study protein–protein interactions can be used to analyze RING complexes. For example, we routinely use qualitative methods, such as GST pulldown assays and analytical SEC, as well as more quantitative techniques, including surface plasmon resonance (SPR) and isothermal titration calorimetry (ITC).

3.1.1 Pulldown assays

In pulldown assays, the RING domain of interest is usually expressed with an N-terminal GST-tag, and purified from the cell lysate using glutathione sepharose affinity chromatography. To evaluate binding ~50–100 μg of glutathione resin-immobilized GST-fused RING domain, as well as a GST only control, are incubated with 20–50 μg of purified E2 ~ Ub conjugate for 1 h at 4 °C in a 200 μL reaction. We routinely use PBS (10 m*M* Na_2HPO_4, 1.8 m*M* KH_2PO_4, 140 m*M* NaCl, 2.7 m*M* KCl, pH 7.4) supplemented with 0.2% Tween-20 and 1–2 m*M* DTT as the binding buffer. However, the choice of buffers has to be optimized for each RING–E2 ~ Ub pair and when disulfide-linked E2 ~ Ub conjugate is used, reducing agents, such as DTT and TCEP, must be excluded from all buffers (Nakatani et al., 2013). During binding, the samples are placed on a rotator at 4 °C to ensure that the resin is dispersed and to maximize interactions with the E2 ~ Ub conjugate. Following extensive washing to remove unbound protein, the samples are analyzed using SDS PAGE and Coomassie blue staining (Fig. 10.4A). Often conjugate binding can be visualized directly, however, if the interaction is weak, immunoblot analysis may be preferable. A number of monoclonal antibodies for different E2 enzymes as well as ubiquitin are available, and can be used to probe for the conjugate. Anti-His_6 antibody can also be used for visualization if the E2 enzyme or ubiquitin is His_6-tagged (Feltham et al., 2011).

Pulldown assays can also be performed using proteins fused to other tags, and these may be preferable in some settings because the dimeric nature of GST can interfere or alter the binding properties of proteins (Sims, Haririnia, Dickinson, Fushman, & Cohen, 2009). For instance, we utilized a maltose binding protein (MBP) fusion to study the effect of disrupting RING dimerization on the interaction between the RING domain from cIAP2 and UbcH5b ~ Ub (Feltham et al., 2011). In this case, the fusion protein was immobilized to amylose resin (New England Biolabs) and because MBP does not bind amylose resin tightly, mild washing steps were required. Others have also employed MBP-fused proteins to study the interactions between RING domains and E2 ~ Ub conjugates (Plechanovova et al., 2011). Other tags, such as His_6-tag, have also been used in pulldown assays (Page, Pruneda, Amick, Klevit, & Misra, 2012). However, in our hands, the purity of Ni^{2+} resin-bound His_6-tagged proteins is often inferior to the

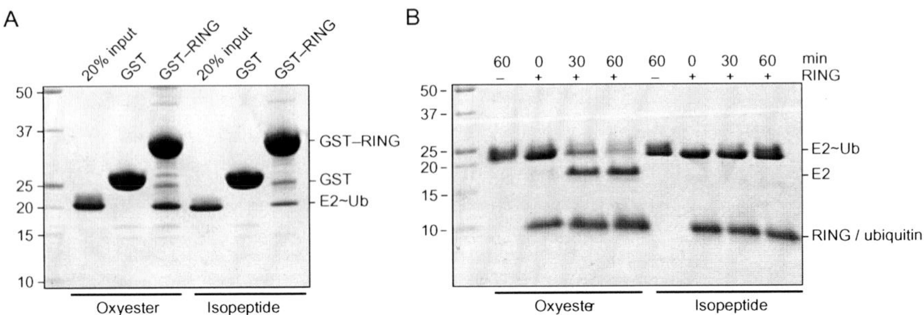

Figure 10.4 Characterization of RING-E2 ~ Ub conjugate interactions. (A) GST-RING (cIAP2) fusion protein (lanes 3 and 6) and GST alone (lanes 2 and 5) were subjected to a pulldown reaction in the presence of either the oxyester- (lanes 1–3) or the isopeptide- (lanes 4–6)-linked conjugate. Lanes 1 and 4 show 20% of the input conjugate. (B) Discharge assay in the presence of 10 μ*M* RING (cIAP1) performed in 50 m*M* Tris–HCl pH 7.2, 50 m*M* NaCl. Lanes 1 and 5 are no-RING controls of the oxyester- and isopeptide-linked conjugates, respectively. The oxyester conjugate is almost fully discharged (lanes 1–4), while the isopeptide shows no discharging (lanes 5–8). Note the RING of cIAP1 and ubiquitin migrates at the same position.

GST-fused counterparts, and nonspecific binding with Ni^{2+} resin is increased compared to glutathione resin.

3.1.2 Analytical SEC

Binding events observed in pulldown assays often require validation using untagged proteins. Analytical SEC is a convenient method to monitor RING:E2 ~ Ub conjugate complex formation because the larger molecular mass of the complex means it elutes before the individual proteins. We have used this technique to monitor the binding between UbcH5b and the RING domain from XIAP (Nakatani et al., 2013). Similarly, others have utilized analytical SEC to show that the first zinc finger of TRAF6 strengthens the binding between its RING domain and Ubc13 (Yin et al., 2009). The affinity between a RING domain and its cognate E2 enzyme is typically in the μ*M* range (i.e., 10–100 μ*M*) (Deshaies & Joazeiro, 2009). Therefore, to observe complex formation, high concentrations of purified E2 and RING are required and we typically load 200–500 μL of 50–300 μ*M* proteins onto a 24-mL Superdex 75 10/300 column (GE Healthcare) (Nakatani et al., 2013).

A number of studies have shown that many RING domains preferentially bind the E2 ~ Ub conjugate (Plechanovova et al., 2011; Spratt, Wu, Kovacev, Pan, & Shaw, 2012). In principle, all E2 ~ Ub conjugates can be used in analytical SEC analyses although if the RING requires reducing agents for stability, then the oxyester may be a better option than the disulfide-linked conjugate. However, oxyester conjugates hydrolyze in the presence of some RING domain and the rate of discharge must be predetermined, and mutation of the conserved Asn may be required. The isopeptide conjugate is stable in the presence of a RING domain (Fig. 10.4B), and can be employed for analytical SEC studies. However, pulldown assays suggest that for some RING domains the affinity for the isopeptide-linked conjugate may be reduced (Fig. 10.4A). This may be because the closed conformation is destabilized by the increased length of the Lys side chain compared to Cys/Ser (Fig. 10.1).

3.1.3 SPR and ITC

To compare binding affinities, quantitative analyses such as SPR and ITC are required. Each of these methods allows different binding parameters to be determined: ITC calculates the overall dissociation constant (K_D), stoichiometry and thermodynamic properties of an interaction at equilibrium, whereas SPR can provide kinetic information (k_{on} and k_{off}) as well as K_D measurements (Hartmann-Petersen & Gordon, 2005; Velazquez-Campoy, Ohtaka, Nezami, Muzammil, & Freire, 2004). We determined that the K_D for the interaction between UbcH5b and the RING domains from cIAP2 and XIAP to be in the range of 20–40 μM using ITC (Mace et al., 2008; Nakatani et al., 2013). In the case of XIAP, a similar affinity was obtained by SPR, highlighting the utility of both approaches.

For ITC, we typically dialyze 200–500 μM of E2 and 20–50 μM of purified RING domain into the same PBS stock. The E2 enzyme is usually loaded into the syringe of a VP-ITC instrument (Microcal), while the RING domain is placed in the sample cell. The temperature is kept constant at 25 °C and the reference power is set to 15 μCal/s. Typically, the instrument is programed to perform up to 30 10-μL injections of the E2 enzyme solution into the sample cell. The data are analyzed using Origin 7 software and fit to a single site-binding model (Mace et al., 2008; Nakatani et al., 2013). Similar experimental conditions have been used to characterize the interactions between E2 enzymes and other RING/U-box domains (Benirschke et al., 2010; Zhang et al., 2005). This method works well with uncharged E2 enzymes. However, the E2~Ub conjugate tends to precipitate and we have found SPR better suited to the analysis of RING-E2~Ub interactions (Nakatani et al., 2013).

For SPR measurements, we frequently utilize a Biacore X100 instrument (GE Healthcare) and a CM5 chip (GE Healthcare), which has been previously coated with anti-GST antibody according to the manufacturer's protocol. In this case, the purified GST-fused E2 enzyme is eluted from the resin by incubation with 10 m*M* reduced glutathione dissolved in 50 m*M* Tris–HCl pH 8.0, and then together with untagged RING, dialyzed into SPR running buffer (PBS supplemented with 0.1% Tween-20). Up to 1000 response unit of GST, or the GST-E2, are captured on the chip by the anti-GST antibody. Different concentrations of RING domain are then injected into the flow cells to yield a binding curve that can be analyzed using either the Biacore X100 evaluation software (GE Healthcare) or Prism (GraphPad) (Nakatani et al., 2013). The interaction between a RING domain and its E2 enzyme (or E2 ~ Ub conjugate) often exhibits fast k_{on} and k_{off}, precluding kinetic analysis and the K_D value is calculated from equilibrium binding analyses (Dou, Buetow, Sibbet, Cameron, & Huang, 2013; Nakatani et al., 2013).

Similar strategies have been used to characterize other RING-E2 or RING-E2 ~ Ub conjugate interactions. For example, to probe the interaction between Ubc13 and TRAF6, amine coupling chemistry was used to directly immobilize the E3 ligase (Yin et al., 2009). Immobilized GST-fused RING domains have also been used to study the interactions with UbcH5b and E2 ~ Ub conjugate (Dou et al., 2012). Importantly, this study provides evidence that the RING domains from cIAP2 and ML-IAP preferentially bind the UbcH5b-C85S ~ Ub conjugate relative to uncharged UbcH5b.

3.2. Discharge assays

Discharge assays are a simple experiment that can be used to measure the ability of an E3 ligase to promote hydrolysis of E2 ~ Ub oxyester conjugate. A number of studies have shown that this activity correlates well with the ability of a RING domain to promote ubiquitin chain formation. It is therefore assumed that this assay measures the ability of the RING domain to stabilize the conformation of the conjugate that is susceptible to nucleophilic attack (Feltham et al., 2011; Plechanovova et al., 2011, 2012; Pruneda et al., 2012). Discharge assays are frequently performed in 50 m*M* NaCl and 50 m*M* MES at pH 6.5, or a higher pH (e.g., 7.5); however, it is essential to include a control with no E3 ligase and incubate at 37 °C for the full experiment in order to account for any discharge activity in the absence of a RING. Typically, we use final concentrations of 20 μ*M* conjugate and 5–10 μ*M* RING E3 ligase. Choosing the time points depends on the E3, but at low pH, 3 and 6 h are often suitable. At higher pH, shorter times (30 and 60 min) are often appropriate.

After completion of the assay, samples are analyzed by SDS PAGE (Fig. 10.4B). The activity of the ligase can be determined by monitoring the reduction in intensity of the conjugate band, as well as the increased intensity of the E2 and ubiquitin bands over time. This assay can also be used to compare the activity of different ligases or mutants. It is also possible to use the E2∼Ub thioester conjugate in this assay, however the labile bond results in rapid discharge making prior purification of the conjugate difficult. Other variations of discharge assays have also been reported, for example inclusion of small molecules that harbor primary amine groups, such as L-lysine, hydroxylamine, or Tris, accelerate discharge (Dou et al., 2012; Plechanovova et al., 2011; Saha, Lewis, Kleiger, Kuhlman, & Deshaies, 2011).

3.3. Structural studies of RING-E2∼Ub complexes

Recent advances in our understanding of RING domain function have been revealed by analyzing the isolated RINGs together with purified E2∼Ub conjugate. Most recently, three structures of RING domains bound to a stable UbcH5b∼Ub conjugate have been reported (Dou et al., 2012, 2013; Plechanovova et al., 2012). Two of these studies report the complexes formed between the conjugate and the related dimeric RING domains from ML-IAP and RNF4 (Fig. 10.5A). The other study reports the structure of the conjugate bound to the monomeric RING domain from Cbl-b (Dou et al., 2013). In all three structures, the conjugate adopts the same closed conformation even though the isopeptide-linked and the oxyester-linked conjugate, that also included mutation of the conserved Asn to Ala, were used. This shows that for UbcH5b introduction of a Lys at the active site only results in small changes at the active site (Fig. 10.5B). The stability of the isopeptide conjugate makes it the preferred conjugate for crystallization, although the extra length of the Lys side chain may destabilize the closed conformation and reduce its affinity for the RING-E3 (Fig. 10.4A). Therefore, the choice of conjugate for structural studies will be dependent on the E2 and RING under investigation.

Once a RING domain and E2∼Ub of interest have been identified and purified, crystallization of the complex is undertaken in much the same way that one would approach crystallization of other protein complexes (Radaev, Li, & Sun, 2006). For example we typically set up drops (e.g., vapor diffusion) with an equimolar ratio of RING:conjugate. The RING domain and the conjugate should be mixed just before setting up the crystal screen. Use a concentration of 15–20 mg/mL conjugate and 10–15 mg/mL RING domain. If the components bind tightly it is also possible to first

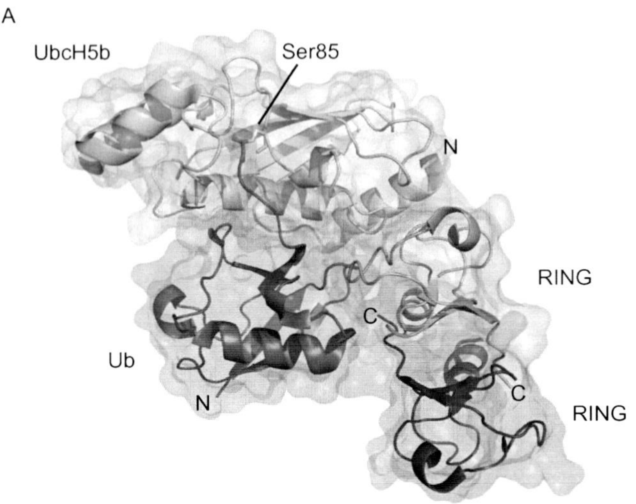

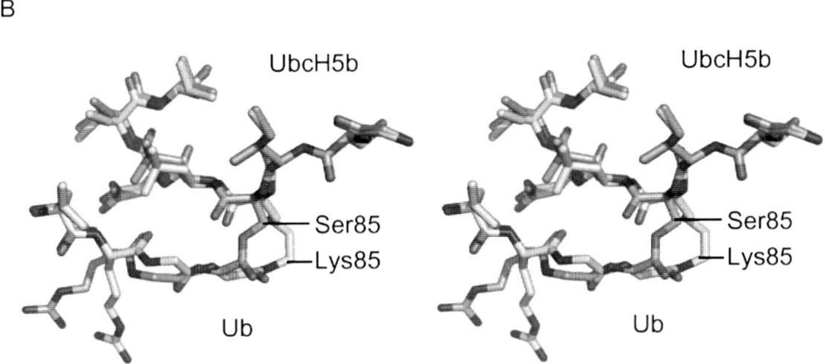

Figure 10.5 Structure of RING-E2 ~ Ub conjugate complexes. (A) Structure of a dimeric RING (BIRC7) in complex with the oxyester-linked UbcH5b ~ Ub conjugate (PDB ID: 4AUQ). E2 is in light blue, ubiquitin in dark blue, and the RING dimer is shown in light and dark gray. (B) Overlay showing the oxyester and isopeptide linkages of closed-conformation conjugates in stereo view. E2 molecules were overlaid from PDB IDs 4AP4 and 4AUQ. (See the color plate.)

purify the complex by SEC prior to setting up the drops. Despite these efforts, E2 ~ Ub crystals may grow on their own, and we have observed clusters of flat, irregular shaped UbcH5b ~ Ub crystals in a number of conditions. These crystals tend to become more prevalent if the drops are left for extended periods (1 + month).

If it is not possible to obtain crystals, then solution studies, such as NMR and cross-linking studies, can also be used to characterize RING–conjugate complexes. Pruneda et al. analyzed interactions between two E3 ligases and the UbcH5c ~ Ub conjugate using NMR (Pruneda et al., 2012). Chemical shift perturbations were used to map interacting surfaces between the E3, UbcH5b, and ubiquitin, while relaxation experiments showed that the closed conformation is more frequently populated in the presence of the RING domain (Pruneda et al., 2012; Soss, Klevit, & Chazin, 2013). In another study, chemical cross-linking together with mass spectrometry was used to identify contacts between the XIAP RING and ubiquitin (Nakatani et al., 2013). Cross-linking experiments were undertaken because overlaying the XIAP RING domain and the UBE2K ~ Ub conjugate onto the structure of cIAP2 bound to UbcH5b (Hamilton et al., 2001; Mace et al., 2008) revealed Lys residues in close proximity. This study correctly identified the contact site even though it was undertaken prior to the availability of the crystal structures. Similar approaches may prove useful for the analysis of other RING-E2 ~ Ub conjugate complexes that resist crystallization efforts.

4. CONCLUSION

In the last year, significant advances in our understanding of ubiquitin transfer by IAPs and related E3 ligases have been made. The three crystal structures of the RING-bound E2 ~ Ub show that the conjugate is bound by the RING domain in the closed conformation, where the I44-centered face of ubiquitin makes extensive contacts with the E2 (Dou et al., 2012, 2013; Plechanovova et al., 2012). This conformation is stabilized by RING–ubiquitin interactions; either an aromatic residue on the C terminus of the dimerized RING, or a region N-terminal to the ligase domain. These interactions place ubiquitin in a conformation that is thought to prime the thioester bond for nucleophilic attack by a substrate Lys/N-terminal Met. This model accounts for much of the prior biochemical data, and explains why RING dimerization and an aromatic residue at the dimer interface are required for ubiquitin transfer of IAPs (Dou et al., 2012; Nakatani et al., 2013).

Despite advances in our understanding of the overall topology of the RING-E2 ~ Ub complex (Dou et al., 2012, 2013; Plechanovova et al., 2012), and the importance of reducing conformational freedom for catalysis (Berndsen et al., 2013), a detailed mechanistic understanding of ubiquitin transfer from the E2 to a substrate Lys remains uncertain. In addition, although many features of the mode of binding and mechanism of transfer are likely to be conserved in most E2s, the RING-bound structure of a conjugate is only available for highly similar UbcH5 family members. Do other less similar E2 ~ Ub conjugates adopt an analogous configuration when they are bound by their cognate RING domains? Likewise, only a handful of RING proteins have been characterized, and it seems likely that residues outside the core RING domain will be important for stabilization of the closed conformation in some cases (Dou et al., 2013). A molecular understanding of the complexes formed by other E2 ~ Ub conjugates and RING E3s is eagerly awaited.

Over the last 5 years, it has become clear that proper regulation of apoptosis is dependent on appropriate ubiquitylation of various cellular components (Vucic, Dixit, & Wertz, 2011). Here, we have focused on IAP proteins that contain a RING domain, but other E3 ligases such as MDM2, which is the primary E3 that regulates the abundance of the tumor suppressor protein p53, also have critical roles in regulating apoptosis. The RING domains from MDM2 and IAPs form similar dimers (Kostic, Matt, Martinez-Yamout, Dyson, & Wright, 2006; Linke et al., 2008; Mace et al., 2008), and it is anticipated that their mechanism of action will be comparable. However, other structurally distinct E3s, such as linear ubiquitin chain assembling complex (Stieglitz et al., 2013) also play important roles in regulating apoptosis and a detailed understanding of their interactions with the E2 ~ Ub conjugate will be of considerable interest.

Recent studies have also uncovered additional roles for E2 enzymes and E2 ~ Ub conjugates in modulating the activity of deubiquitinating enzymes (DUBs), such as OTUB1 (Juang et al., 2012; Wiener et al., 2013; Wiener, Zhang, Wang, & Wolberger, 2012). Interestingly, OTUB1 has also been shown to associate and modulate the stability of cIAP1 (Goncharov et al., 2013). Together, these studies highlight the central role of E2 ~ Ub conjugates in regulating cell death and it will be of considerable interest to understand the factors that influence the balance between E3-E2 ~ Ub and DUB-E2 ~ Ub interactions. We anticipate that biochemically defined assays that require preparations of pure reagents, such as those described here, will be essential for this understanding.

ACKNOWLEDGMENTS

The authors thank the members of the Day laboratory for their helpful discussions, and Bodhi Bettjeman and Bronwyn Carlisle for help with preparation of the figures. Our research is supported by grants from the Marsden Fund (NZ), the Health Council of New Zealand, and the University of Otago.

REFERENCES

Benirschke, R. C., Thompson, J. R., Nominé, Y., Wasielewski, E., Juranić, N., Macura, S., et al. (2010). Molecular basis for the association of human E4B U box ubiquitin ligase with E2-conjugating enzymes UbcH5c and Ubc4. *Structure*, *18*, 955–965.

Berndsen, C. E., Wiener, R., Yu, I. W., Ringel, A. E., & Wolberger, C. (2013). A conserved asparagine has a structural role in ubiquitin-conjugating enzymes. *Nature Chemical Biology, 9*, 154–156.

Bertrand, M. J. M., Milutinovic, S., Dickson, K. M., Ho, W. C., Boudreault, A., Durkin, J., et al. (2008). cIAP1 and cIAP2 facilitate cancer cell survival by functioning as E3 ligases that promote RIP1 ubiquitination. *Molecular Cell, 30*, 689–700.

Brzovic, P. S., Lissounov, A., Christensen, D. E., Hoyt, D. W., & Klevit, R. E. (2006). A UbcH5/ubiquitin noncovalent complex is required for processive BRCA1-directed ubiquitination. *Molecular Cell, 21*, 873–880.

Budhidarmo, R., Nakatani, Y., & Day, C. L. (2012). RINGs hold the key to ubiquitin transfer. *Trends in Biochemical Sciences, 37*, 58–65.

Clem, R. J., Fechheimer, M., & Miller, L. K. (1991). Prevention of apoptosis by a baculovirus gene during infection of insect cells. *Science, 254*, 1388–1390.

Crook, N., Clem, R., & Miller, L. (1993). An apoptosis-inhibiting baculovirus gene with a zinc finger-like motif. *Journal of Virology, 67*, 2168–2174.

Damgaard, R. B., Fiil, B. K., Speckmann, C., Yabal, M., Stadt, U. Z., Bekker-Jensen, S., et al. (2013). Disease-causing mutations in the XIAP BIR2 domain impair NOD2-dependent immune signalling. *EMBO Molecular Medicine, 5*, 1278–1295.

Damgaard, R. B., Nachbur, U., Yabal, M., Wong, W. W.-L., Fiil, B. K., Kastirr, M., et al. (2012). The ubiquitin ligase XIAP recruits LUBAC for NOD2 signaling in inflammation and innate immunity. *Molecular Cell, 46*, 746–758.

Deshaies, R. J., & Joazeiro, C. A. P. (2009). RING domain E3 ubiquitin ligases. *Annual Review of Biochemistry, 78*, 399–434.

Dou, H., Buetow, L., Sibbet, G. J., Cameron, K., & Huang, D. T. (2012). BIRC7–E2 ubiquitin conjugate structure reveals the mechanism of ubiquitin transfer by a RING dimer. *Nature Structural and Molecular Biology, 19*, 876–883.

Dou, H., Buetow, L., Sibbet, G. J., Cameron, K., & Huang, D. T. (2013). Essentiality of a non-RING element in priming donor ubiquitin for catalysis by a monomeric E3. *Nature Structural and Molecular Biology, 20*, 982–986.

Dueber, E. C., Schoeffler, A. J., Lingel, A., Elliott, J. M., Fedorova, A. V., Giannetti, A. M., et al. (2011). Antagonists induce a conformational change in cIAP1 that promotes autoubiquitination. *Science, 334*, 376–380.

Eddins, M. J., Carlile, C. M., Gomez, K. M., Pickart, C. M., & Wolberger, C. (2006). Mms2-Ubc13 covalently bound to ubiquitin reveals the structural basis of linkage-specific polyubiquitin chain formation. *Nature Structural and Molecular Biology, 13*, 915–920.

Feltham, R., Bettjeman, B., Budhidarmo, R., Mace, P. D., Shirley, S., Condon, S. M., et al. (2011). Smac mimetics activate the E3 ligase activity of cIAP1 protein by promoting RING domain dimerization. *Journal of Biological Chemistry, 286*, 17015–17028.

Goncharov, T., Niessen, K., de Almagro, M. C., Izrael-Tomasevic, A., Fedorova, A. V., Varfolomeev, E., et al. (2013). OTUB1 modulates c-IAP1 stability to regulate signalling pathways. *EMBO Journal, 32*, 1103–1114.

Hamilton, K., Ellison, M., Barber, K., Williams, R., Huzil, J., McKenna, S., et al. (2001). Structure of a conjugating enzyme-ubiquitin thiolester intermediate reveals a novel role for the ubiquitin tail. *Structure, 9*, 897–904.

Hartmann-Petersen, R., & Gordon, C. (2005). Quantifying protein-protein interactions in the ubiquitin pathway by surface plasmon resonance. *Methods in Enzymology, 399*, 164–177.

Juang, Y.-C., Landry, M.-C., Sanches, M., Vittal, V., Leung, C. C. Y., Ceccarelli, D. F., et al. (2012). OTUB1 co-opts Lys48-linked ubiquitin recognition to suppress E2 enzyme function. *Molecular Cell, 45*, 384–397.

Kee, Y., & Huibregtse, J. M. (2007). Regulation of catalytic activities of HECT ubiquitin ligases. *Biochemical and Biophysical Research Communications, 354*, 329–333.

Kostic, M., Matt, T., Martinez-Yamout, M., Dyson, H., & Wright, P. (2006). Solution structure of the Hdm2 C2H2C4 RING, a domain critical for ubiquitination of p53. *Journal of Molecular Biology, 363*, 433–450.

Lee, I., & Schindelin, H. (2008). Structural insights into E1-catalyzed ubiquitin activation and transfer to conjugating enzymes. *Cell, 134*, 268–278.

Linke, K., Mace, P. D., Smith, C. A., Vaux, D. L., Silke, J., & Day, C. L. (2008). Structure of the MDM2/MDMX RING domain heterodimer reveals dimerization is required for their ubiquitylation in trans. *Cell Death and Differentiation, 15*, 841–848.

Lorick, K. L., Jensen, J. P., & Weissman, A. M. (2005). Expression, purification, and properties of the Ubc4/5 family of E2 enzymes. *Methods in Enzymology, 398*, 54–68.

Mace, P. D., Linke, K., Feltham, R., Schumacher, F.-R., Smith, C. A., Vaux, D. L., et al. (2008). Structures of the cIAP2 RING domain reveal conformational changes associated with ubiquitin-conjugating enzyme (E2) recruitment. *Journal of Biological Chemistry, 283*, 31633–31640.

Mace, P. D., Shirley, S., & Day, C. L. (2010). Assembling the building blocks: Structure and function of inhibitor of apoptosis proteins. *Cell Death and Differentiation, 17*, 46–53.

Merkley, N. (2005). Ubiquitin manipulation by an E2 conjugating enzyme using a novel covalent intermediate. *Journal of Biological Chemistry, 280*, 31732–31738.

Nakatani, Y., Kleffmann, T., Linke, K., Condon, S. M., Hinds, M. G., & Day, C. L. (2013). Regulation of ubiquitin transfer by XIAP, a dimeric RING E3 ligase. *Biochemical Journal, 450*, 629–638.

Page, R. C., Pruneda, J. N., Amick, J., Klevit, R. E., & Misra, S. (2012). Structural insights into the conformation and oligomerization of E2 ~ ubiquitin conjugates. *Biochemistry, 51*, 4175–4187.

Pickart, C. M. (2001). Mechanisms underlying ubiquitination. *Annual Review of Biochemistry, 70*, 503–533.

Plechanovova, A., Jaffray, E. G., McMahon, S. A., Johnson, K. A., Navrátilová, I., Naismith, J. H., et al. (2011). Mechanism of ubiquitylation by dimeric RING ligase RNF4. *Nature Structural and Molecular Biology, 18*, 1052–1059.

Plechanovova, A., Jaffray, E. G., Tatham, M. H., Naismith, J. H., & Hay, R. T. (2012). Structure of a RING E3 ligase and ubiquitin-loaded E2 primed for catalysis. *Nature, 489*, 115–120.

Pruneda, J. N., Littlefield, P. J., Soss, S. E., Nordquist, K. A., Chazin, W. J., Brzovic, P. S., et al. (2012). Structure of an E3:E2 ~ Ub complex reveals an allosteric mechanism shared among RING/U-box ligases. *Molecular Cell, 47*, 933–942.

Pruneda, J. N., Stoll, K. E., Bolton, L. J., Brzovic, P. S., & Klevit, R. E. (2011). Ubiquitin in motion: Structural studies of the ubiquitin-conjugating enzyme ~ ubiquitin conjugate. *Biochemistry, 50*, 1624–1633.

Raasi, S., & Pickart, C. M. (2005). Ubiquitin chain synthesis. *Methods in Molecular Biology, 301*, 47–55.

Radaev, S., Li, S., & Sun, P. D. (2006). A survey of protein-protein complex crystallizations. *Acta Crystallographica. Section D, Biological Crystallography, 62*, 605–612.

Rothe, M., Pan, M. G., Henzel, W. J., Ayres, T. M., & Goeddel, D. V. (1995). The TNFR2-TRAF signaling complex contains two novel proteins related to baculoviral inhibitor of apoptosis proteins. *Cell, 83*, 1243–1252.

Saha, A., Lewis, S., Kleiger, G., Kuhlman, B., & Deshaies, R. J. (2011). Essential role for ubiquitin-ubiquitin-conjugating enzyme interaction in ubiquitin discharge from Cdc34 to substrate. *Molecular Cell, 42*, 75–83.

Sakata, E., Satoh, T., Yamamoto, S., Yamaguchi, Y., Yagi-Utsumi, M., Kurimoto, E., et al. (2010). Crystal structure of UbcH5b ~ ubiquitin intermediate: Insight into the formation of the self-assembled E2 ~ Ub conjugates. *Structure, 18*, 138–147.

Sato, Y., Yoshikawa, A., Yamagata, A., Mimura, H., Yamashita, M., Ookata, K., et al. (2008). Structural basis for specific cleavage of Lys 63-linked polyubiquitin chains. *Nature, 455*, 358–362.

Serniwka, S. A., & Shaw, G. S. (2009). The structure of the UbcH8–ubiquitin complex shows a unique ubiquitin interaction site. *Biochemistry*, *48*, 12169–12179.

Sims, J. J., Haririnia, A., Dickinson, B. C., Fushman, D., & Cohen, R. E. (2009). Avid interactions underlie the Lys63-linked polyubiquitin binding specificities observed for UBA domains. *Nature Structural and Molecular Biology*, *16*, 883–889.

Soss, S. E., Klevit, R. E., & Chazin, W. J. (2013). Activation of the UbcH5c ~ Ub is the result of a shift in interdomain motions of the conjugate bound to U-Box E3 ligase E4B. *Biochemistry*, *52*, 2991–2999.

Spratt, D. E., Wu, K., Kovacev, J., Pan, Z. Q., & Shaw, G. S. (2012). Selective recruitment of an E2 ubiquitin complex by an E3 ubiquitin ligase. *Journal of Biological Chemistry*, *287*, 17374–17385.

Stieglitz, B., Rana, R. R., Koliopoulos, M. G., Morris-Davies, A. C., Schaeffer, V., Christodoulou, E., et al. (2013). Structural basis for ligase-specific conjugation of linear ubiquitin chains by HOIP. *Nature*, *503*, 422–426.

Uldrijan, S., Pannekoek, W. J., & Vousden, K. H. (2007). An essential function of the extreme C-terminus of MDM2 can be provided by MDMX. *EMBO Journal*, *26*, 102–112.

Uren, A. G., Pakusch, M., Hawkins, C. J., Puls, K. L., & Vaux, D. L. (1996). Cloning and expression of apoptosis inhibitory protein homologs that function to inhibit apoptosis and/or bind tumor necrosis factor receptor-associated factors. *Proceedings of the National Academy of Sciences of the United States of America*, *93*, 4974–4978.

Vandenabeele, P., & Bertrand, M. J. M. (2012). The role of the IAP E3 ubiquitin ligases in regulating pattern-recognition receptor signalling. *Nature Reviews Immunology*, *12*, 833–834.

Varfolomeev, E., Blankenship, J. W., Wayson, S. M., Fedorova, A. V., Kayagaki, N., Garg, P., et al. (2007). IAP antagonists induce autoubiquitination of c-IAPs, NF-kappaB activation, and TNFα-dependent apoptosis. *Cell*, *131*, 669–681.

Varshavsky, A. (2012). The ubiquitin system, an immense realm. *Annual Review of Biochemistry*, *81*, 167–176.

Vaux, D. L., & Silke, J. (2005). IAPs, RINGs and ubiquitylation. *Nature Reviews. Molecular Cell Biology*, *6*, 287–297.

Velazquez-Campoy, A., Ohtaka, H., Nezami, A., Muzammil, S., & Freire, E. (2004). Isothermal titration calorimetry. *Current Protocols in Cell Biology*, *23*, 35–54.

Vince, J. E., Wong, W. W.-L., Khan, N., Feltham, R., Chau, D., Ahmed, A. U., et al. (2007). IAP antagonists target cIAP1 to induce TNFα-dependent apoptosis. *Cell*, *131*, 682–693.

Vucic, D., Dixit, V. M., & Wertz, I. E. (2011). Ubiquitylation in apoptosis: A post-translational modification at the edge of life and death. *Nature Reviews. Molecular Cell Biology*, *12*, 439–452.

Wenzel, D. M., Stoll, K. E., & Klevit, R. E. (2010). E2s: Structurally economical and functionally replete. *Biochemical Journal*, *433*, 31–42.

Wenzel, D. M., Lissounov, A., Brzovic, P. S., & Klevit, R. E. (2011). UBCH7 reactivity profile reveals parkin and HHARI to be RING/HECT hybrids. *Nature*, *474*, 105–108.

Wiener, R., DiBello, A. T., Lombardi, P. M., Guzzo, C. M., Zhang, X., Matunis, M. J., et al. (2013). E2 ubiquitin-conjugating enzymes regulate the deubiquitinating activity of OTUB1. *Nature Structural and Molecular Biology*, *20*, 1033–1039.

Wiener, R., Zhang, X., Wang, T., & Wolberger, C. (2012). The mechanism of OTUB1-mediated inhibition of ubiquitination. *Nature*, *483*, 618–622.

Yin, Q., Lin, S.-C., Lamothe, B., Lu, M., Lo, Y.-C., Hura, G., et al. (2009). E2 interaction and dimerization in the crystal structure of TRAF6. *Nature Structural and Molecular Biology*, *16*, 658–666.

Zhang, M., Windheim, M., Roe, S., Peggie, M., Cohen, P., Prodromou, C., et al. (2005). Chaperoned ubiquitylation—Crystal structures of the CHIP U box E3 ubiquitin ligase and a CHIP-Ubc13-Uev1a complex. *Molecular Cell*, *20*, 525–538.

CHAPTER ELEVEN

Multidimensional Profiling in the Investigation of Small-Molecule-Induced Cell Death

Adam J. Wolpaw[*], **Brent R. Stockwell**[†,‡,§,1]

[*]Residency Program in Pediatrics, The Children's Hospital of Philadelphia, Philadelphia, Pennsylvania, USA
[†]Department of Biological Sciences, Columbia University, New York, USA
[‡]Department of Chemistry, Columbia University, New York, USA
[§]Howard Hughes Medical Institute, Columbia University, New York, USA
[1]Corresponding author: e-mail address: bstockwell@columbia.edu

Contents

Methods in Enzymology, Volume 545
ISSN 0076-6879
http://dx.doi.org/10.1016/B978-0-12-801430-1.00011-1

Abstract

Numerous morphological variations of cell death have been described. These processes depend on a complex and overlapping cellular signaling network, making molecular definition of the pathways challenging. This review describes one solution to this problem for small-molecule-induced death, the creation of high-dimensionality profiles for compounds that can be used to define and compare pathways. Such profiles have been assembled from gene expression measurements, protein quantification, chemical–genetic interactions, chemical combination interactions, cancer cell line sensitivity profiling, quantitative imaging, and modulatory profiling. We discuss the advantages and limitations of these techniques in the study of cell death.

1. INTRODUCTION

While descriptions of active cell death processes can be traced as far back as the nineteenth century (Virchow & Chance, 1860), the modern era of cell death research was firmly established by the description and coining of apoptosis in 1972 (Kerr, Wyllie, & Currie, 1972). The authors described the consistent nuclear, cytoplasmic, and organellar changes in cells dying in a variety of physiological and pathological settings. Their description was entirely morphological for obvious reasons: the molecular tools to further characterize the phenomena were not available.

Such tools began to be developed in the late 1980s and early 1990s. For example, Robert Horvitz and colleagues uncovered the genetic basis of apoptosis in *Caenorhabiditis elegans* and showed that these pathways were largely conserved in mammalian cells (Ellis & Horvitz, 1986; Hengartner, Ellis, & Horvitz, 1992; Hengartner & Horvitz, 1994; Miura, Zhu, Rotello, Hartwieg, & Yuan, 1993; Yuan & Horvitz, 1990; Yuan, Shaham, Ledoux, Ellis, & Horvitz, 1993). These and subsequent studies allowed a transition from purely morphological descriptions of cell death processes to biochemical descriptions. However, this transition has been incomplete, and morphological descriptors remain prominent, if not predominant, in the study of cell death. There is a growing push to move away from morphological characterizations, however, given their dependence on subjective criteria and the recognition that morphology is not always a marker of unique underlying biochemistry. A panel of cell death experts recently published formal recommendations to transition to fully biochemical descriptions of cell death and provided recommended biochemical descriptors of a number of cell death processes (Galluzzi et al., 2012).

In the past decade, however, not only have the identified morphological varieties of cell death expanded significantly (Fig. 11.1), but the biochemical pathways underlying these processes have been shown to be complex and interconnected. Calling a form of cell death "caspase dependent," for example, does little to clarify if the signaling was conducted through the intrinsic, extrinsic, or granzyme-mediated pathway (Taylor, Cullen, & Martin, 2008), or even if the resultant morphology is consistent with apoptosis or with pyroptosis, an inflammatory form of cell death dependent on the activity of caspase 1 (Fernandes-Alnemri et al., 2007). Necroptosis, a well-accepted form of regulated necrosis that involves signaling through the RIP family proteins, can be activated by binding of the same death receptor ligands that can initiate extrinsic pathway apoptosis (Degterev et al., 2008, 2005). Other forms of caspase-independent death can be initiated via mitochondrial outer membrane permeabilization (MOMP), the stimulus that typically initiates intrinsic pathway apoptosis (Colell et al., 2007).

How can we fully characterize, and distinguish between, complex, interconnected processes that can be difficult to distinguish either morphologically or biochemically? One solution is to vastly increase the dimensionality of

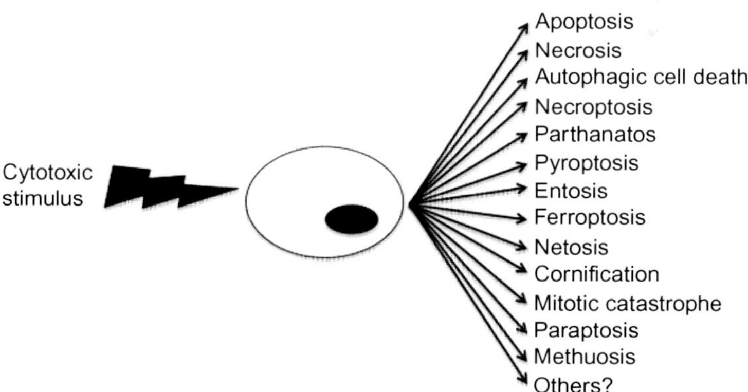

Figure 11.1 Diversity of cell death pathways. The number of characterized cell death pathways has expanded significantly in the past decade. In a given context, the death pathway is determined by the cytotoxic stimulus, the cell type, the microenvironment, and the presence of cotreatments, among others. The total number of death pathways accessible to cells remains unknown and an important question for investigation.

the measurements taken. These high-dimensionality profiles can then be compared to each other in order to relate and distinguish between lethal processes.

Such systems have been developed and primarily implemented in the study of the bioactivity of small molecules. Small molecules are versatile tools for studying a range of biological processes (Stockwell, 2004) and are particularly useful in the study of cell death. They can easily be applied to different cellular contexts and a variety of organisms and potentially translated into *in vivo* studies. Concentrations can be varied to investigate the thresholds for processes and intermediate effects of inhibiting protein function. Compounds can be applied and removed with precise temporal control, allowing for the investigation of the kinetics of events. Small molecules can inhibit single functions of multifunctional enzymes, allowing for more detailed investigation of processes. The utility of small molecules in cell death is demonstrated by the widespread use of small-molecule-induced cell death as a model for studying apoptosis. More recently, small-molecule screens have identified compounds that are essential for defining alternative cell death processes (Degterev et al., 2005; Dixon et al., 2012).

This review summarizes a number of the systems that have been developed to create high-dimensionality profiles for small molecules (see Fig. 11.2 and Table 11.1) and focuses on their utilization or potential utilization in the study of cell death.

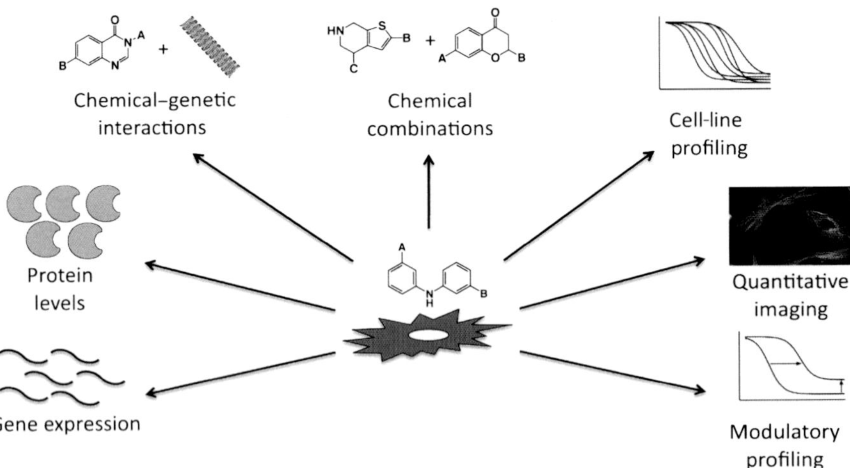

Figure 11.2 Small-molecule profiling technologies. Overview of the different methodologies that have been used to create profiles for small molecules. Profiles are based on quantitative measurements of the effect of a small molecule on cells. These profiles can be used to define a small-molecule-induced process and to compare to other processes.

Table 11.1 Small-molecule profiling modalities

Measurement type	Throughput	Functional measurement	Application to mammalian cells	Application to diverse cell type	Single-cell resolution	Accessibility to scientific community	Reproducibility	References
Gene expression	High	No	Yes	Yes	No	Very high	Low	Hughes et al. (2000), Lamb et al. (2006)
Proteomic	Intermediate	No	Yes	Yes	Sometimes	Low	Low	Dix, Simon, and Cravatt (2008), Mahrus et al. (2008), Muroi et al. (2010), Sevecka and MacBeath (2006)
Gene–small-molecule interaction	High	Yes	Limited	No	No	Intermediate	Intermediate	Parsons et al. (2004, 2006)
Small-molecule combinations	Low	Yes	Yes	Yes	No	Low	High	Farha and Brown (2010), Lehar et al. (2007), Yeh, Tschumi, and Kishony (2006)

Continued

Table 11.1 Small-molecule profiling modalities—cont'd

Measurement type	Throughput	Functional measurement	Application to mammalian cells	Application to diverse cell type	Single-cell resolution	Accessibility to scientific community	Reproducibility	References
Cell line profiling	Low	Yes	Yes	No	No	High	Low	Barretina et al. (2012), Basu et al. (2013), Garnett et al. (2012), Paull et al. (1989), Weinstein et al. (1997)
Quantitative imaging	Low	No	Yes	Yes	Yes	Low	Low	Perlman et al. (2004), Young et al. (2008)
Modulatory profiling	Intermediate	Yes	Yes	Yes	No	Low	High	Wolpaw et al. (2011)

2. GENE EXPRESSION PROFILING

Gene expression profiling is a powerful tool to explore cellular states, development, and disease. Investigation of small-molecule mechanisms of action was among the first applications of gene expression profiling (Schena et al., 1996; Stockwell, Hardwick, Tong, & Schreiber, 1999). Given the informational richness of gene expression profiles and their widespread availability and relative affordability, this method has developed into the most widely utilized system for profiling and comparing small-molecule bioactivities.

2.1. Comparing small-molecule profiles

The initial study demonstrating the utility of comparing small-molecule-induced gene expression profiles was performed in yeast by Steve Friend and colleagues (Hughes et al., 2000). This landmark study was largely focused on gene expression changes induced by genetic deletion, but it also measured genome-wide profiles for 13 well-characterized small molecules. They clustered the compounds and the gene knockouts based on their expression profiles and found that knockouts of genes with similar cellular function clustered together and that the small molecules clustered with knockouts of their characterized targets. Additionally, they found that the changes induced by the small-molecule dyclonine clustered with the knockout of *erg2*. The target of dyclonine was unknown at this time, and the authors provided evidence that dyclonine in fact inhibited Erg2p.

Inspired by the success of this approach, Todd Golub and colleagues developed the "Connectivity Map," a compendium of gene expression profiles generated after treatment with 164 different small molecules, largely in two different human cell lines (Lamb et al., 2006). They applied a nonparametric, rank-based pattern matching approach to analyze the data. A unitless "query signature" was generated based on genes up- and downregulated in a biological process of interest. This signature was compared to the database of gene expression profiles; the compounds in the database were ranked based on how well they correlated to the query signature (see Fig. 11.3).

The authors first demonstrated that their system had platform independence by showing that signatures for compounds derived from gene expression experiments on other platforms had high "connectivity scores" to the compounds in the database with the same mechanism of action. They then demonstrated the utility in investigating the mechanism of action of an uncharacterized compound. They queried the database with a signature based on the gene expression

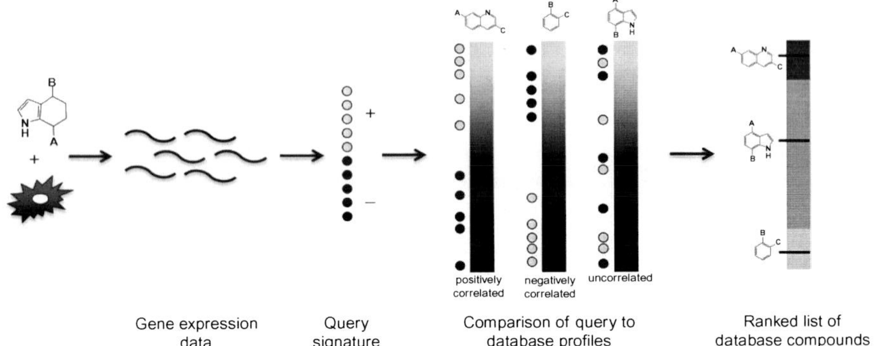

Figure 11.3 Connectivity Map schematic. An outline of the process of querying the Connectivity Map is shown. Gene expression information is generated from an independent source, shown here as small-molecule treatment of cells. Those data are used to generate a unitless query signature which is then compared to the database of gene expression profiles. Small molecules in the database are ranked based on how well their profiles are correlated to the query signature. (See the color plate.)

changes induced by treatment with gedunin, a small-molecule natural product without a characterized mechanism of action, and found that it was highly correlated to HSP90 pathway inhibitors. They subsequently demonstrated that gedunin did in fact inhibit this pathway (Hieronymus et al., 2006). Lastly, the authors demonstrated that signatures generated based on disease states, such as obesity and Alzheimer's disease, could be used to identify compounds capable of inducing or reversing these states.

2.2. Protocol for the use of the Connectivity Map database

All of the data generated by the Connectivity Map project were made publicly available on their Web site (www.broadinstitute.org/cmap) with software tools allowing for the uploading of a user-generated query profile and interrogation of the database. Since the initial publication, the database has been updated to include profiles generated from treatment with 1309 compounds and a much larger expansion is planned, likely to be released prior to publication of this article. A stepwise protocol for querying the database is described below. Further instruction and guidance are available at the project Web site.

I. *Generate a query signature*

A "query signature" is a unitless list of up- and downregulated genes representing a biological process of interest, recommended to involve anywhere between 10 and 500 genes depending on the knowledge of the process. There are a number of ways to generate a signature. It can be derived from independent gene expression profiling experiments, it can be manually generated based on prior biological knowledge or experiments, or it can be generated from the profiles contained within the Connectivity Map database itself. The Web site software allows for the use of up to three profiles for generation of a signature; however, the raw or partially processed data files can be downloaded and used independently to generate signatures.

II. *Upload signature*

In order to upload a custom-generated signature, up- and down-regulated genes must be converted to the corresponding probeset name used in the Affymetrix HG-U133A array. This can be done using tools available at http://www.affymetrix.com/analysis/netaffx/index.affx. The list of probesets is then converted to a .grp file (this can be done using Microsoft Excel).

III. *Query the Connectivity Map database*

Separate .grp files for up- and downregulated probesets are uploaded and selected for use in the query.

IV. *Analysis of the results*

The results of a query can be viewed as a ranked list of either the separate instances (specific compound, cell line, concentration, time point) or as a ranked list of all instances of a specific compound, all instances of a compound/cell line combination, or all instances of compounds classified under the same ATC code. The "connectivity score" is calculated based on the enrichment of the upregulated query genes among the most overexpressed genes in the database instances (or groups of instances) and the enrichment of the downregulated query genes among the most underexpressed genes in the database instances (or groups of instances). The instances or groups of instances are ranked from those with the highest "connectivity score" (correlated) to those with the lowest score (anticorrelated) (see Fig. 11.3).

V. *Independent validation*

The authors make clear that the Connectivity Map is best used as a hypothesis generator. Finding based on a query must be independently validated in separate assays.

2.3. Applications in cell death

In the 7 years since the initial publication of the Connectivity Map database, it has been widely utilized, cited by nearly 800 scientific articles. These include a number of interesting new insights into cell death, a selection of which are summarized below.

One of the initial studies that utilized the Connectivity Map found that modulation of the antiapoptotic protein MCL1 was important for restoring glucocorticoid-induced apoptosis in acute lymphoblastic leukemia (ALL) cells (Wei et al., 2006). The authors first generated a query signature of genes whose expression distinguished between glucocorticoid-sensitive and glucocorticoid-resistant ALL samples. They used this signature to query the Connectivity Map database and found that multiple instances of the compound rapamycin were among the most highly ranked instances, suggesting that rapamycin may induce a glucocorticoid-sensitive state. They then showed in multiple cell lines derived from lymphoid malignancies that rapamycin sensitized to glucocorticoid-induced apoptosis, but not to other cell-death-inducing agents. They went on to show that rapamycin down-regulated MCL1, that overexpression of MCL1 conferred glucocorticoid resistance, and that MCL1 suppression conferred sensitivity. This study was thus able to use an insight gleaned from the Connectivity Map to demonstrate the specific dependence of glucocorticoid-induced apoptosis on MCL1 and not on other BCL2-family member proteins. A subset of the authors of this study used the Connectivity Map further in a subsequent study investigating MCL1 (Wei et al., 2012). They used the Connectivity Map to identify compounds that induce profiles similar to triptolide, a compound that they had shown represses MCL1 expression, but not the expression of proapoptotic proteins. They found a number of related compounds that also repress MCL1 and that all of these compounds were acting as transcriptional inhibitors. MCL1 repression was essential for these compounds' activity, thus suggesting a biochemical basis for the mechanism of cell death induced by such compounds.

One of the essential challenges in cancer therapeutic development is the identification of compounds that can induce cancer-cell-selective cell death. Hassane and colleagues used the Connectivity Map to identify compounds that selectively kill acute myeloid leukemia (AML) cells, and particularly AML stem cells (Hassane et al., 2008). The authors identified a gene expression signature from the treatment of cells with parthenolide, which had previously been shown to selectively induce death in AML cells (Guzman et al., 2005). They used this signature as a query to search multiple gene expression databases, including the Connectivity Map, and identified other compounds with similar profiles. They subsequently showed that these compounds were also able to kill AML cells, including AML stem cells, and that this death was dependent on inhibition of NF-κB and on the generation of oxidative stress. In a different study, Stumpel and colleagues used the Connectivity Map to find compounds to target the particularly aggressive form of ALL harboring MLL rearrangements (Stumpel et al., 2012). They derived a query signature of genes selectively overexpressed in MLL-rearranged cell lines and found that a number of HDAC inhibitors produced profiles anticorrelated with the query. They went on to show that these compounds were selectively lethal in MLL-rearranged cells compared to MLL wild-type cells and that the compounds decreased expression of many of the signature genes and increased methylation at their promoters.

Endoplasmic reticulum (ER) stress plays a role in activating cell death (Rasheva & Domingos, 2009). Two recent studies have used the Connectivity Map to further investigate the role of ER stress in small-molecule-induced cell death. The first study investigated the difference in the mechanisms of action of two structurally related procaspase-activating compounds and found that at high concentrations one compound (PAC-1) activated ER-stress-mediated apoptosis, while the other (SPAC-1) did not (West et al., 2012). They created a signature query out of the 50 most highly up- and downregulated genes after treatment with a high concentration of PAC-1 and found that thapsigargin, a known inducer of ER stress, was the compound with the highest connectivity score. They then showed that high concentration of PAC-1 caused similar ultrastructural changes as thapsigargin and caused a similar increase in cytosolic calcium and a decrease in ER calcium, elucidating a side effect of this compound when used at high concentrations. A different study identified a novel, reversible ER stress response that is regulated in part by MCL1. This process was initially observed in response to treatment with apogossypol, a putative BCL-2 inhibitor. The authors created a query signature based on expression changes induced by apogossypol and used the Connectivity Map to identify 20 diverse compounds capable of initiating the same process, demonstrating its widespread occurrence.

2.4. Advantages and limitations in the study of cell death

Advantages of the use of gene expression profiles include their widespread availability and accessibility and the high dimensionality and information richness of the data. Gene expression profiling is already a widely available technology. With the recent rapid decrease in cost and increase in speed of sequencing technology, RNA-seq has replaced microarrays as the preferred method for gene expression measurement (McGettigan, 2013). This transition is likely to continue to decrease costs and increase the quality of gene expression measurements. The ubiquity of these measurements has driven the development of a host of readily available tools for the processing and analysis of gene expression data. The Connectivity Map project made an explicit goal to allow for platform independence, validated this aspect in their initial publication, made all of their data publicly available, and developed Web-based tools for querying their data. Multiple successful independent studies, including some of those described above, have further confirmed the utility of applying independently generated expression data. This accessibility is particularly valuable to the broader scientific community and unique among the technologies described within this review. The high dimensionality of a genome-wide expression profile allows the opportunity to accurately characterize such a highly complex process such as cell death. In addition, the content of the individual profiles are information-rich and can be mined to identify specific genes pathways involved in a death process.

Notable disadvantages of the use of expression data in studying cell death include difficulty detecting time-dependent, cell type-dependent, and concentration-dependent phenomena; the high barrier to reproducing a similar system focused on cell death; and the difficulty distinguishing on-target from off-target transcriptional effects of small molecules. As noted by the authors of the Connectivity Map study, for reasons of feasibility they were forced to limit the time points of treatment, the number of different cell types, and the concentrations used. The profiles are therefore snapshots of a cellular state that may or may not successfully represent prior and future states. While the relatively early time that they chose (6 h) is appropriate for many cell death processes, others can occur more rapidly (Newman, Crown, Leppla, & Moayeri, 2010) or much more slowly (Turmaine et al., 2000). In addition, rapid enzymatic cascades, such as activation of preformed zymogens, may occur too quickly to be accurately represented by transcriptional changes. While the authors suggest that insights can be gained across cell types and species, cell death processes can be active only in specific cellular or genetic contexts and therefore not likely accessible with the use of only two cancer cell lines. Given the cost of producing the Connectivity Map database, it is not feasible to reproduce the system in a relevant cellular context to study a phenomena poorly accessed in the chosen cell lines. The Connectivity Map largely incorporated a single dose of compounds. This limitation may mask effects only activated at higher or lower concentrations. It also can exacerbate the problem of distinguishing off-target effects. Even reportedly specific small molecules can have pleiotropic actions on cells (Campillos, Kuhn, Gavin, Jensen, & Bork, 2008; Keiser et al., 2009).

It is not obvious or trivial to distinguish the transcriptional effects of a compound that are related to the process of interest from those that are related to an off-target effect. Thus, despite the proven value of Connectivity Map, additional tools are needed to probe cell death mechanisms with small molecules, using measurements with high dimensionality.

3. PROTEIN QUANTIFICATION

One potential improvement over gene expression measurements involves the direct detection of changes in protein abundance and protein modifications. While mRNA levels are often used as a surrogate for protein level, changes in mRNA can correlate poorly to changes in protein level (Haider & Pal, 2013). A number of methods have been developed for the widespread measurement of protein levels and modifications to those proteins, basally and in response to small-molecule treatments. Changes in protein levels and posttranslational modifications can be monitored with two-dimensional difference gel electrophoresis (DIGE; Cecconi et al., 2007; Unlu, Morgan, & Minden, 1997), sandwiched antibody microarrays (Schweitzer et al., 2002), antibody microarrays analysis of dual color-labeled proteomes (Haab, Dunham, & Brown, 2001; MacBeath, 2002), lysate microarrays (Nishizuka et al., 2003), fluorescence-based flow cytometry (Krutzik & Nolan, 2006), and more recently multiplexed mass cytometry (Bodenmiller et al., 2012). The activity and small-molecule binding to individual enzyme classes can be monitored in some cases with activity-based protein profiling (Cravatt, Wright, & Kozarich, 2008; Leung, Hardouin, Boger, & Cravatt, 2003). Cell-wide proteolytic events can be tracked using labeling and mass spectrometry-based techniques (Dix et al., 2008; Mahrus et al., 2008). All of these methods assess the effects of small molecule at the protein level, instead of the mRNA level, which is likely more relevant to the final phenotypic effects of small molecules.

3.1. Comparing small-molecule profiles

A limited number of studies have compared compounds based on signatures created from protein profiles. Using lysate microarrays, Sevecka and MacBeath compared 84 kinase and phosphatase inhibitors based on their ability to change the phosphorylation state of 12 proteins in the EGF receptor pathway (Sevecka & MacBeath, 2006). Osada and colleagues created profiles for 19 compounds based on changes in protein levels detected by DIGE after compound treatment of HeLa cells (Muroi et al., 2010).

Clustering of these profiles accurately grouped compounds based on their known mechanisms of action, similar to some of the results obtained using gene expression data. No direct comparison to gene expression data was made. Such a comparison would be highly valuable, and to our knowledge, no such study has been conducted.

3.2. Application in cell death

One protein-profiling technique, proteolytic profiling, deserves further attention due to its particular focus on cell death. Two studies introducing this technique were published simultaneously in 2008. James Wells' group used an engineered enzyme to biotinylate and subsequently enrich and identify proteins with free N-termini (Mahrus et al., 2008). They applied this technique to analyze 333 cleavage sites in 292 proteins identified in apoptosis induced by etoposide in Jurkat cells. They found that many of these cleavage sites were poorly predicted by *in vitro* caspase cleavage site selectivity measurements and that cleavage sites were enriched within interacting proteins. The Wells lab subsequently applied this technique to more comprehensively identify cleavage events in apoptotic cells in culture (Crawford et al., 2013) and to characterize the substrates of inflammatory caspases (Agard, Maltby, & Wells, 2010). They also compared the profiles generated in three different cell lines treated with three different small molecules to attempt to identify unique fingerprints for compound mechanisms of action (Shimbo et al., 2012).

In the second study, Cravatt and colleagues combined SDS-PAGE with LC–MS–MS in a technique they named Protein Topography and Migration Analysis Platform (PROTOMAP) (Dix et al., 2008). They used this technique to analyze proteolytic events induced by staurosporine treatment of Jurkat cells, identifying 91 characterized and 170 previously uncharacterized cleavage events. Analysis of these fragments demonstrated that many were persistent and preserved intact protein domains, raising the possibility of cleavage of proteins results in the generation of active fragments. More recently, the Cravatt group updated their PROTOMAP technology to also identify phosphorylation events (Dix et al., 2012). By identifying more than 700 cleaved proteins and 5000 sites of phosphorylation during apoptosis in Jurkat cells, they were able to demonstrate that these two modifications are intricately linked with phosphorylation-driving cleavage and vice versa.

3.3. Advantages and limitations in the study of cell death

Advantages of protein-profiling methods include high dimensionality, the direct measurement of the effector proteins rather precursor changes, the ability to distinguish and quantify posttranslational modifications of proteins, and the ability of some techniques to make single-cell measurements. While the dimensionality of protein profiles is typically lower than that of gene expression profiles, techniques are available for global monitoring of large numbers of proteins. These profiles have the advantage that they are direct measurements of the level of what is typically the functional entity (the protein) rather than the precursor mRNA, which is not always well correlated to the protein level (Nishizuka et al., 2003). They also can detect modifications of proteins, which increase the dimensionality of the data and allow for dissection of these critical events. This is particularly notable in cell death where changes in signaling cascades and cleavage events are essential (Kurokawa & Kornbluth, 2009). While flow and mass cytometry are limited to fewer simultaneous measurements (mass cytometry can detect up to 34) (Bendall et al., 2011), they have the ability to make single-cell measurements. Such measurements can be highly valuable in cell death, where there can be significant cell-to-cell variability in time-to-response and response to lethal stimuli among cells in a clonal population (Spencer, Gaudet, Albeck, Burke, & Sorger, 2009).

Protein profiling shares a number of the limitations of gene expression profiling, including the difficulty in identifying changes over time and concentration ranges and the difficulty separating primary from off-target effects of small molecules. Additional disadvantages include relatively high cost, low throughput, difficulty in detecting low-abundance proteins, and reliance on antibodies. While improvements in technology, particularly mass spectrometry and labeling techniques, have improved proteome coverage, costs are still high relative to gene expression profiling and detection of low-abundance remains problematic (Carragher, Brunton, & Frame, 2012). A number of techniques for the simultaneous measurement of proteins, including lysate arrays and flow and mass cytometry, require the use of antibodies and are therefore clearly limited by the availability and quality of the antibodies. Many of the protein-profiling technologies have had limited applications to date in cell death. An exception is the proteolytic profiling techniques employed by the Wells and Cravatt groups. These techniques are promising and it will be interesting to see the result of their application to larger numbers of small molecules, particularly of inducers of non-apoptotic cell death pathways.

4. GENE–SMALL-MOLECULE INTERACTIONS

Improvements in the generation and monitoring of gene knockouts in model organisms and the availability of RNAi technology in mammalian cells have led to their application in the investigation of small-molecule mechanisms of action (Brummelkamp et al., 2006; Hoon et al., 2008; Lum et al., 2004; Luo et al., 2008). These techniques have also been applied to create and compare profiles for small molecules.

4.1. Chemical–genetic profiling in yeast

Charlie Boone's lab, Guri Giaever's lab, and Cori Nislow's lab have developed and utilized a system called "chemical–genetic profiling" that measures the hypersensitivity to small molecules conferred by the full collection of viable haploid deletion mutants in yeast. The Boone lab initially compared 12 small-molecule profiles (Parsons et al., 2004) and then extended their analysis to 82 compounds (Parsons et al., 2006). Using hierarchical clustering and sparse matrix factorization, they showed that compounds with similar mechanisms of action had similar profiles and were able to identify that the estrogen modulator tamoxifen can cause increases in intracellular calcium concentrations. In their initial as well as a subsequent study (Costanzo et al., 2010), they showed that by integrating chemical–genetic profiles with genetic interaction profiles, they could identify the mechanism of action of previously uncharacterized compounds based on the similarity of the profiles of a compound and the profile from the deletion of its target.

4.2. Applications of yeast profiling in cell death

Subsequent studies have used this technology in the investigation of cytotoxic compounds. The first examined the mechanism of action of two previously uncharacterized and structurally similar antifungal compounds (Yu et al., 2008). By comparing their chemical–genetic profiles to those from the studies published above, they found that one compound clustered with mitochondrial inhibitors while the other clustered with DNA-damaging agents. They subsequently validated these mechanistic predictions. Spitzer and colleagues compared chemical–genetic profiles for compounds found to potentiate the activity of the antifungal fluconazole (Spitzer et al., 2011). Comparing these profiles led them to identify two principle mechanisms of synergy and allowed them to identify additional synergistic compounds. The Nislow and Giaever group took an analogous approach to investigating the mechanisms of action of DNA-damaging agents (Lee et al., 2005). They created chemical–genetic profiles for 12 compounds against ~4700 homozygous deletion strains. As a part of a broader analysis, they clustered the compounds and found that compounds with similar mechanisms of DNA damage clustered together.

4.3. Chemical–genetic profiling in mammalian cells

Citing the difficulty in applying the yeast system to study cancer chemotherapeutics due to the lack of conservation of certain drug targets, Hemann and colleagues developed a profiling system using RNAi in mammalian cells (Jiang, Pritchard, Williams, Lauffenburger, & Hemann, 2011). They measured the ability of 29 shRNAs (targeting either the Bcl2 family or p53 and its activating kinases) to alter the lethality in a murine lymphoma cell line of 15 chemotherapeutic compounds, each used at a single dose. Clustering of these profiles accurately grouped compounds according to their known mechanism of action. They went on to show that a subset of eight shRNAs was sufficient to accurately classify the compounds and that profiles generated with those eight shRNAs for additional compounds were able to successfully classify both compounds with mechanisms already represented within their database as well as compounds with unique mechanisms.

4.4. Advantages and limitations in the study of cell death

Chemical–genetic profiling systems can generate large amounts of high-quality, functional information about the mechanism of action of a compound. Each profile is information-rich and can be individually mined for mechanistic data as well as used as a fingerprint of compound action that can be compared to other profiles. The use of a functional assay is a key advantage over gene expression and protein-profiling methods. Although efflux pumps and multidrug resistance genes can be complicating (Parsons et al., 2004), the use of a functional assay removes the difficulty in distinguishing relevant from off-target effects. Functional assays are able to give information about what took place over the time course of the experiment, rather than taking a snapshot of one point in the process. The yeast system is well established and robust and the pooled barcoding systems allow for genome-wide coverage with good throughput. There are large available data sets including extensive gene–gene interaction data to which chemical–genetic profiles can be compared (Costanzo et al., 2010; Hillenmeyer et al., 2008). The RNAi system is less widely used but is promising for its applicability in mammalian cells. Hemann and colleagues used only 29 genes, but pooled RNAi approaches may make it feasible to create large numbers of genome-wide profiles (Luo et al., 2008). Their approach is also appealing for the ability to readily generate mini-profiles focused on a process of interest. For example, Hemann and colleagues used a set of established apoptosis-related genes, but other choices could be made to create a systems concentrated on a different cell death process.

The limitation of the yeast system is primarily that the technology can only be applied in yeast. This is a particular drawback in studying non-apoptotic cell death, which is poorly conserved even between mammals and other metazoans (Tait & Green, 2008). While the shRNA system is applicable in mammalian cells, it may be difficult to apply to specialized cell types that are not easily transfected. There are also off-target concerns with RNAi, exemplified in the Hemann study by their inability to fully reproduce the clusters when each clone in their eight-gene set was replaced with a different clone targeting the same gene. Compounds in both systems are generally used at a single dose, masking concentration-dependent changes and raising the likelihood of using too high a concentration and increasing the likelihood of pleiotropic effects.

5. SMALL-MOLECULE COMBINATION INTERACTIONS

Combinations of drugs are the foundation of treatment for a number of diseases including HIV, tuberculosis, and multiple types of cancer. There is a long history of the analysis and interpretation of the interactions between small molecules (Keith, Borisy, & Stockwell, 2005). More recently, these effects have been quantified and compared to help understand compound mechanisms of action.

5.1. Profiles based on small-molecule interactions

Small-molecule interactions have been used as fingerprints both in model organisms and in mammalian cells. Kishony and colleagues scored pairwise interactions between 21 antibiotics as antagonistic, additive, or synergistic, based on their combined effect on the growth of *E. coli* (Yeh et al., 2006). These data allowed them to accurately group the compounds according to their known mechanisms of action and suggest a novel mechanism for one poorly characterized compound. Farha and Brown took a similar approach, screening ~200 compounds capable of inhibiting *Escherichia coli* growth for synergy with 14 well-characterized antibiotics (Farha & Brown, 2010). They compared these profiles and were able to make mechanistic inferences about novel compounds. Lehar and colleagues created dose matrices for combinations of 10 sterol inhibitors in yeast (Lehar et al., 2007). Analysis of the shape of these interaction maps allowed distinction between compounds acting in the same pathway and compounds acting in different pathways. They performed a similar analysis on data from a screen of all combinations of 90 characterized compounds in a human colon cancer cell line and found that compounds with similar mechanism of action were more likely to have similar interactions with the rest of the set.

5.2. Advantages and limitations in the study of cell death

Analysis of small-molecule combinations is valuable for its versatility and widespread applicability. Unlike genetic changes or RNAi, it can be applied across species and in specialized cell types. It has the potential to generate high-dimensionality data, it typically uses a functional assay as the output, and the profiles can be information-rich and mined for further insights. While some studies used single doses, larger dose matrices like those used by Lehar et al. can be used to capture dose-dependent effects. Additionally, as demonstrated by Lehar and colleagues in their yeast experiments with sterol inhibitors, focused, information-dense mini-profiles can be readily generated for specific processes of interest.

Using combinations of small molecules has several drawbacks, primarily centered on throughput and coverage. Larger systems in which all pairwise interactions are tested are limited by the exponential increase in the number of required experiments. The same problem prevents the use of multiple doses and higher-order combinations of small molecules. While with genetic techniques it is possible to barcode and pool experiments, this is not theoretically possible with small-molecule combinations. Additionally, genetic approaches can achieve genome-wide coverage while chemical interaction approaches are limited by the availability of relevant small molecules. It can require the testing of large numbers of combinations to identify statistically significant interactions. For example, Farha and Brown detected an interaction with only 45 of the 186 compounds tested against their 14-compound reference set.

6. CELL LINE VIABILITY PROFILING

George Gey and colleagues first successfully cultured human tumor cells (HeLa cells) in 1952 (Gey, Coffman, & Kubicek, 1952). Since that time, cultured tumor cells have been a central pillar of biological investigation in general and cell death in particular. While attempting to identify potential chemotherapeutic agents, researchers at the NCI noted that differential cell line toxicity was a useful marker of compound mechanism of action (Shoemaker, 2006). Cell line viability profiling has continued today as one of the most frequently utilized profiling methodologies.

6.1. NCI60 screen

In the late 1980s, the National Cancer Institute developed a screen of 60 human tumor cell lines from 9 different tumor types (NCI60) with the intention of identifying disease-specific lethal compounds. The NCI60 screen was instituted as a primary drug screening platform in the 1990s and then transitioned to a research tool in 2000. By 2005, over 350,000 compounds had been screened and compounds continue to be screened at a reduced rate of approximately 3000 per year (Shoemaker, 2006). Early on in the screening program, it was noted that the patterns of relative potency across the cell lines could be used as a marker of compound mechanism of action. They subsequently developed their "COMPARE" algorithm to further analyze these patterns and allow for querying of their large database of compound sensitivities (Paull et al., 1989). In this initial work, they demonstrated that the algorithm was able to accurately link alkylating agents, topoisomerase inhibitors, and antimetabolites to other compounds from the same mechanistic class. See Fig. 11.4 for an example of the display format and use of the COMPARE algorithm from a recent study (Yang et al., 2012).

6.1.1 Notable applications of the NCI60 screen

One of the first successful applications of the COMPARE algorithm was in the investigation of halichondrin B, a cytotoxic natural product without a known mechanism of action. The NCI team found that halichondrin B had a sensitivity profile similar to known microtubule destabilizers and subsequently demonstrated that halichondrin B was also a microtubule destabilizer (Bai et al., 1991). Later the database was used to investigate the mechanism of action of recombinant anthrax lethal factor. The

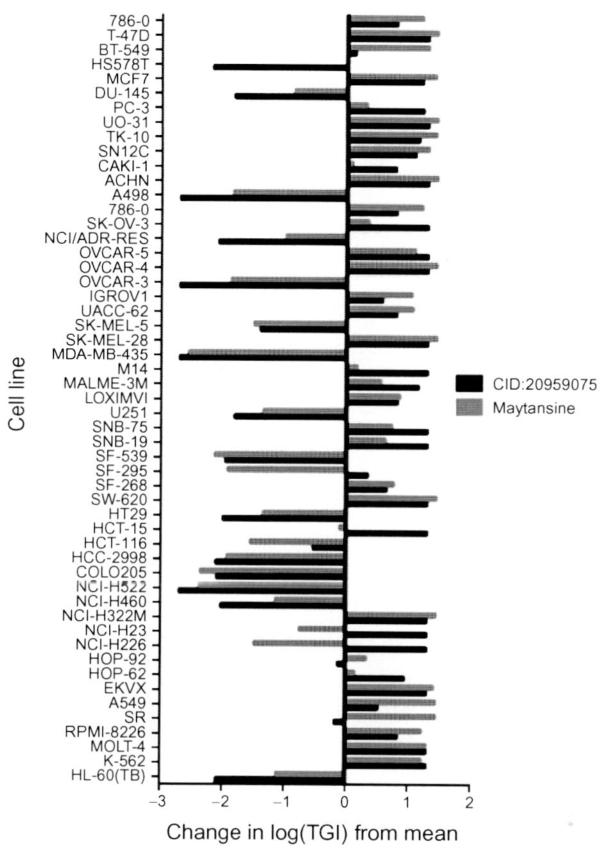

Figure 11.4 NCI60 data display and COMPARE analysis. Yang and colleagues (Yang et al., 2012) identified the highly potent cytotoxic compound 20959075. Its profile across the NCI60 cell lines is shown in the black bars. The bars represent the distance of the total growth inhibition (TGI) from the mean of all cells lines tested. COMPARE analysis identified maytansine, a known microtubule destabilizer, as the most highly correlated compound in the database with a correlation of 0.808. Its profile is shown in the red bars. The authors went on to show that 20959075 is in fact a microtubule destabilizer. (See the color plate.)

sensitivity profile was similar to the profile of PD98059, an MEK inhibitor. Anthrax lethal factor was then shown to act as a protease that cleaves MEK1 and MEK2 (Duesbery et al., 1998).

The NCI60 screen was also valuable in identifying compounds with unique profiles in the database, therefore suggesting that such compounds acted through a novel mechanism of action. This was notable the case for the proteasome inhibitor bortezomib (Holbeck & Sausville, 2004).

Bortezomib and several analogues were tested in the NCI60 screen. Initial COMPARE analysis demonstrated that the compounds were similar to each other but had profiles distinct from the database compounds, supporting the novel mechanism of action of bortezomib and its analogues. In addition, the potency of the analogues in the cells lines correlated well to the *in vitro* potency of the compounds against the proteasome, supporting proteasome inhibition as the lethal mechanism of action of the compounds.

6.2. Use of molecularly characterized cell lines

The use of multiple cell lines can be conceptualized as a chemical–genetic profile, akin to testing an unknown combination of an unknown number of genetic (and epigenetic) mutations. Researchers at the NCI noted the value in defining those mutations in order to then be able to link molecular characteristics of the cell lines to compound activity. The advent of genome-wide profiling techniques has enhanced this ability as it has become possible to molecularly characterize larger numbers of cell lines. Programs have since been initiated to greatly expand on the approach taken by the NCI60 and profile compounds in hundreds or thousands of well-characterized cell lines.

6.2.1 Molecular characterization of NCI60 cell lines

In the mid-1990s, the NCI60 group began annotating their cell lines with "molecular targets"—genetic mutations, mRNA levels, protein levels, and enzymatic activity (Weinstein, 2006). In a pioneering study, Jonathan Weinstein and colleagues developed visualization and computational tools to analyze the relationship between cell sensitivity profiles and molecular characteristics (Weinstein et al., 1997). In this study, they showed that they could identify compounds that were likely multidrug resistance transporter substrates. They also identified compounds whose activity was either dependent on or independent of wild-type p53. Further studies have integrated genome-wide expression data (Ross et al., 2000; Scherf et al., 2000), proteomic profiles (Nishizuka et al., 2003; Park et al., 2010; Shankavaram et al., 2007), and mutation data (Ikediobi et al., 2006). The NCI has made all of their data and analysis tools available on their web site (http://dtp.nci.nih.gov/mtargets/mt_index.html).

6.2.2 Expanded cell line databases

Three recent projects have tested small molecules in large numbers of molecularly characterized cell lines. Garnett and colleagues tested 130 compounds in an average of 368 cell lines that had been characterized by sequencing of 64 commonly mutated cancer genes, evaluation of 7 commonly rearranged cancer genes, genome-wide copy number analysis, and genome-wide expression analysis (Garnett et al., 2012). The Cancer Cell Line Encyclopedia (CCLE) project tested 24 anticancer drugs against 479 cell lines that were characterized genetically by sequencing of >1600 genes and testing of 392 recurrent mutations in 33 known cancer genes, DNA copy number evaluation, and genome-wide expression analysis (Barretina et al., 2012). Using a subset of 242 of the cell lines in the CCLE, Basu and colleagues profiled 354 highly selective small molecules (Basu et al., 2013). The data from all three projects are publicly available. Basu and colleagues designed a Web portal for querying their database (http://www.broadinstitute.org/ctrp).

All three of these studies were primarily focused on identifying genomic markers that could be used to predict responsiveness to a chemotherapeutic drug. While characterizing compound mechanism was not the central goal, they do provide rich data sets to investigate compound activity. Garnett and colleagues and the CCLE project did cluster their compounds based on the cell line responses. The CCLE had a small number of compounds that shared a common target, and these groups did cluster together. Garnett and colleagues had a larger number of such compounds, many of which clustered together (MEK1/2 inhibitors, EGFR, IGFR1) although others did not (SRC inhibitors, microtubule stabilizers). Basu et al. clustered the compounds based on the most significant genomic features correlated with compound activity and highlighted several compounds with common targets that clustered together.

6.3. Advantages and limitations in the study of cell death

Cell line profiling of small molecules is the first established profiling technique and has a number of attractive features. The primary advantage is the collection of a large amount of publicly accessible, information-rich data. The lifespan of the NCI60 project has allowed for the profiling of an unparalleled number of compounds. The three more recent projects described above have profiled fewer compounds but in an impressive number of cell lines with extensive molecular characterization. Similar to chemical–genetic and chemical interaction profiling, the assays are functional and therefore more likely to detect relevant effects, even when used at a single time point. A recent study suggests that the concentration–response data produced by such studies is even more information-rich than previously appreciated and that additional features can be mined for mechanistic information (Fallahi-Sichani, Honarnejad, Heiser, Gray, & Sorger, 2013). The data dimensionality depends on the number of cell lines tested, but even given similar dimensionality, the data are likely to be more information-dense than a chemical–genetic profile. Single genetic changes often produce no effect on a compound, but the complex combination of alterations represented by each cancer cell line is much more likely to alter compound response, resulting in fewer null values. Lastly, genomic characterization of cell lines allows for the mining of the profiles for mechanistic information.

Limitations of cell line profiling include dependence on available cancer cell lines and the requirement for the use of large numbers of cell lines to achieve representation of uncommon targets and pathways. One of the driving motivations for the assembly of the larger cell line profiling projects was the realization that 60 cell lines was insufficient to allow for the statistical detection of some selective agents, particularly those dependent on mutations found in only of fraction of cancers of a particular lineage (Sharma, Haber, & Settleman, 2010). Unlike in chemical–genetic profiling in which genome-wide coverage can be achieved, cancer cell line profiling is likely to only appreciably access changes relevant to cancer. The dependence on cancer cell lines is particularly a limitation in the study of cell death, since inactivation of cell death is one of the hallmarks of cancer (Hanahan & Weinberg, 2011). It also limits the ability to study highly specialized processes only accessible in specific cell types.

7. QUANTITATIVE IMAGING

As described earlier, morphology has long formed the foundation of cell death characterization. While qualitative descriptions of morphology are problematic, advances in microscopy and computational analysis have allowed for the extraction of quantitative parameters from images that can be used to create profiles to compare compounds.

7.1. High-content imaging in cell culture

The first large-scale imaging-based compound profiling system was developed by the Altschuler and Wu lab. They generated profiles for 100 compounds by testing a range of concentrations in HeLa cells and after 20 h staining with 11 fluorescent probes and imaging to capture up to 8000 cells per well (Perlman et al., 2004). They derived 93 descriptors from the images and used them to compare the 61 compounds for which they detected a significant phenotype. They showed that replicates of compounds and compounds with similar known mechanisms of action clustered together (with some notable exceptions including protein synthesis inhibitors). They were also able to predict the mechanism of one poorly characterized compound and showed that their profiles were more informative than the analysis of only the intensity of the staining. A second study used 3 cellular stains and measured 36 cytological features from which they derived 6 factors by combining highly correlated features (Young et al., 2008). Using this system, they screened a library of over 6000 compounds at a single dose and time point and did further analysis of the 211 compounds that caused the largest perturbations of the 6 factors. Clustering of these compounds accurately classified compounds with known biological activity.

7.2. Advantages and limitations of image-based profiles in studying cell death

Imaging systems continue to decrease in cost and increase in throughput and quality. They are generally accessible across species and cell types that can be cultured. Such systems are valuable for studying cell death because links can be made between quantitative metrics and traditional cell death phenotypes. For example, one of the factors defined by Young and colleagues combined chromatin condensation and decrease in nuclear size and was thought to correspond to apoptosis. Another appealing feature of imaging systems is the possibility of obtaining single-cell data, which could make it feasible to identify and characterize a death process that occurs in only a subset of a cell population.

Image-based profiles are limited in that they collect observational and not functional data, which raises the possibility that the changes observed are not driven by the same process that leads to cell death. While factor analysis like that performed by Young and colleagues attempts to connect biological meaning to the extracted parameters, image-based data do not readily correlate to underlying biochemistry. In fact, in cell death, morphological changes do not necessarily represent unique biochemistry. While the additions of cellular stains increase the dimensionality and sensitivity of the data, many compounds may not cause a significant morphological change, particularly at a chosen time point. This was demonstrated in all of the studies described above where only a minority of compounds displayed an appreciable phenotype.

8. MODULATORY PROFILING

To address some of the deficiencies in the above-described methods and assist in the characterization of small-molecule-induced cell death, we recently developed a system called "modulatory profiling" that creates information-dense, functional profiles for small molecules based on quantification of the degree to which various agents are able to perturb the death induced by lethal compounds (Wolpaw et al., 2011).

8.1. Design and validation

To understand a process, one needs a way to perturb it. From this principle, it follows that a rigorous description of a process could be generated from cataloging various agents capable of inhibiting or enhancing it. To apply this concept to cell death, we assembled a collection of 32 chemical and genetic "modulators" of known cell death pathways and quantified their ability to change the extent of death induced by lethal compounds in two cell lines. We did this by testing a single, literature-derived concentration of the modulator against a 12-point dilution series of the lethal compounds and extracted two parameters, the change in potency and the change in efficacy (see Fig. 11.5). As a proof-of-principle, we tested 28 well-characterized lethal compounds and showed that clustering based on their "modulatory profiles" correctly grouped together compound replicates and compounds with the same mechanism of action. Clustering based on modulatory profiles was shown to more accurately group compounds according to their established mechanisms compared to clustering using a similar algorithm applied to gene expression profiles or chemical structure.

To further demonstrate the value of this approach, we generated profiles for 25 poorly characterized or uncharacterized lethal compounds and clustered them with the characterized compounds (see Fig. 11.6). Three compounds had profiles that clustered with known microtubule destabilizers, and we subsequently showed that these compounds do in fact destabilize microtubules in cultured cells. Other compounds grouped together and based on those groupings were determined to act through nonspecific methods, either compound reactivity or detergent-like membrane disruption. Lastly, poorly characterized and uncharacterized compounds in a cluster without any well-characterized compounds were shown to act through a nonapoptotic process that involved the mitochondria.

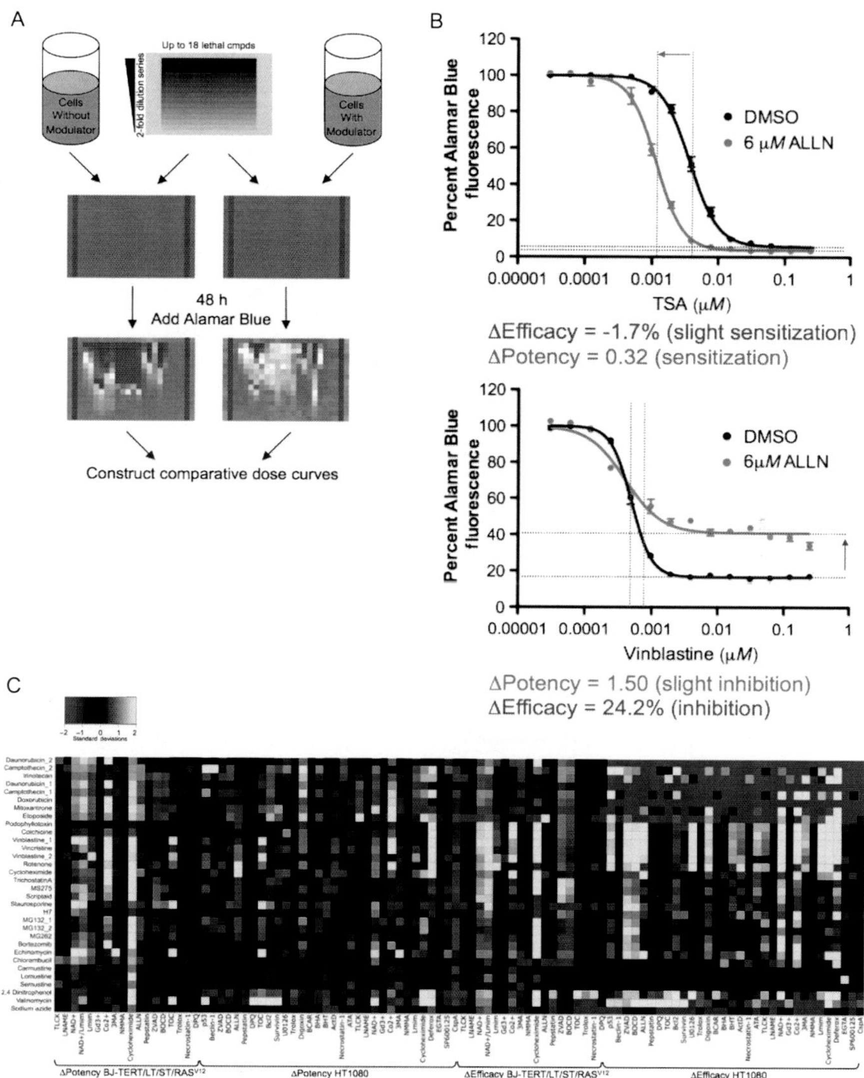

Figure 11.5 Creating modulatory profiles. (A) Cells with or without modulator were seeded into 384-well plates and lethal compounds were added in a dilution series. Viability was measured after 48 h with Alamar Blue and comparative concentration–response curves were constructed from the data. (B) Two examples of comparative dose curves. These illustrate the two parameters extracted from each pair of curves, the change in potency and the change in efficacy. (C) Heat map illustrating the modulatory profiles for a number of characterized lethal compounds. Each row represents the modulatory profiles for a different compound. *Reproduced with permission from Wolpaw et al. (2011).* (See the color plate.)

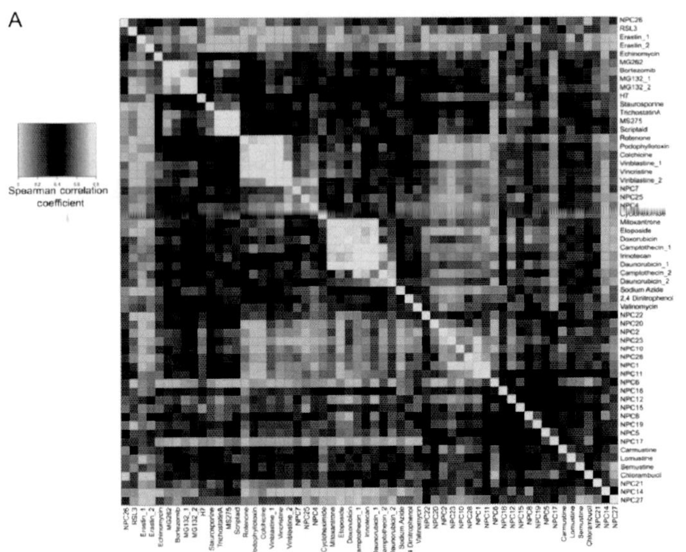

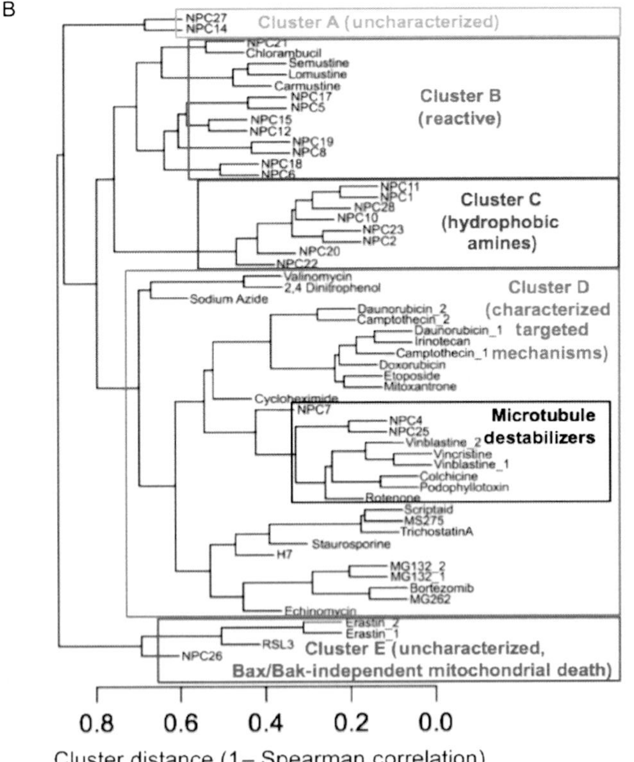

Figure 11.6 Comparing and clustering modulatory profiles. (A) Heat map of the similarity matrix showing the Spearman correlation between modulatory profiles of both characterized and uncharacterized lethal compounds. (B) Dendrogram derived from

8.2. Modulatory profiling protocol

A detailed protocol for performing modulatory profiling was published previously (Wolpaw et al., 2011). Below we will describe a summary of our protocol and identify opportunities for repurposing the system to study specific cell death processes.

I. *Selection of modulators, reference lethal compounds, and cell lines*

We initially selected literature-reported modulators of established cell death pathways including reactive oxygen species scavengers, calcium channel blockers, protein synthesis inhibitors, and protease inhibitors, among others. A different set of modulators could in principle be chosen to focus more clearly on a specific process of interest. We used four genetic modifiers (two shRNAs and two cDNAs), which left significant opportunity for expansion. We used a diverse group of well-characterized lethal compounds, but alternative reference lethal compounds could be chosen. For quality control purposes, it is advantageous to select multiple compounds from the same mechanistic class. We chose two cell lines, the human fibrosarcoma HT-1080 line and an engineered tumorigenic line, derived from human foreskin fibroblasts (BJ-TERT/LT/ST/RAS^{V12}). Both cell lines grow rapidly in and uniformly in culture, HT-1080 cells are easily manipulated with genetic tools, and BJ-TERT/LT/ST/RAS^{V12} have defined genetic changes and therefore do not have mutations in poorly characterized death pathways that may be found in a cancer cell line. Different cell lines could be chosen in order to focus on a cell death process specific to a certain cell type.

II. *Cell plating and addition of reagents*

Cells were cultured as previously described (Yang & Stockwell, 2008). Cells were trypsinized, counted, and seeded in 384-well plates with or without a specific modulator. Lethal compounds were added from a separate 384-well plate containing 12-point, twofold dilutions of the compounds and DMSO-only control wells (see Fig. 11.5A).

clustering the similarity matrix shown in (A). Five broad clusters are highlighted and lettered. In addition, microtubule destabilizers are shown in black, a cluster that includes three previously uncharacterized compounds. Other features that are not highlighted include clustering of characterized compounds according to their known mechanisms of action—alkylating agents, mitochondrial poisons, topoisomerase inhibitors, histone deacetylase inhibitors, and proteasome inhibitors. *Reproduced with permission from Wolpaw et al. (2011) and slightly altered.* (See the color plate.)

After 48 h, the cell viability dye Alamar Blue (Nociari, Shalev, Benias, & Russo, 1998) was added and plates were read 16 h later on a Victor 3 plate reader (Perkin Elmer). All liquid transfers were performed using a Biomek FX AP384 module (Beckman Coulter). All assays were performed at least in triplicate.

III. *Calculation of changes in potency and efficacy*

Background fluorescence (no cells with modulator or no cells with lethal compound) was subtracted and values were normalized to vehicle or modulator-only controls. GraphPad Prism was used to calculate logistic best-fit curves based on four parameters—EC_{50}, Top, Bottom, and Hill slope. Curves with and without modulator were compared. Changes in the Top (change in survival in very low concentrations of compound) were appropriately not observed and therefore the Top was set as equal to one (100% survival). We found that changes in the Hill slope were particularly error prone and uninformative (Kenichi Shimada and Brent Stockwell, unpublished). We therefore used a modification of the changes in the EC_{50} (potency) and the Bottom (efficacy) of these curves as the parameters with which to create the profiles, giving each parameter equal weight (see Fig. 11.5B). While most studies have used potency alone or a combination of potency and efficacy (area under the curve), a recent study validated our approach, demonstrating the utility of efficacy as a separate marker of compound activity (Fallahi-Sichani et al., 2013). These data-processing steps were performed primarily using MatLab.

IV. *Comparing, visualizing, and clustering modulatory profiles*

Potency and efficacy changes were individually normalized so that each had a standard deviation of one (for all compounds, not for each compound). This removed the units and gave each parameter equal weight. Spearman correlations were then calculated for each pair of compounds to produce a similarity matrix like the one shown in Fig. 11.6A. This was done within the R programming environment. The R function hclust was then used to cluster the similarity matrix, using the group average method for defining new clusters (Kaufman & Rousseeuw, 1990). This produced a dendrogram such as the one shown in Fig. 11.6B.

V. *Further validation and exploration of findings*

Similar to other profiling methods, modulatory profiling suggests connections and mechanisms that then must be confirmed through independent experimental validation.

8.3. Application of modulatory profiling to the investigation of ferroptosis

Our lab recently described a nonapoptotic cell death process, "ferroptosis," induced by the small-molecule erastin and dependent on intracellular iron (Dixon et al., 2012). This study used two focused variations of modulatory profiling to demonstrate the uniqueness of ferroptosis. Six chemical modulators of ferroptosis were tested for their ability to alter the death induced by 16 diverse lethal compounds, including two small-molecule inducers of ferroptosis. This analysis showed that ferroptosis modulators were largely inactive against other lethal compounds, and clustering showed that the two ferroptosis inducers were similar to each other and distinct from other lethal compounds. A similar analysis was performed using seven shRNAs, targeting genes that were found in a screen to be required for ferroptosis. These hairpins were tested against seven lethal compounds (including one ferroptosis inducer, erastin) and were only active against erastin. Clustering also placed erastin in a unique cluster, demonstrating the uniqueness of the genetic network required for ferroptosis.

A further study from our lab investigated the role of glutathione peroxidase 4 (Gpx4) in ferroptosis (Yang et al., 2014). This study created a modulatory profile for Gpx4 siRNA and showed that it was similar to the profiles of the ferroptosis-inducing compound RSL3, providing evidence that Gpx4 is the relevant cellular target of RSL3. This demonstrates the ability of modulatory profiling to link small molecules to their targets based on the similarity of their modulatory profiles, similar to what has been done in yeast by combining genetic interaction and chemical–genetic interaction profiles.

8.4. Advantages and limitations in the study of cell death

Modulatory profiling was specifically designed for the study of small-molecule-induced cell death and has some distinct advantages. Perhaps its most important feature is its flexibility, as illustrated in its application to ferroptosis. Modulatory profiles can be readily applied to a specific cell death process of interest, including in specialized cell types. In addition, it uses a functional assay (changes in cell death) and therefore is more likely to identify the relevant mechanism of action of a compound. It uses a broad concentration range, allowing for the detection of dose-dependent effects. It utilizes human cell lines, allowing for the study of processes not found in lower organisms. Modulatory profiling uses a combination of genetic and chemical modulators. Combining both modalities improves resolution, allows investigation in cell types that are difficult to access with genetic methods, and allows access to processes not well targeted currently by small molecules. The chemicals used in modulatory profiling are selected and optimized prior to the profiling experiment. This eliminates the need for a dose matrix like those used in small-molecule interaction profiling and reduces the potentially enormous number of experiments required to test all pairwise interactions among a large collection of small molecules.

Modulatory profiling is limited by the availability of appropriate modulators. If relevant modulators for a given process are unknown or unavailable, a broad-based system can demonstrate the uniqueness of that process, but it will not be able to offer any detailed mechanistic information or differentiate between inducers of that process. In addition, the use of chemical modulators requires individual wells, which makes it difficult to increase throughput. The ability to create profiles for genetic lethal agents, as described above for Gpx4, is promising but has only been demonstrated in specific circumstances. Its widespread applicability remains to be demonstrated.

9. CONCLUSIONS

Over the past 15 years, technological advances and miniaturization have decreased cost and allowed for the implementation and expansion of high-dimensionality profiling systems for the analysis of biological processes. As we have described, these systems take advantage of the ability to quantify transcript and protein levels, create and test large numbers of genetic mutants, perform large numbers of assays required for cancer cell line profiling or chemical combination experiments, and extract quantifiable features from high-content imaging data.

Cell death is an essential and diverse process with a highly complex and interconnected underlying cellular signaling network and significant species and cell-type specificity. High-dimensionality profiling systems are appealing in the study of cell death for their potential to create fingerprints for processes that can be used both for comparison to other pathways and definition of a specific pathway. While unfocused systems like genome-wide expression profiling are valuable for their universality, there is also an important role for smaller, process-focused profiling modalities. We have used modulatory profiling both broadly and specifically, to try to create a map of all of the death pathways available to the cell, but also to create focused systems to analyze specific pathways or contexts. Such systems should continue to play an important role in future cell death research.

REFERENCES

Agard, N. J., Maltby, D., & Wells, J. A. (2010). Inflammatory stimuli regulate caspase substrate profiles. *Molecular and Cellular Proteomics*, *9*(5), 880–893.

Bai, R. L., Paull, K. D., Herald, C. L., Malspeis, L., Pettit, G. R., & Hamel, E. (1991). Halichondrin B and homohalichondrin B, marine natural products binding in the vinca domain of tubulin. Discovery of tubulin-based mechanism of action by analysis of differential cytotoxicity data. *The Journal of Biological Chemistry*, *266*(24), 15882–15889.

Barretina, J., Caponigro, G., Stransky, N., Venkatesan, K., Margolin, A. A., Kim, S., et al. (2012). The Cancer Cell Line Encyclopedia enables predictive modelling of anticancer drug sensitivity. *Nature*, *483*(7391), 603–607.

Basu, A., Bodycombe, N. E., Cheah, J. H., Price, E. V., Liu, K., Schaefer, G. I., et al. (2013). An interactive resource to identify cancer genetic and lineage dependencies targeted by small molecules. *Cell*, *154*(5), 1151–1161.

Bendall, S. C., Simonds, E. F., Qiu, P., Amir el, A. D., Krutzik, P. O., Finck, R., et al. (2011). Single-cell mass cytometry of differential immune and drug responses across a human hematopoietic continuum. *Science*, *332*(6030), 687–696.

Bodenmiller, B., Zunder, E. R., Finck, R., Chen, T. J., Savig, E. S., Bruggner, R. V., et al. (2012). Multiplexed mass cytometry profiling of cellular states perturbed by small-molecule regulators. *Nature Biotechnology*, *30*(9), 858–867.

Brummelkamp, T. R., Fabius, A. W., Mullenders, J., Madiredjo, M., Velds, A., Kerkhoven, R. M., et al. (2006). An shRNA barcode screen provides insight into cancer cell vulnerability to MDM2 inhibitors. *Nature Chemical Biology*, *2*(4), 202–206.

Campillos, M., Kuhn, M., Gavin, A. C., Jensen, L. J., & Bork, P. (2008). Drug target identification using side-effect similarity. *Science*, *321*(5886), 263–266.

Carragher, N. O., Brunton, V. G., & Frame, M. C. (2012). Combining imaging and pathway profiling: An alternative approach to cancer drug discovery. *Drug Discovery Today*, *17*(5–6), 203–214.

Cecconi, D., Donadelli, M., Rinalducci, S., Zolla, L., Scupoli, M. T., Scarpa, A., et al. (2007). Proteomic analysis of pancreatic endocrine tumor cell lines treated with the histone deacetylase inhibitor trichostatin A. *Proteomics*, 7(10), 1644–1653.

Colell, A., Ricci, J. E., Tait, S., Milasta, S., Maurer, U., Bouchier-Hayes, L., et al. (2007). GAPDH and autophagy preserve survival after apoptotic cytochrome c release in the absence of caspase activation. *Cell*, *129*(5), 983–997.

Costanzo, M., Baryshnikova, A., Bellay, J., Kim, Y., Spear, E. D., Sevier, C. S., et al. (2010). The genetic landscape of a cell. *Science*, *327*(5964), 425–431.

Cravatt, B. F., Wright, A. T., & Kozarich, J. W. (2008). Activity-based protein profiling: From enzyme chemistry to proteomic chemistry. *Annual Review of Biochemistry*, *77*, 383–414.

Crawford, E. D., Seaman, J. E., Agard, N., Hsu, G. W., Julien, O., Mahrus, S., et al. (2013). The DegraBase: A database of proteolysis in healthy and apoptotic human cells. *Molecular and Cellular Proteomics*, *12*(3), 813–824.

Degterev, A., Hitomi, J., Germscheid, M., Ch'en, I. L., Korkina, O., Teng, X., et al. (2008). Identification of RIP1 kinase as a specific cellular target of necrostatins. *Nature Chemical Biology*, *4*(5), 313–321.

Degterev, A., Huang, Z., Boyce, M., Li, Y., Jagtap, P., Mizushima, N., et al. (2005). Chemical inhibitor of nonapoptotic cell death with therapeutic potential for ischemic brain injury. *Nature Chemical Biology*, *1*(2), 112–119.

Dix, M. M., Simon, G. M., & Cravatt, B. F. (2008). Global mapping of the topography and magnitude of proteolytic events in apoptosis. *Cell, 134*(4), 679–691.

Dix, M. M., Simon, G. M., Wang, C., Okerberg, E., Patricelli, M. P., & Cravatt, B. F. (2012). Functional interplay between caspase cleavage and phosphorylation sculpts the apoptotic proteome. *Cell, 150*(2), 426–440.

Dixon, S. J., Lemberg, K. M., Lamprecht, M. R., Skouta, R., Zaitsev, E. M., Gleason, C. E., et al. (2012). Ferroptosis: An iron-dependent form of nonapoptotic cell death. *Cell, 149*(5), 1060–1072.

Duesbery, N. S., Webb, C. P., Leppla, S. H., Gordon, V. M., Klimpel, K. R., Copeland, T. D., et al. (1998). Proteolytic inactivation of MAP-kinase-kinase by anthrax lethal factor. *Science, 280*(5364), 734–737.

Ellis, H. M., & Horvitz, H. R. (1986). Genetic control of programmed cell death in the nematode C. elegans. *Cell, 44*(6), 817–829.

Fallahi-Sichani, M., Honarnejad, S., Heiser, L. M., Gray, J. W., & Sorger, P. K. (2013). Metrics other than potency reveal systematic variation in responses to cancer drugs. *Nature Chemical Biology, 9*(11), 708–714.

Farha, M. A., & Brown, E. D. (2010). Chemical probes of Escherichia coli uncovered through chemical-chemical interaction profiling with compounds of known biological activity. *Chemistry & Biology, 17*(8), 852–862.

Fernandes-Alnemri, T., Wu, J., Yu, J. W., Datta, P., Miller, B., Jankowski, W., et al. (2007). The pyroptosome: A supramolecular assembly of ASC dimers mediating inflammatory cell death via caspase-1 activation. *Cell Death and Differentiation, 14*(9), 1590–1604.

Galluzzi, L., Vitale, I., Abrams, J. M., Alnemri, E. S., Baehrecke, E. H., Blagosklonny, M. V., et al. (2012). Molecular definitions of cell death subroutines: Recommendations of the Nomenclature Committee on Cell Death 2012. *Cell Death and Differentiation, 19*(1), 107–120.

Garnett, M. J., Edelman, E. J., Heidorn, S. J., Greenman, C. D., Dastur, A., Lau, K. W., et al. (2012). Systematic identification of genomic markers of drug sensitivity in cancer cells. *Nature, 483*(7391), 570–575.

Gey, G. O., Coffman, W. D., & Kubicek, M. C. (1952). Tissue culture studies of the proliferative capacity of the cervical carcinoma and normal epithelium. *Cancer Research, 12*(4), 264–265.

Guzman, M. L., Rossi, R. M., Karnischky, L., Li, X., Peterson, D. R., Howard, D. S., et al. (2005). The sesquiterpene lactone parthenolide induces apoptosis of human acute myelogenous leukemia stem and progenitor cells. *Blood, 105*(11), 4163–4169.

Haab, B. B., Dunham, M. J., & Brown, P. O. (2001). Protein microarrays for highly parallel detection and quantitation of specific proteins and antibodies in complex solutions. *Genome Biology, 2*(2), RESEARCH0004.

Haider, S., & Pal, R. (2013). Integrated analysis of transcriptomic and proteomic data. *Current Genomics, 14*(2), 91–110.

Hanahan, D., & Weinberg, R. A. (2011). Hallmarks of cancer: The next generation. *Cell, 144*(5), 646–674.

Hassane, D. C., Guzman, M. L., Corbett, C., Li, X., Abboud, R., Young, F., et al. (2008). Discovery of agents that eradicate leukemia stem cells using an in silico screen of public gene expression data. *Blood, 111*(12), 5654–5662.

Hengartner, M. O., Ellis, R. E., & Horvitz, H. R. (1992). Caenorhabditis elegans gene ced-9 protects cells from programmed cell death. *Nature, 356*(6369), 494–499.

Hengartner, M. O., & Horvitz, H. R. (1994). C. elegans cell survival gene ced-9 encodes a functional homolog of the mammalian proto-oncogene bcl-2. *Cell, 76*(4), 665–676.

Hieronymus, H., Lamb, J., Ross, K. N., Peng, X. P., Clement, C., Rodina, A., et al. (2006). Gene expression signature-based chemical genomic prediction identifies a novel class of HSP90 pathway modulators. *Cancer Cell, 10*(4), 321–330.

Hillenmeyer, M. E., Fung, E., Wildenhain, J., Pierce, S. E., Hoon, S., Lee, W., et al. (2008). The chemical genomic portrait of yeast: Uncovering a phenotype for all genes. *Science, 320*(5874), 362–365.

Holbeck, S. L., & Sausville, E. A. (2004). The proteasome and the COMPARE algorithm. In J. Adam (Ed.), *Proteasome inhibitors in cancer therapy* (pp. 99–107). Totowa, NJ: Humana Press.

Hoon, S., Smith, A. M., Wallace, I. M., Suresh, S., Miranda, M., Fung, E., et al. (2008). An integrated platform of genomic assays reveals small-molecule bioactivities. *Nature Chemical Biology, 4*(8), 498–506.

Hughes, T. R., Marton, M. J., Jones, A. R., Roberts, C. J., Stoughton, R., Armour, C. D., et al. (2000). Functional discovery via a compendium of expression profiles. *Cell, 102*(1), 109–126.

Ikediobi, O. N., Davies, H., Bignell, G., Edkins, S., Stevens, C., O'Meara, S., et al. (2006). Mutation analysis of 24 known cancer genes in the NCI-60 cell line set. *Molecular Cancer Therapeutics, 5*(11), 2606–2612.

Jiang, H., Pritchard, J. R., Williams, R. T., Lauffenburger, D. A., & Hemann, M. T. (2011). A mammalian functional-genetic approach to characterizing cancer therapeutics. *Nature Chemical Biology*, 7(2), 92–100.

Kaufman, L., & Rousseeuw, P. J. (1990). *Finding groups in data: An introduction to cluster analysis.* New York: Wiley.

Keiser, M. J., Setola, V., Irwin, J. J., Laggner, C., Abbas, A. I., Hufeisen, S. J., et al. (2009). Predicting new molecular targets for known drugs. *Nature, 462*(7270), 175–181.

Keith, C. T., Borisy, A. A., & Stockwell, B. R. (2005). Multicomponent therapeutics for networked systems. *Nature Reviews. Drug Discovery, 4*(1), 71–78.

Kerr, J. F., Wyllie, A. H., & Currie, A. R. (1972). Apoptosis: A basic biological phenomenon with wide-ranging implications in tissue kinetics. *British Journal of Cancer, 26*(4), 239–257.

Krutzik, P. O., & Nolan, G. P. (2006). Fluorescent cell barcoding in flow cytometry allows high-throughput drug screening and signaling profiling. *Nature Methods, 3*(5), 361–368.

Kurokawa, M., & Kornbluth, S. (2009). Caspases and kinases in a death grip. *Cell, 138*(5), 838–854.

Lamb, J., Crawford, E. D., Peck, D., Modell, J. W., Blat, I. C., Wrobel, M. J., et al. (2006). The Connectivity Map: Using gene-expression signatures to connect small molecules, genes, and disease. *Science, 313*(5795), 1929–1935.

Lee, W., St Onge, R. P., Proctor, M., Flaherty, P., Jordan, M. I., Arkin, A. P., et al. (2005). Genome-wide requirements for resistance to functionally distinct DNA-damaging agents. *PLoS Genetics, 1*(2), e24.

Lehar, J., Zimmermann, G. R., Krueger, A. S., Molnar, R. A., Ledell, J. T., Heilbut, A. M., et al. (2007). Chemical combination effects predict connectivity in biological systems. *Molecular Systems Biology, 3*, 80.

Leung, D., Hardouin, C., Boger, D. L., & Cravatt, B. F. (2003). Discovering potent and selective reversible inhibitors of enzymes in complex proteomes. *Nature Biotechnology, 21*(6), 687–691.

Lum, P. Y., Armour, C. D., Stepaniants, S. B., Cavet, G., Wolf, M. K., Butler, J. S., et al. (2004). Discovering modes of action for therapeutic compounds using a genome-wide screen of yeast heterozygotes. *Cell, 116*(1), 121–137.

Luo, B., Cheung, H. W., Subramanian, A., Sharifnia, T., Okamoto, M., Yang, X., et al. (2008). Highly parallel identification of essential genes in cancer cells. *Proceedings of the National Academy of Sciences of the United States of America, 105*(51), 20380–20385.

MacBeath, G. (2002). Protein microarrays and proteomics. *Nature Genetics, 32*(Suppl.), 526–532.

Mahrus, S., Trinidad, J. C., Barkan, D. T., Sali, A., Burlingame, A. L., & Wells, J. A. (2008). Global sequencing of proteolytic cleavage sites in apoptosis by specific labeling of protein N termini. *Cell, 134*(5), 866–876.

McGettigan, P. A. (2013). Transcriptomics in the RNA-seq era. *Current Opinion in Chemical Biology, 17*(1), 4–11.

Miura, M., Zhu, H., Rotello, R., Hartwieg, E. A., & Yuan, J. (1993). Induction of apoptosis in fibroblasts by IL-1 beta-converting enzyme, a mammalian homolog of the C. elegans cell death gene ced-3. *Cell, 75*(4), 653–660.

Muroi, M., Kazami, S., Noda, K., Kondo, H., Takayama, H., Kawatani, M., et al. (2010). Application of proteomic profiling based on 2D-DIGE for classification of compounds according to the mechanism of action. *Chemistry & Biology, 17*(5), 460–470.

Newman, Z. L., Crown, D., Leppla, S. H., & Moayeri, M. (2010). Anthrax lethal toxin activates the inflammasome in sensitive rat macrophages. *Biochemical and Biophysical Research Communications, 398*(4), 785–789.

Nishizuka, S., Charboneau, L., Young, L., Major, S., Reinhold, W. C., Waltham, M., et al. (2003). Proteomic profiling of the NCI-60 cancer cell lines using new high-density reverse-phase lysate microarrays. *Proceedings of the National Academy of Sciences of the United States of America, 100*(24), 14229–14234.

Nociari, M. M., Shalev, A., Benias, P., & Russo, C. (1998). A novel one-step, highly sensitive fluorometric assay to evaluate cell-mediated cytotoxicity. *Journal of Immunological Methods, 213*(2), 157–167.

Park, E. S., Rabinovsky, R., Carey, M., Hennessy, B. T., Agarwal, R., Liu, W., et al. (2010). Integrative analysis of proteomic signatures, mutations, and drug responsiveness in the NCI 60 cancer cell line set. *Molecular Cancer Therapeutics, 9*(2), 257–267.

Parsons, A. B., Brost, R. L., Ding, H., Li, Z., Zhang, C., Sheikh, B., et al. (2004). Integration of chemical-genetic and genetic interaction data links bioactive compounds to cellular target pathways. *Nature Biotechnology, 22*(1), 62–69.

Parsons, A. B., Lopez, A., Givoni, I. E., Williams, D. E., Gray, C. A., Porter, J., et al. (2006). Exploring the mode-of-action of bioactive compounds by chemical-genetic profiling in yeast. *Cell, 126*(3), 611–625.

Paull, K. D., Shoemaker, R. H., Hodes, L., Monks, A., Scudiero, D. A., Rubinstein, L., et al. (1989). Display and analysis of patterns of differential activity of drugs against human tumor cell lines: Development of mean graph and COMPARE algorithm. *Journal of the National Cancer Institute, 81*(14), 1088–1092.

Perlman, Z. E., Slack, M. D., Feng, Y., Mitchison, T. J., Wu, L. F., & Altschuler, S. J. (2004). Multidimensional drug profiling by automated microscopy. *Science, 306*(5699), 1194–1198.

Rasheva, V. I., & Domingos, P. M. (2009). Cellular responses to endoplasmic reticulum stress and apoptosis. *Apoptosis, 14*(8), 996–1007.

Ross, D. T., Scherf, U., Eisen, M. B., Perou, C. M., Rees, C., Spellman, P., et al. (2000). Systematic variation in gene expression patterns in human cancer cell lines. *Nature Genetics, 24*(3), 227–235.

Schena, M., Shalon, D., Heller, R., Chai, A., Brown, P. O., & Davis, R. W. (1996). Parallel human genome analysis: Microarray-based expression monitoring of 1000 genes. *Proceedings of the National Academy of Sciences of the United States of America, 93*(20), 10614–10619.

Scherf, U., Ross, D. T., Waltham, M., Smith, L. H., Lee, J. K., Tanabe, L., et al. (2000). A gene expression database for the molecular pharmacology of cancer. *Nature Genetics, 24*(3), 236–244.

Schweitzer, B., Roberts, S., Grimwade, B., Shao, W., Wang, M., Fu, Q., et al. (2002). Multiplexed protein profiling on microarrays by rolling-circle amplification. *Nature Biotechnology, 20*(4), 359–365.

Sevecka, M., & MacBeath, G. (2006). State-based discovery: A multidimensional screen for small-molecule modulators of EGF signaling. *Nature Methods*, *3*(10), 825–831.
Shankavaram, U. T., Reinhold, W. C., Nishizuka, S., Major, S., Morita, D., Chary, K. K., et al. (2007). Transcript and protein expression profiles of the NCI-60 cancer cell panel: An integromic microarray study. *Molecular Cancer Therapeutics*, *6*(3), 820–832.
Sharma, S. V., Haber, D. A., & Settleman, J. (2010). Cell line-based platforms to evaluate the therapeutic efficacy of candidate anticancer agents. *Nature Reviews. Cancer*, *10*(4), 241–253.
Shimbo, K., Hsu, G. W., Nguyen, H., Mahrus, S., Trinidad, J. C., Burlingame, A. L., et al. (2012). Quantitative profiling of caspase-cleaved substrates reveals different drug-induced and cell-type patterns in apoptosis. *Proceedings of the National Academy of Sciences of the United States of America*, *109*(31), 12432–12437.
Shoemaker, R. H. (2006). The NCI60 human tumour cell line anticancer drug screen. *Nature Reviews. Cancer*, *6*(10), 813–823.
Spencer, S. L., Gaudet, S., Albeck, J. G., Burke, J. M., & Sorger, P. K. (2009). Non-genetic origins of cell-to-cell variability in TRAIL-induced apoptosis. *Nature*, *459*(7245), 428–432.
Spitzer, M., Griffiths, E., Blakely, K. M., Wildenhain, J., Ejim, L., Rossi, L., et al. (2011). Cross-species discovery of syncretic drug combinations that potentiate the antifungal fluconazole. *Molecular Systems Biology*, 7, 499.
Stockwell, B. R. (2004). Exploring biology with small organic molecules. *Nature*, *432*(7019), 846–854.
Stockwell, B. R., Hardwick, J. S., Tong, J. K., & Schreiber, S. L. (1999). Chemical genetic and genomic approaches reveal a role for copper in specific gene activation. *Journal of the American Chemical Society*, *121*(45), 10662–10663.
Stumpel, D. J., Schneider, P., Seslija, L., Osaki, H., Williams, O., Pieters, R., et al. (2012). Connectivity mapping identifies HDAC inhibitors for the treatment of t(4;11)-positive infant acute lymphoblastic leukemia. *Leukemia*, *26*(4), 682–692.
Tait, S. W., & Green, D. R. (2008). Caspase-independent cell death: Leaving the set without the final cut. *Oncogene*, *27*(50), 6452–6461.
Taylor, R. C., Cullen, S. P., & Martin, S. J. (2008). Apoptosis: Controlled demolition at the cellular level. *Nature Reviews. Molecular Cell Biology*, *9*(3), 231–241.
Turmaine, M., Raza, A., Mahal, A., Mangiarini, L., Bates, G. P., & Davies, S. W. (2000). Nonapoptotic neurodegeneration in a transgenic mouse model of Huntington's disease. *Proceedings of the National Academy of Sciences of the United States of America*, *97*(14), 8093–8097.
Unlu, M., Morgan, M. E., & Minden, J. S. (1997). Difference gel electrophoresis: A single gel method for detecting changes in protein extracts. *Electrophoresis*, *18*(11), 2071–2077.
Virchow, R. L. K., & Chance, F. (1860). *Cellular pathology, as based upon physiological and pathological histology: Twenty lectures delivered in the pathological institute of Berlin during the months of February, March and April, 1858*. New York: Robert M. De Witt.
Wei, G., Margolin, A. A., Haery, L., Brown, E., Cucolo, L., Julian, B., et al. (2012). Chemical genomics identifies small-molecule MCL1 repressors and BCL-xL as a predictor of MCL1 dependency. *Cancer Cell*, *21*(4), 547–562.
Wei, G., Twomey, D., Lamb, J., Schlis, K., Agarwal, J., Stam, R. W., et al. (2006). Gene expression-based chemical genomics identifies rapamycin as a modulator of MCL1 and glucocorticoid resistance. *Cancer Cell*, *10*(4), 331–342.
Weinstein, J. N. (2006). Spotlight on molecular profiling: "Integromic" analysis of the NCI-60 cancer cell lines. *Molecular Cancer Therapeutics*, *5*(11), 2601–2605.
Weinstein, J. N., Myers, T. G., O'Connor, P. M., Friend, S. H., Fornace, A. J., Jr., Kohn, K. W., et al. (1997). An information-intensive approach to the molecular pharmacology of cancer. *Science*, *275*(5298), 343–349.

West, D. C., Qin, Y., Peterson, Q. P., Thomas, D. L., Palchaudhuri, R., Morrison, K. C., et al. (2012). Differential effects of procaspase-3 activating compounds in the induction of cancer cell death. *Molecular Pharmaceutics*, *9*(5), 1425–1434.

Wolpaw, A. J., Shimada, K., Skouta, R., Welsch, M. E., Akavia, U. D., Pe'er, D., et al. (2011). Modulatory profiling identifies mechanisms of small molecule-induced cell death. *Proceedings of the National Academy of Sciences of the United States of America*, *108*(39), E771–E780.

Yang, W. S., Shimada, K., Delva, D., Patel, M., Ode, E., Skouta, R., et al. (2012). Identification of simple compounds with microtubule-binding activity that inhibit cancer cell growth with high potency. *ACS Medicinal Chemistry Letters*, *3*(1), 35–38.

Yang, W. S., SriRamaratnam, R., Welsch, M. E., Shimada, K., Skouta, R., Viswanathan, V. S., et al. (2014). Regulation of ferroptotic cancer cell death by GPX4. *Cell*, *156*(1–2), 317–331.

Yang, W. S., & Stockwell, B. R. (2008). Synthetic lethal screening identifies compounds activating iron-dependent, nonapoptotic cell death in oncogenic-RAS-harboring cancer cells. *Chemistry & Biology*, *15*(3), 234–245.

Yeh, P., Tschumi, A. I., & Kishony, R. (2006). Functional classification of drugs by properties of their pairwise interactions. *Nature Genetics*, *38*(4), 489–494.

Young, D. W., Bender, A., Hoyt, J., McWhinnie, E., Chirn, G. W., Tao, C. Y., et al. (2008). Integrating high-content screening and ligand-target prediction to identify mechanism of action. *Nature Chemical Biology*, *4*(1), 59–68.

Yu, L., Lopez, A., Anaflous, A., El Bali, B., Hamal, A., Ericson, E., et al. (2008). Chemical-genetic profiling of imidazo[1,2-a]pyridines and -pyrimidines reveals target pathways conserved between yeast and human cells. *PLoS Genetics*, *4*(11), e1000284.

Yuan, J. Y., & Horvitz, H. R. (1990). The Caenorhabditis elegans genes ced-3 and ced-4 act cell autonomously to cause programmed cell death. *Developmental Biology*, *138*(1), 33–41.

Yuan, J., Shaham, S., Ledoux, S., Ellis, H. M., & Horvitz, H. R. (1993). The C. elegans cell death gene ced-3 encodes a protein similar to mammalian interleukin-1 beta-converting enzyme. *Cell*, *75*(4), 641–652.

AUTHOR INDEX

Note: Page numbers followed by "*f*" indicate figures and "*t*" indicate tables.

C

D

E

H

I

J

K

M

N

O

P

S

T

U

X

SUBJECT INDEX

Note: Page numbers followed by "*f*" indicate figures and "*t*" indicate tables.

D

I

K

L

M

N

P

R

S

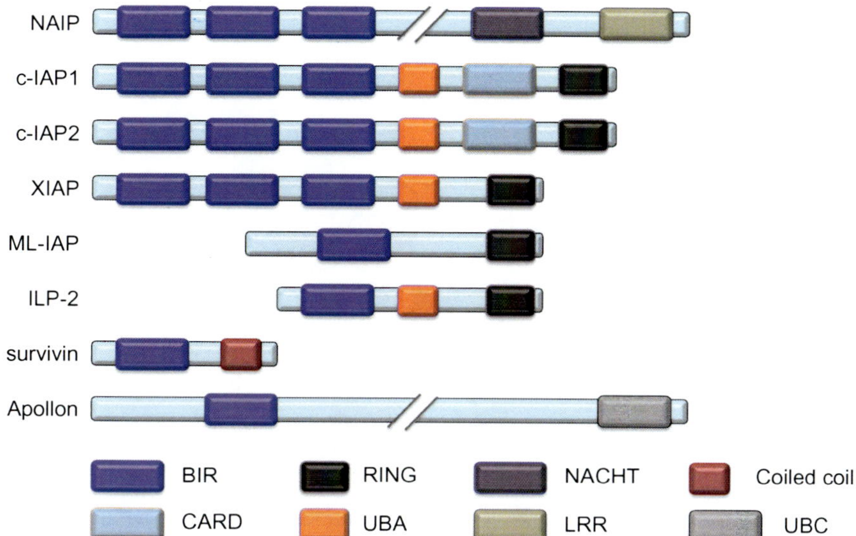

John Silke and Domagoj Vucic, Figure 2.1 Schematic representation of human IAP proteins. BIR: baculovirus IAP repeat; CARD: caspase recruitment domain; LRR: leucine-rich repeat; NACHT: NAIP, CIITA, HET-E, and TP1; RING: really interesting gene; UBA: ubiquitin-associated domain; UBC: ubiquitin-conjugating domain.

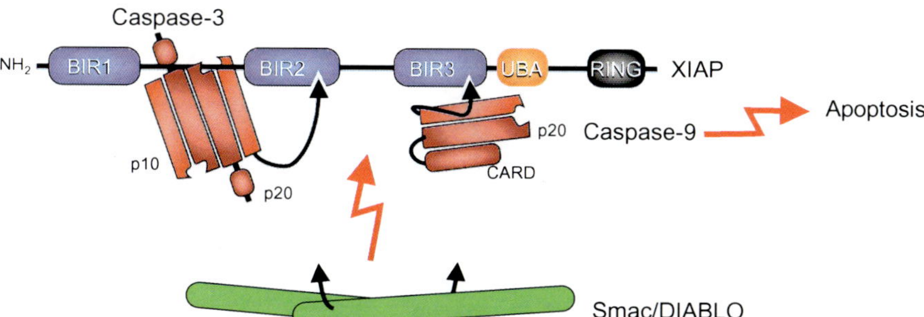

John Silke and Domagoj Vucic, Figure 2.2 XIAP binds and inhibits processed caspases-3 and -9 and possibly functions as a safety catch to block limited activation of the apoptosome. If mitochondria are permeabilized, they release IAP-antagonist molecules including Smac/DIABLO and HtrA2, which liberate caspases from inhibition allowing apoptosis to proceed.

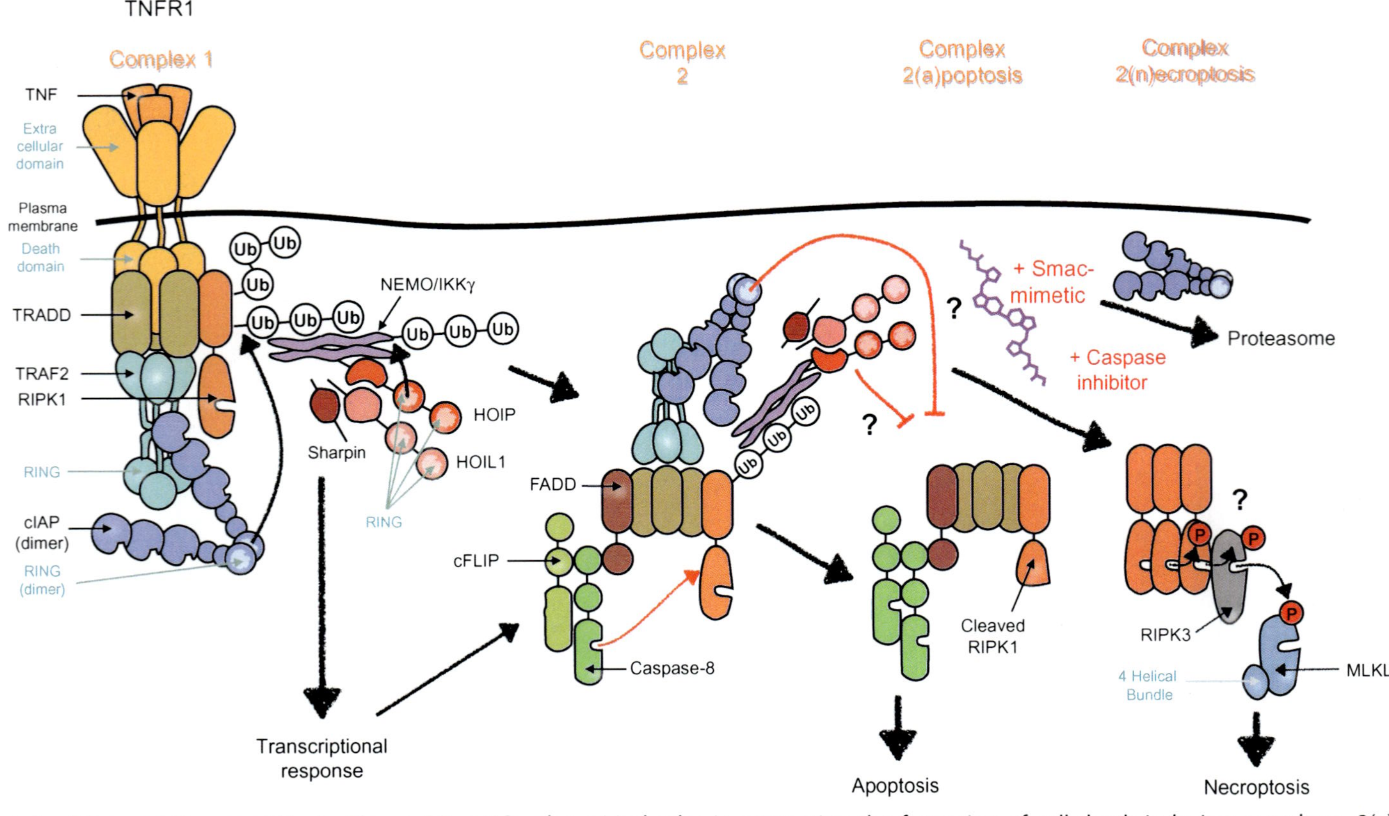

John Silke and Domagoj Vucic, Figure 2.3 c-IAPs play critical roles in preventing the formation of cell death inducing complexes 2(a)poptosis and 2(n)ecroptosis.

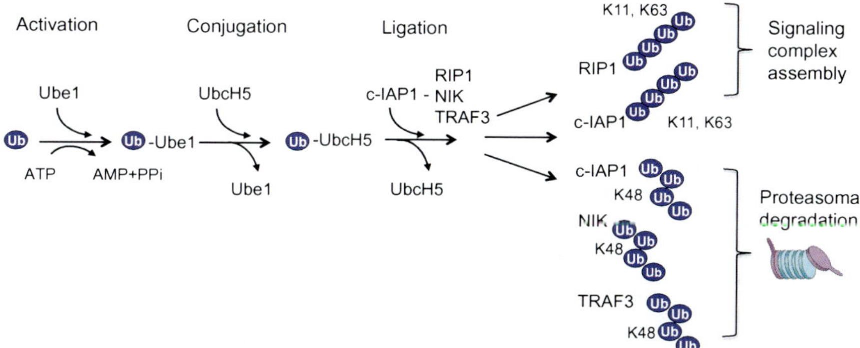

John Silke and Domagoj Vucic, Figure 2.4 The enzymes and reactions of the IAP-mediated ubiquitylation. Activation reaction involves transfer of ubiquitin to an E1 enzyme (Ube1) in an ATP-dependent fashion and leads to transfer of activated ubiquitin to an E2 enzyme (UbcH5) in the conjugation reaction. The E2 with ubiquitin binds E3 ubiquitin ligase (c-IAP1), which can also bind a substrate—commonly via a different protein interaction domain—and thus allows the ubiquitin ligation to occur. When polyubiquitin chains are assembled, this process will be repeated with a lysine (K) residue of the ubiquitin molecule itself serving as a substrate. The assembly of K63- or K11-link polyubiquitin chains on RIP1 or c-IAP1 itself promotes the formation of signaling complexes, while K48-linked ubiquitylation of NIK, TRAF3, or c-IAP1 targets them for proteasomal degradation.

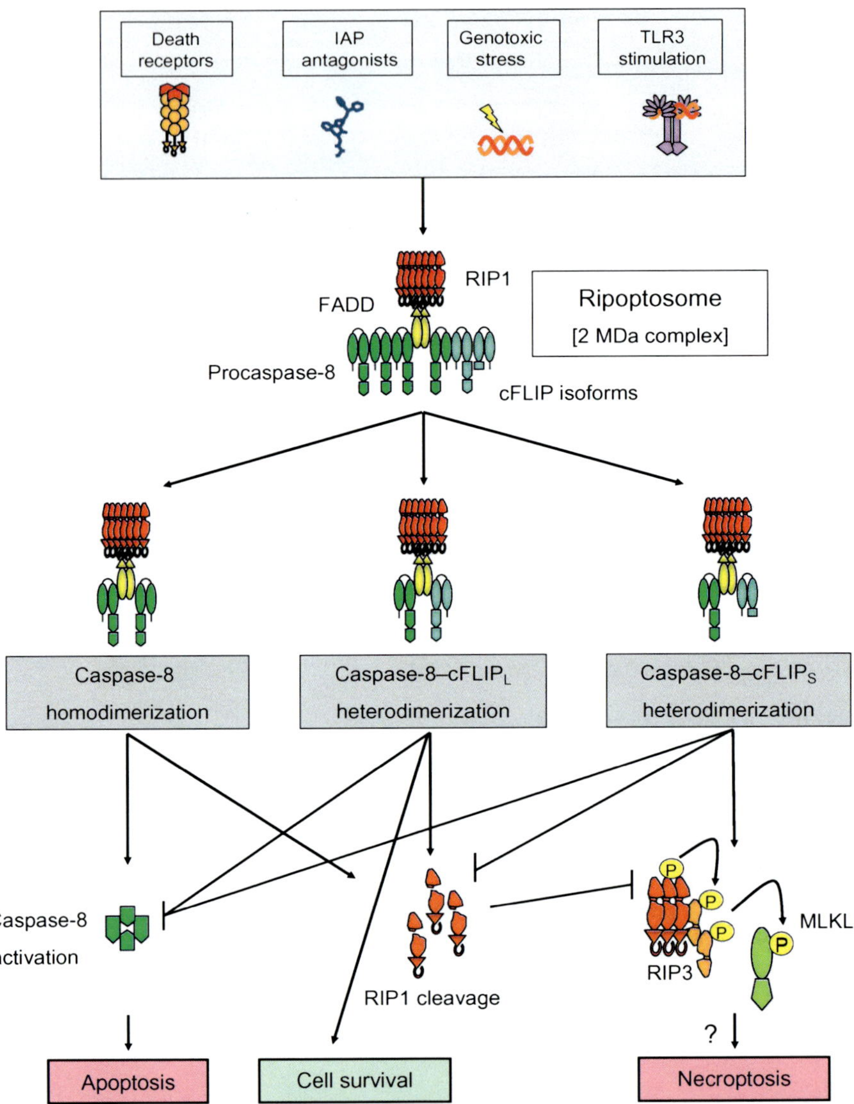

Ramon Schilling *et al.*, Figure 4.1 The stoichiometric composition of the ripoptosome determines the mode of cell death. Depletion of cIAPs, genotoxic stress, death receptor activation, or TLR3 stimulation can lead to the formation of the ripoptosome, consisting of the proteins FADD, caspase-8, RIP1, and cFLIP isoforms. Within the ripoptosome RIP1 and caspase-8 assemble in chains. Thereby, the stoichiometric composition determines the downstream signaling. Caspase-8 homodimerization results in the propagation of apoptosis. cFLIP$_L$ caspase-8 heterodimerization leads to RIP1 cleavage and thereby promotes cell survival. The heterodimerization of caspase-8 and cFLIP$_S$ blocks apoptotic cell death and leads to increased level of active RIP1, which can then phosphorylate RIP3, resulting in MLKL-dependent necroptotic cell death.

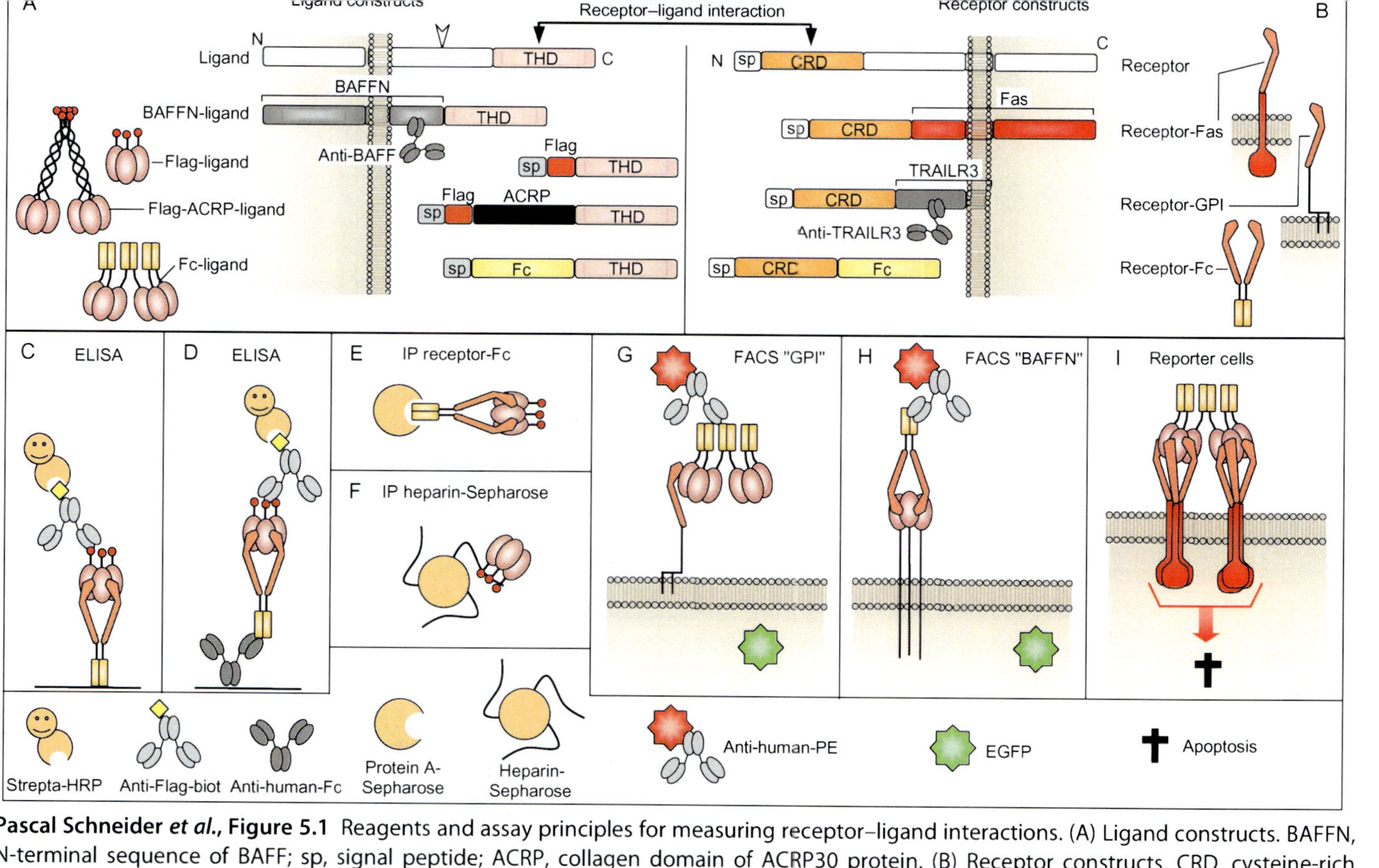

Pascal Schneider *et al.*, Figure 5.1 Reagents and assay principles for measuring receptor–ligand interactions. (A) Ligand constructs. BAFFN, N-terminal sequence of BAFF; sp, signal peptide; ACRP, collagen domain of ACRP30 protein. (B) Receptor constructs. CRD, cysteine-rich domains; Sp, signal peptide. (C and D) ELISA-based assays to detect receptor–ligand interactions using purified or nonpurified receptors-Fc, respectively. (E) Coimmunoprecipitation assay to monitor receptor–ligand interactions. (F) Precipitation assay to detect interactions of ligands with proteoglycans or heparin. (G) FACS-based assay with GPI-anchored receptors. (H) FACS-based assay with membrane-bound BAFFN-ligands. (I) Cell-based reporter assay in which apoptosis is induced by engagement of receptor-Fas chimeric receptors by multimerized ligands.

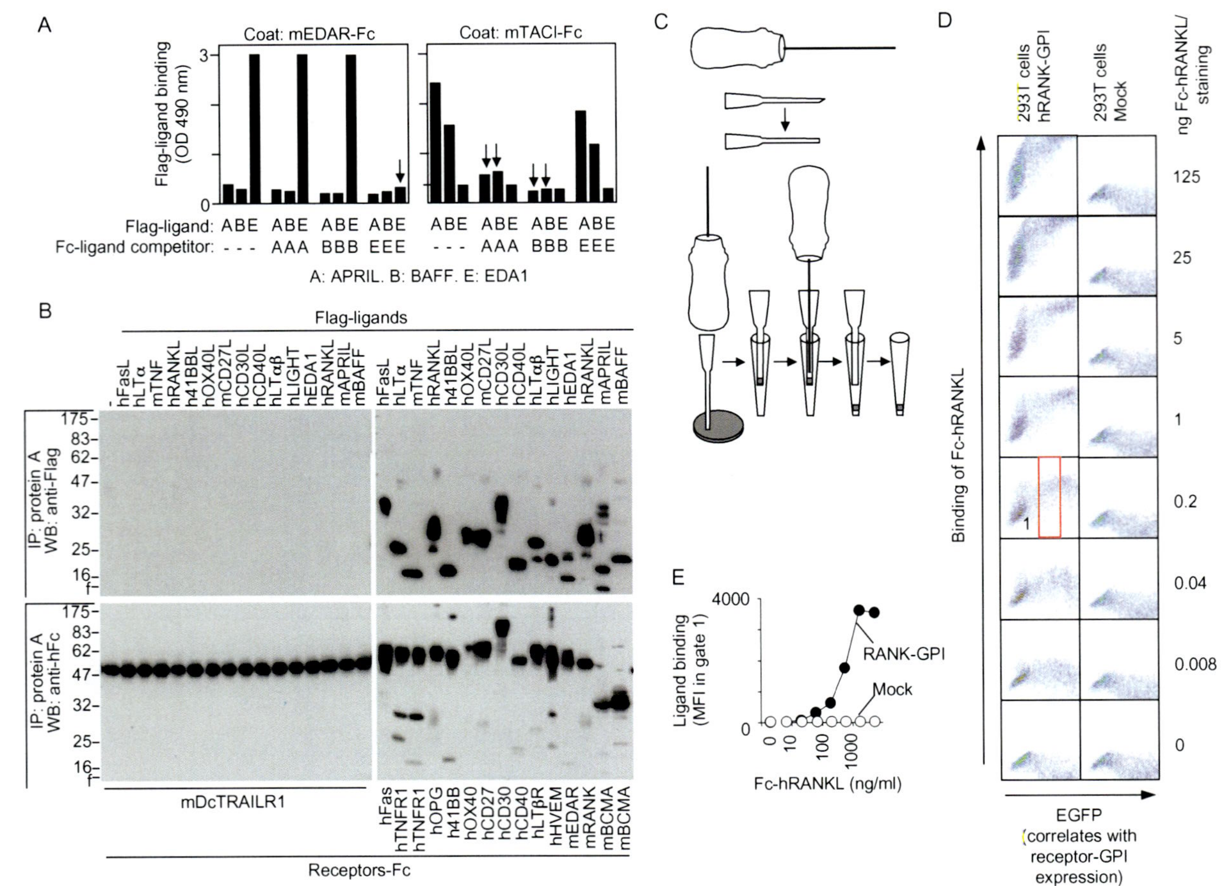
A
Coat: mEDAR-Fc
Coat: mTACI-Fc
Flag-ligand binding (OD 490 nm)
3
0
Flag-ligand: ABE ABE ABE ABE ABE ABE ABE ABE
Fc-ligand competitor: - - - AAA BBB EEE - - - AAA BBB EEE
A: APRIL. B: BAFF. E: EDA1
B
Flag-ligands
- hFasL hLTα mTNF hRANKL h41BBL hOX40L mCD27L hCD30L hCD40L hLTαβ hLIGHT hEDA1 hRANKL mAPRIL mBAFF
IP: protein A WB: anti-Flag
IP: protein A WB: anti-hFc
175- 83- 62- 47- 32- 25- 16- f-
mDcTRAILR1
hFas hTNFR1 hTNFR1 hOPG h41BB hOX40 hCD27 hCD30 hCD40 hLTβR hHVEM mEDAR mRANK mBCMA mBCMA
Receptors-Fc
C
D
293T cells hRANK-GPI
293T cells Mock
ng Fc-hRANKL/ staining
125
25
5
1
0.2
0.04
0.008
0
Binding of Fc-hRANKL
EGFP (correlates with receptor-GPI expression)
E
Ligand binding (MFI in gate 1)
4000
0
RANK-GPI
Mock
0 10 100 1000
Fc-hRANKL (ng/ml)

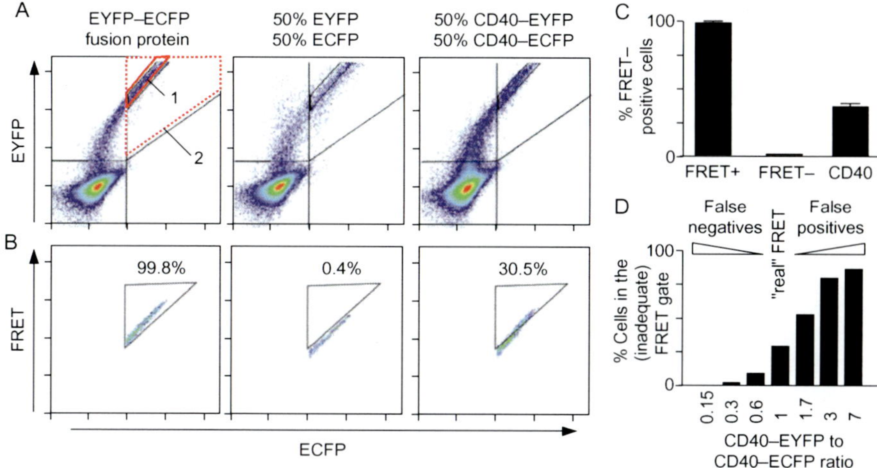

Pascal Schneider *et al.*, Figure 5.4 Ligand-independent interaction of CD40 as measured by FRET. 293T cells were transfected with the indicated plasmids at the indicated ratio. The total amount of plasmid was always constant. (A) Scattergrams showing expression of both ECFP and EYFP in all transfections. (B) FRET signal of cells in gate 1 of panel A. (C) Graphic representation of data from panel (B). Mean ± SEM of 10 independent experiments. (D) 293T cells were transfected with different CD40–EYFP to CD40–ECFP ratios. All cells falling in gate 2 were analyzed. The percentage of cells falling in the FRET gate increases with increasing EYFP to ECFP ratios. This is mainly due to the fact that the FRET gate established with positive and negative controls at a one-to-one ratio is not valid for other ECFP to EYFP ratios.

Pascal Schneider *et al.*, Figure 5.2 Detection of ligand–receptor interactions by ELISA, by coimmunoprecipitation and by FACS. (A) ELISA plates were coated with the indicated receptor-Fc and revealed with the indicated Flag-ligands, in the presence or absence of the indicated Fc-ligands as competitors. Arrows indicate conditions in which competition did occur. (B) A collection of Flag-ligands were immunoprecipitated with an irrelevant receptor (mDcTRAILR1) or with their cognate receptors, as indicated. Western blot was revealed with anti-Flag (blots at the top), then with anti-human Fc (blots at the bottom). Molecular mass markers are in kDa. (C) Mini-column preparation for immunoprecipitations. A blunted gauge 18 needle is used to punch a frit. The needle is introduced in a tip, and the frit is pushed down with a metal wire mounted on a handle. (D) 293T cells transfected with RANK-GPI and EGFP (or cells transfected with EGFP only) were stained with serial dilutions of Fc-RANKL. (E) Quantification of the mean fluorescence in gate 1 of panel (D), which represents the binding of Fc-RANKL to RANK-GPI.

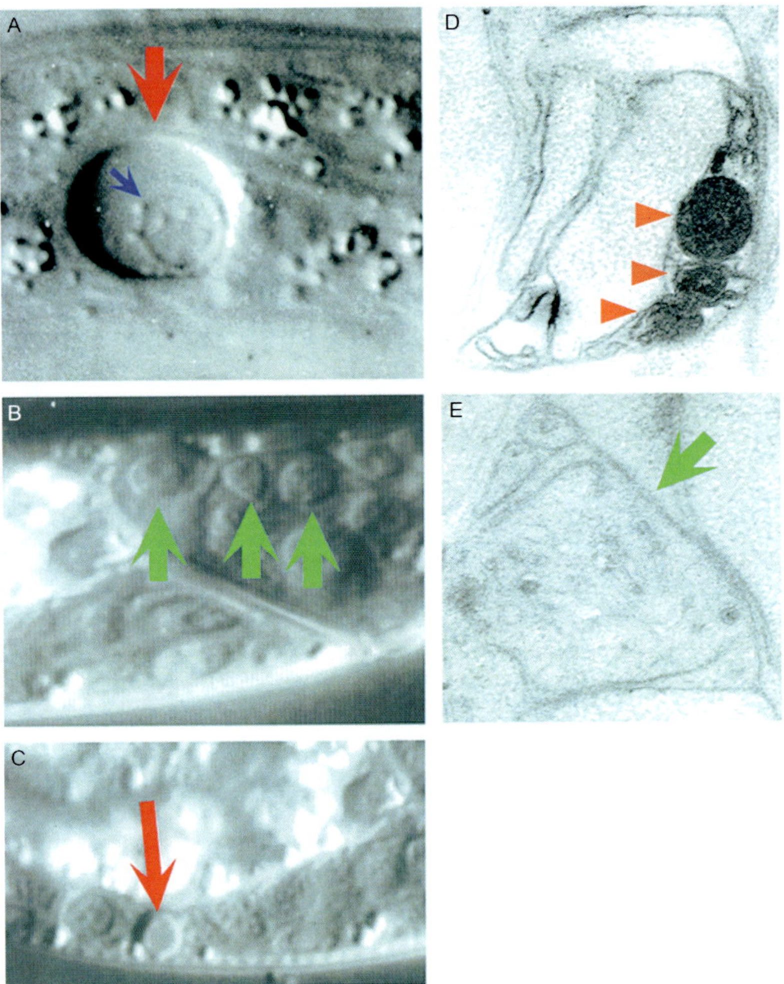

Vassiliki Nikoletopoulou and Nektarios Tavernarakis, Figure 6.2 Necrotic cell death in *C. elegans*. The most prominent morphological characteristic of necrosis is the outstretched swelling of the cell to several times its normal diameter, which is manifested by a hollow, vacuole-like appearance under the optic microscope. For example, a dying PVM (posterior ventral microtubule) touch receptor is shown in (A) by a red arrow. This neuron is expressing a toxic variant of the degenerin MEC-4 (mechanosensory) protein that induces necrosis. The nucleus follows the cellular expansion (A; blue arrow). Healthy cells are indicated by green arrows for comparison (B). In sharp contrast, apoptosis, which normally occurs during nematode development, generates retractile cell corpses, compact in size, with a characteristic button-like appearance (C; red arrow). Under the electron microscope, the same degenerating neuron exhibits dark, electron-dense formations, most likely originating from plasma membrane-internalized material, arranged in onion-like concentric circles (D; arrowheads). At later stages of degeneration, the cytoplasm of the dying cell appears extensively depredated and fragmented. A normal neuron is shown in (E) by a green arrow. *Reprinted from Syntichaki and Tavernarakis (2002), copyright (2002), with permission from Nature Publishing Group.*

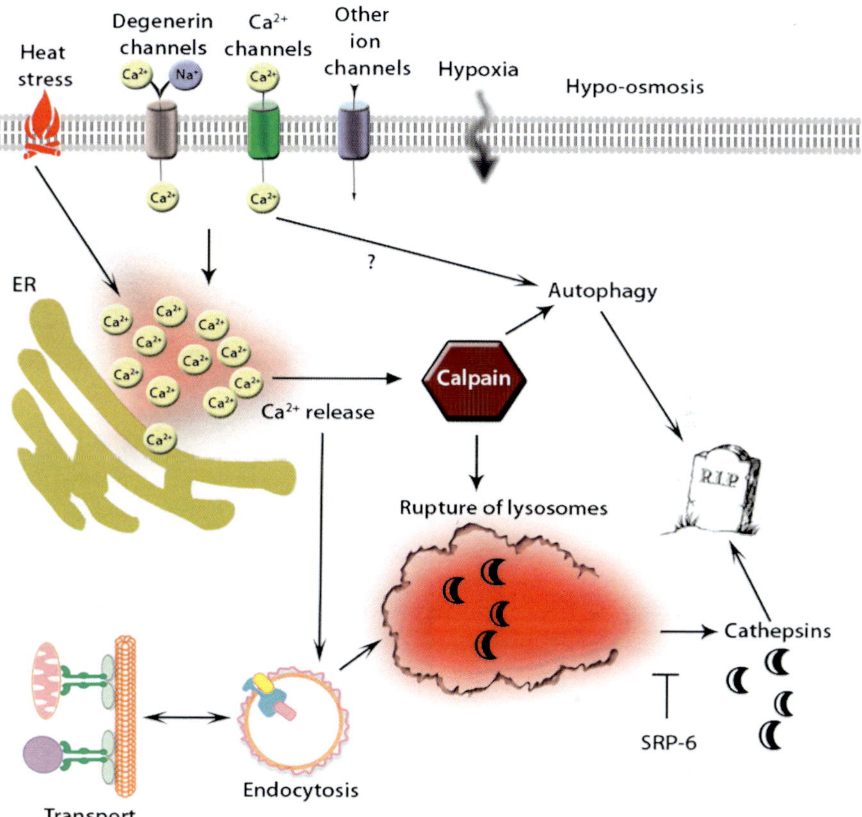

Vassiliki Nikoletopoulou and Nektarios Tavernarakis, Figure 6.3 Necrotic cell death mechanisms. Various types of extrinsic necrotic insults converge on increased intracellular Ca^{2+} levels, caused by increased influx from extracellular pools through plasma membrane channels, or by Ca^{2+} efflux from intracellular stores, such as the endoplasmic reticulum (ER). Ca^{2+} then activates calpain proteases in the cytoplasm that attack lysosomal membrane proteins to compromise lysosomal integrity. Rupture of the lysosomes follows, and release of hydrolytic enzymes such as cathepsin proteases. In addition, autophagy is induced during necrosis, either directly by Ca^{2+} or via calpains and also contributes to cellular destruction. Moreover, both clathrin-mediated endocytosis and intracellular transport are required for necrotic death and are induced by necrosis-triggering insults.

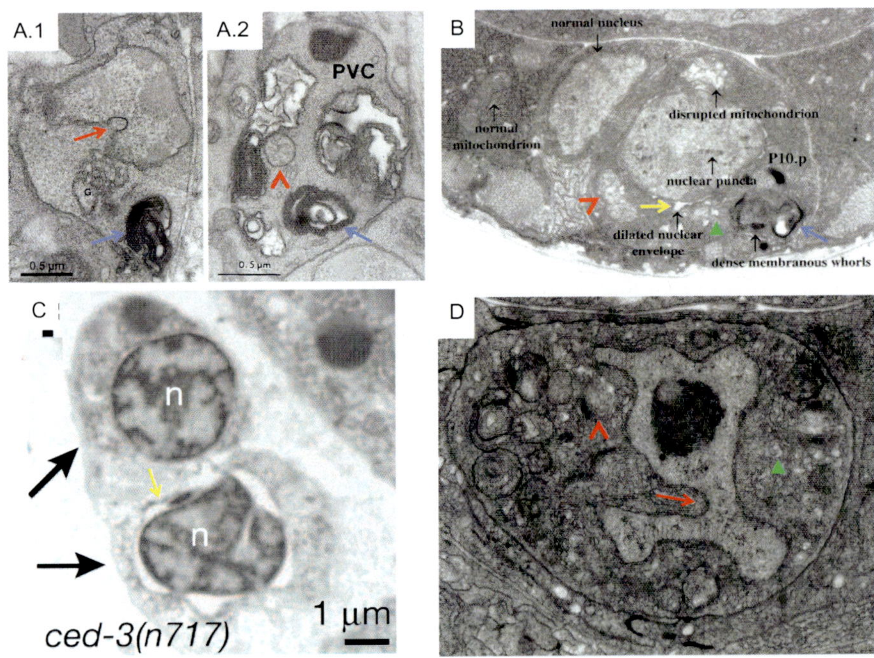

Maxime J. Kinet and Shai Shaham, Figure 7.1 Different cell death pathways share morphological features. (A) PVM neuron (A.1) of a *mec-4*(gf) mutant and PVC neuron (A.2) of a *deg-1*(gf) mutant. (B) P10.p cell in a *lin-33*(gf) animal. (C) Shed cells (arrows) in a *ced-3(n717)* embryo. (D) Dying linker cell. Nuclear indentations, red arrows in A.1, D. Membranous whorls, blue arrows in A.1, A.2, B. Dilated ER, green arrowheads in B, D. Dilated nuclear envelope, yellow arrows in B, C. Dilated mitochondria, red arrowheads in A.2, B, D. Dark intranuclear structure in (D) is the linker cell nucleolus. *Reproduced with permission from (A) Hall et al. (1997), (B) Galvin, Kim, and Horvitz (2008), (C) Denning, Hatch, and Horvitz (2012), and (D) Abraham, Lu, and Shaham (2007).*

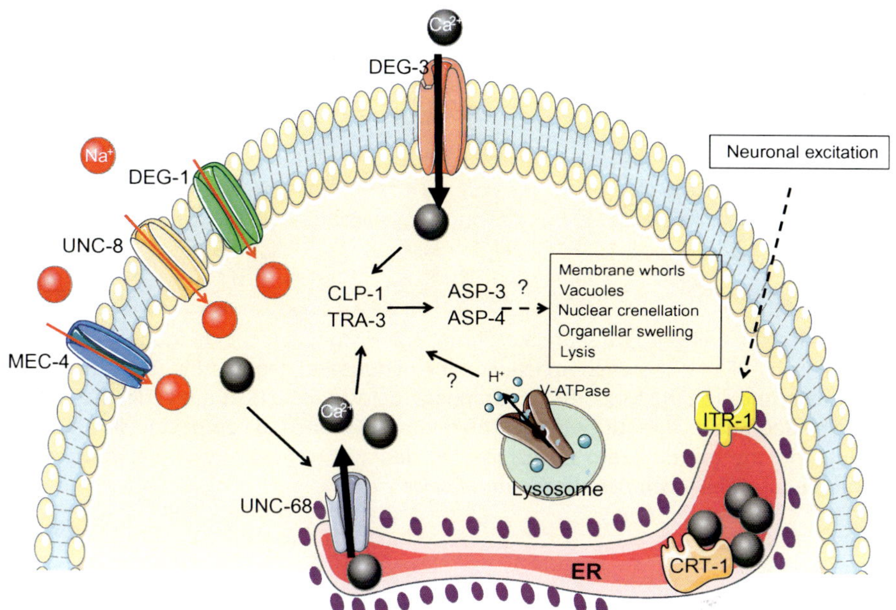

Maxime J. Kinet and Shai Shaham, Figure 7.2 Mechanisms of ion channel mutation induced death in *C. elegans*.

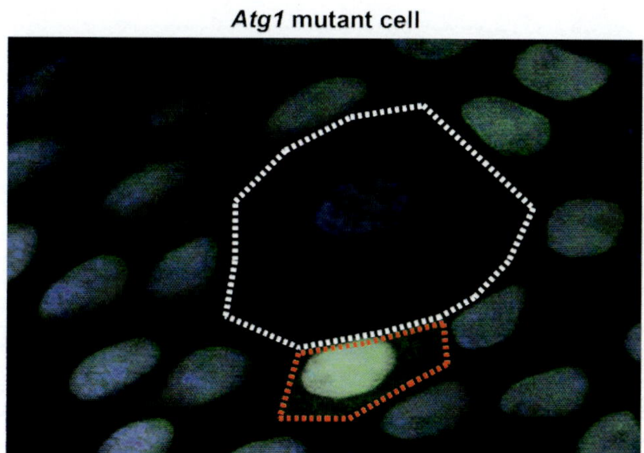

Charles Nelson and Eric H. Baehrecke, Figure 8.1 *Drosophila* genetic techniques allows for the generation of mutant cells in an otherwise heterozygous animal. In this case, the white-outlined *Atg1* homozygous mutant cell was generated using flippase-based mitotic recombination and is marked by the absence of green fluorescent protein (GFP). The greener GFP-marked cell (red outline) is a control homozygous wild-type cell, while the remaining lighter green GFP cells are control heterozygous wild-type/*Atg1* mutant cells. *Reprinted from Chang et al. (2013) with permission from Nature Publishing Group.*

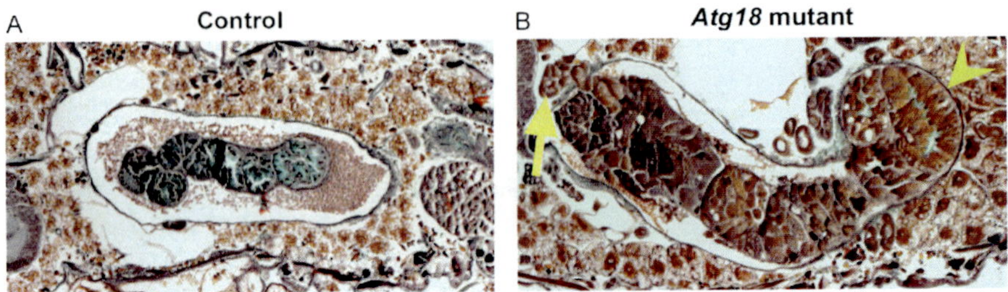

Charles Nelson and Eric H. Baehrecke, Figure 8.2 Histology of the midgut. (A) A control animal 12 h APF that shows compaction of the midgut and degradation of gastric caeca and proventriculus structures. (B) An autophagy-defective *Atg18* mutant animal that has failed to degrade its midgut properly. The midgut has failed to compact (yellow arrowhead), and the gastric caeca have failed to degrade (yellow arrow). *Reprinted from Denton et al. (2009) with permission from Elsevier.*

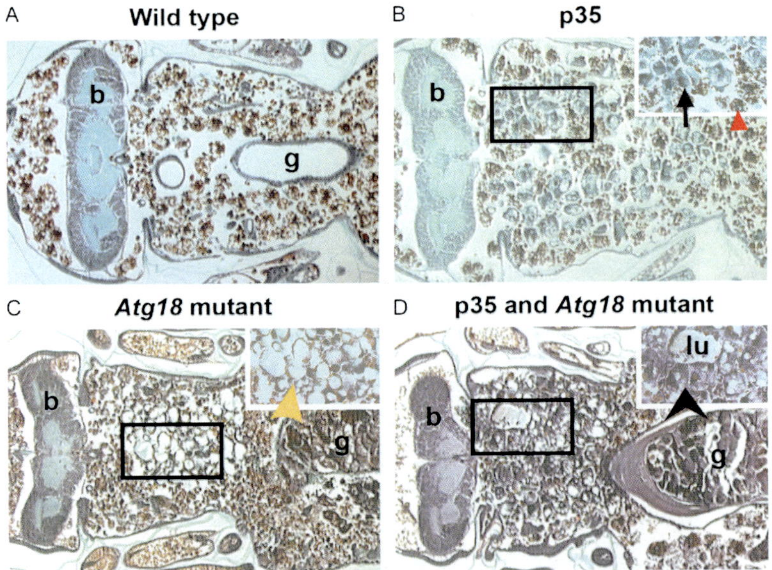

Charles Nelson and Eric H. Baehrecke, Figure 8.3 Histology of the salivary glands. (A) A wild-type animal in which the salivary glands have been properly degraded and cleared. (B) An animal in which the caspase inhibitor p35 has been expressed specifically in the salivary glands. This animal has condensed cell fragments (black arrow) diffused throughout its thorax. The red arrowhead indicates a fat body cell. (C) An autophagy-defective *Atg18* mutant animal that has failed to degrade its salivary glands. This animal has vacuolated cell fragments (yellow arrowhead) diffused throughout its thorax. (D) An *Atg18* mutant animal in which p35 has been expressed specifically in the salivary glands. This animal has gland fragments (black arrowhead) that also display remnants of salivary gland structure such as a lumen. Symbols are (b) brain, (g) midgut, and (lu) salivary gland lumen. *Reprinted from Berry and Baehrecke, 2007 with permission from Elsevier.*

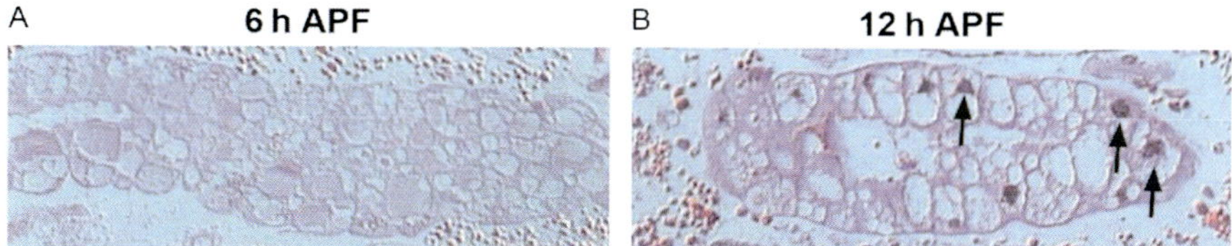

Charles Nelson and Eric H. Baehrecke, Figure 8.4 TUNEL staining. (A) A wild-type salivary gland 6 h APF showing no TUNEL staining indicating that DNA is not nicked and the salivary gland is not undergoing programmed cell death. (B) A wild-type salivary gland 12 h APF showing TUNEL-positive staining (black arrows) indicating nicked DNA and the salivary gland is undergoing programmed cell death. *Reprinted from Lee and Baehrecke (2001) with permission from Company of Biologists.*

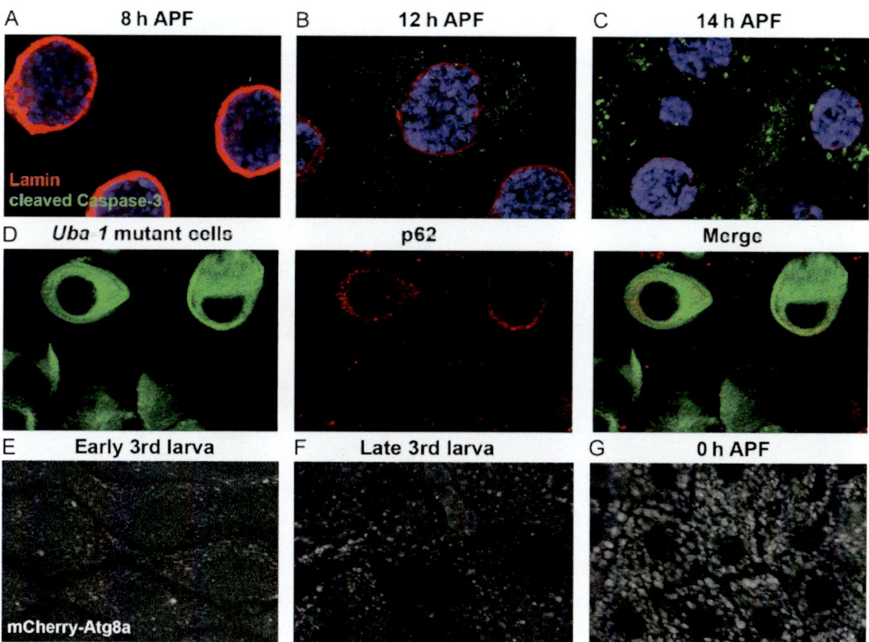

Charles Nelson and Eric H. Baehrecke, Figure 8.6 Immunofluorescence in the salivary gland and midgut. (A–C) Salivary glands of varying stages stained with anti-Lamin (red) and anti-cleaved-Caspase3 (green) antibodies. (A) A salivary gland 8 h APF when caspases are not active so Lamins remain intact (anti-Lamin staining is highly detectable and surrounds the nucleus) and Caspase-3 remains uncleaved. (B) A salivary gland 12 h APF when caspases are first activated resulting in the cleavage of Lamins and Caspase-3 (anti-cleaved-Caspase3 staining appears as puncta). (C) A salivary gland 14 h APF when Lamins are almost undetected and cleaved Caspase-3 is abundant. (D) A 0 h APF midgut in which mitotic recombination has been induced to create *Uba1* mutant clone cells that are marked with GFP (green). In these cells, autophagy is defective and the autophagy substrate p62 accumulates (red). (E–G) Midguts from early third instar (E), late third instar (F), and 0 h APF (G) larvae that express a mCherry-Atg8a reporter. An increase in reporter puncta occurs as the animals pupate indicating an increase in autophagy levels. *Panels (A)–(C): Reprinted from Martin and Baehrecke (2004) with permission from Company of Biologists. Panels (D)–(G): Reprinted from Chang et al. (2013) with permission from Nature Publishing Group.*

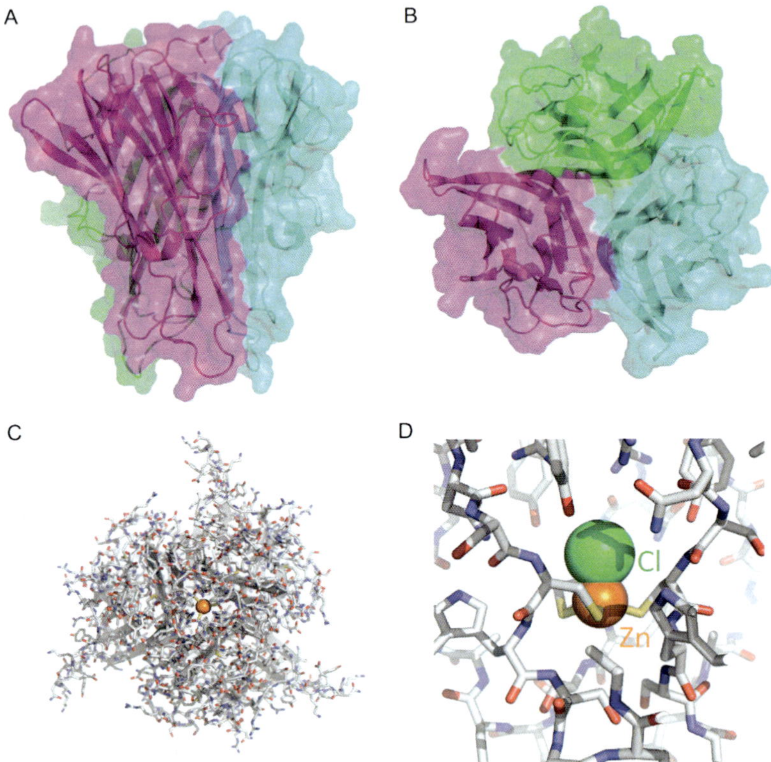

Paul C. Driscoll, Figure 9.1 The 3D structure of death ligands. (A and B) The homotrimer structure of TNFα (PDB 1TNF) (Eck & Sprang, 1989), showing the three chains in transparent van der Waals surface representation, and secondary structure in ribbon format. In the orientation shown in (A), the N-termini are at the top of the structure; the stalk regions connect the N-terminus to the transmembrane regions that are not part of the crystal structure. The structure has been rotated by 90° in (B) to highlight the N-terminal face and the threefold symmetry. (C and D) The structure of TRAIL (PDB 1DG6) (Hymowitz et al., 2000) shown in stick format, with (C) the N-terminal face to the front (C), and the zinc atom highlighted in orange. (D) The coordination of the zinc and the associated chlorine atom in side-on view; the threefold symmetry axis is collinear with the Zn—Cl bond.

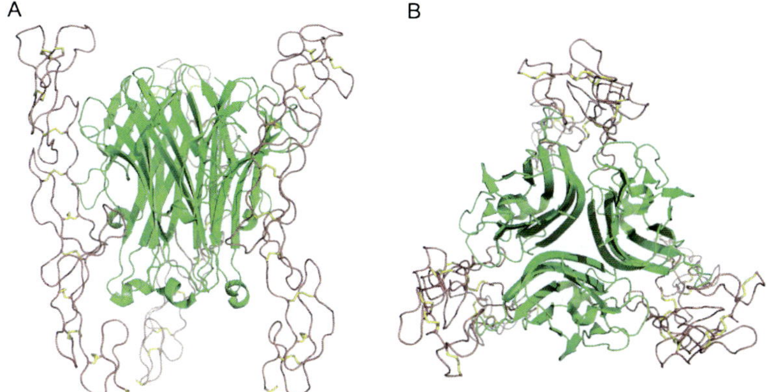

Paul C. Driscoll, Figure 9.2 The structure of the complex formed between lymphotoxin-α (TNFβ) and the ectodomain of TNFR1 (PDB 1TNR) (Banner et al., 1993). The lymphotoxin-α homotrimer is shown in green, and the three chains of TNFR1 in mauve. Disulfide connections are highlighted in yellow. In the orientation shown in (A), the ligand would be attached to a membrane at the top of the picture, and the receptor chains to a membrane at the bottom. The structure has been rotated through 90° in (B) so that the top (N-terminal) face of the complex is to the front, and to highlight that the each receptor chain to two chains of the ligand, but does not contact any other receptor chain.

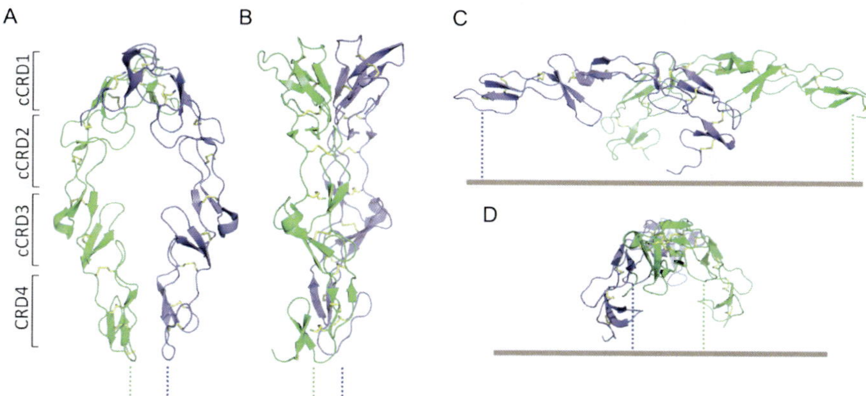

Paul C. Driscoll, Figure 9.3 The 3D structure of the ectodomain of TNFR1 crystallized in the absence of a ligand (PDB 1NCF) (Naismith et al., 1995). The structure is composed of three canonical cysteine-rich domains (cCRD1–3) and a fourth cysteine-rich domain (CRD4) that has different disulfide bond connectivities (A). The disulfide bond linkages are depicted in yellow. Within the crystal lattice two types of inter-chain contacts that suggest the potential for homodimerization can be discerned. (A and B) Two orientations, rotated by 90° about the vertical symmetry axis, of the parallel homodimer, and (C and D) the antiparallel homodimer. The approximate locations of the linking segments to the transmembrane regions are indicated by dashed lines. The transmembrane domains are likely to be much further apart (by up to 100 Å) in the case of the antiparallel homodimer.

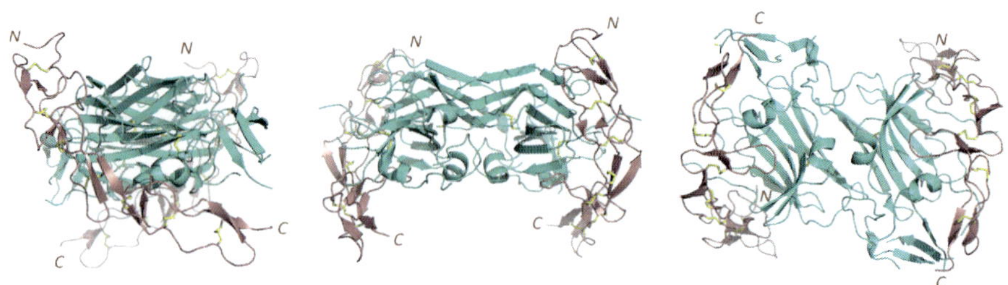

Paul C. Driscoll, Figure 9.4 The 3D structure of the complex formed between the human cytomegalovirus glycoprotein UL141 homodimer (teal) and the ectodomain of TRAIL-R2 (gray-brown) that blocks cell surface expression of the receptor, viewed from three orthogonal directions (PDB 4I9X) (Nemcovicova et al., 2013).

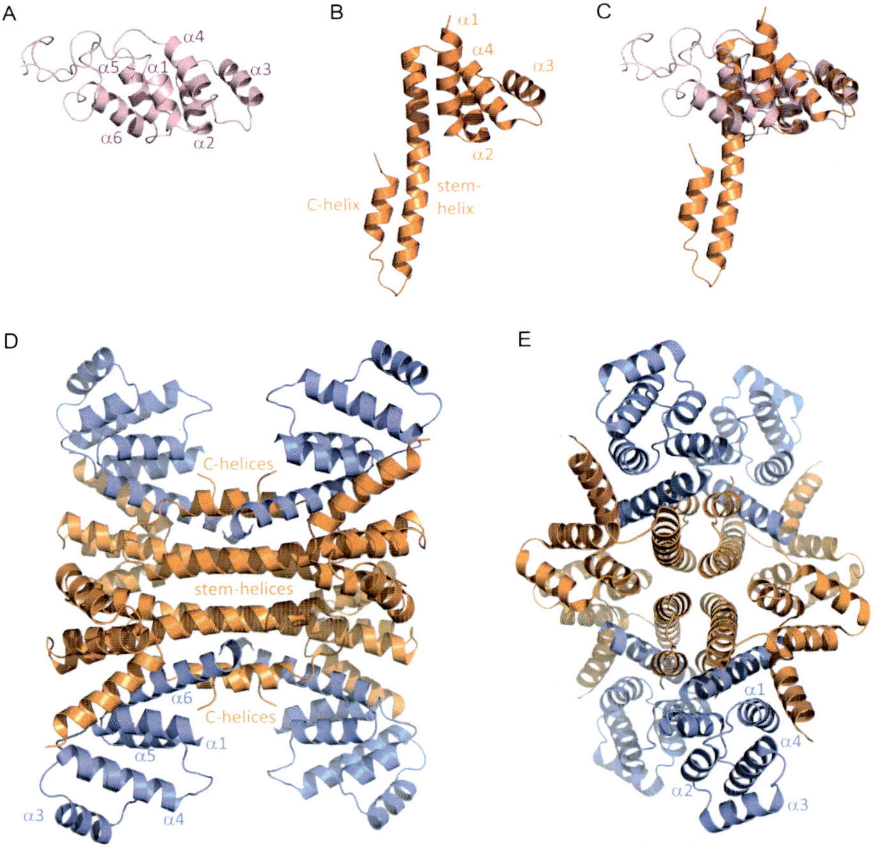

Paul C. Driscoll, Figure 9.5 Solution structure of the DD of CD95 and the crystallographic complex with the DD of the adaptor protein FADD. (A) The NMR-derived solution structure of the CD95-DD (PDB 1DDF) (Huang et al., 1996) illustrating the characteristic six-helix Greek-key topology. (B) The structure of CD95-DD extracted from the heterotetramic complex with FADD-DD (D and E) obtained by Scott et al. (2009) (PDB 3EZQ) showing the restructuring of helices $\alpha5$ and $\alpha6$ into the stem- and C-helix $\alpha\alpha$ hairpin. In (B–E), CD95-DD from 3EZQ is shown in orange; in (D and E) FADD-DD is shown in blue. (C) A superposition of the solution structure of CD95-DD with that observed in the complex with FADD-DD. The orientation shown in (E) is rotated by 90° relative to that in (D).

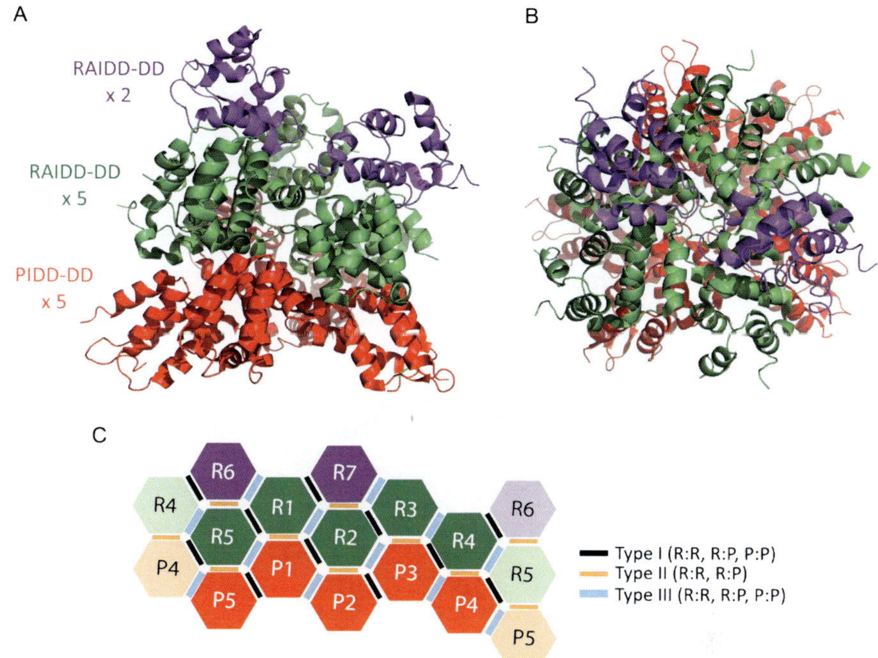

Paul C. Driscoll, Figure 9.6 The 3D structure of the PIDDosome core particle (PDB 2OF5) (Park, Logette, et al., 2007). The structure is comprised of seven RAIDD-DD domains and five PIDD-DD domains arranged approximately as three layers (A) around a screw axis evident in the center of the structure shown in the orientation (B) rotated by 90° relative to (A). The structure lacks formal symmetry but contacts between domains can be characterized in three classes types I–III as summarized schematically in (C) with is adapted from Park, Logette, et al. (2007). In (C), light colors indicate domains that appear twice in the diagram, due to the pseudo-cylindrical nature of the structure.

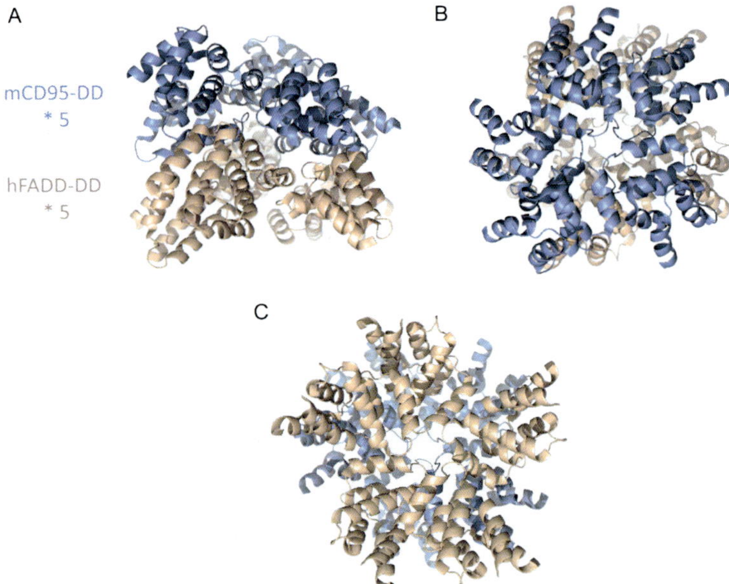

Paul C. Driscoll, Figure 9.7 The low-resolution (6.9 Å) crystal structure of the chimeric complex between mouse CD95-DD and human FADD-DD (PDB 3OQ9) (Wang et al., 2010). (A) The two-layer arrangement of five CD95-DD chains (blue) atop five FADD-DD chains (light brown). The model is based upon the arrangement of the bottom two layers of the PIDDosome core (Fig. 9.6). (B and C) Two different 90° rotations of the structure, relative to that in (A) to highlight the pseudo-helical symmetry axis with either the CD95-DD or FADD-DD domains to the fore, respectively.

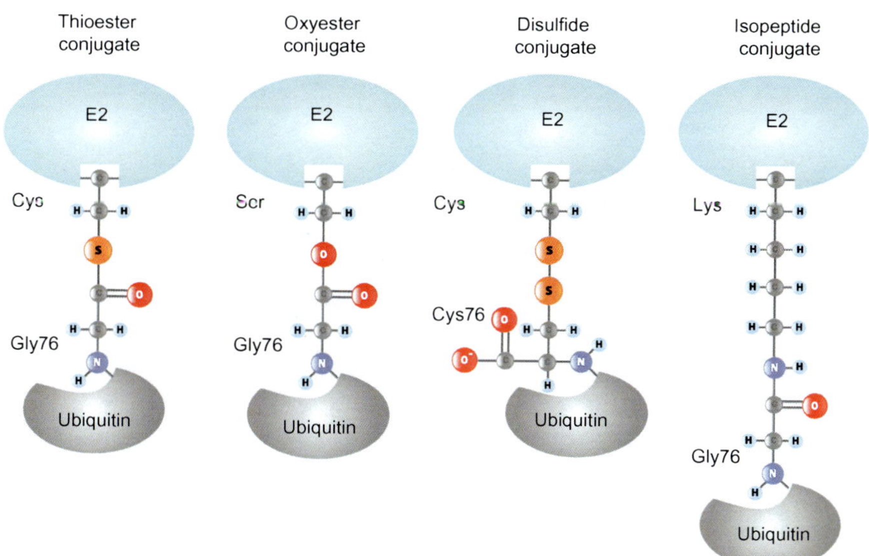

Adam J. Middleton *et al.*, Figure 10.1 Schematic showing the different E2 ~ Ub conjugates discussed.

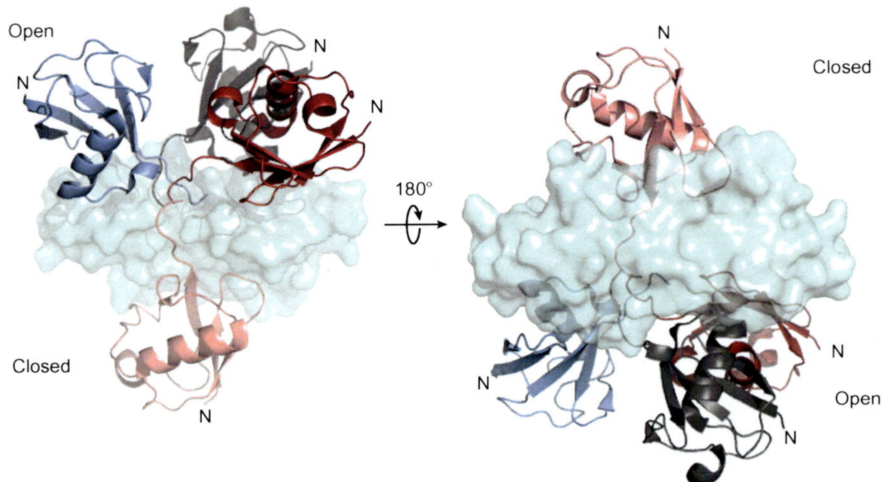

Adam J. Middleton *et al.*, Figure 10.3 E2 ~ Ub conjugate can adopt open and closed conformations. The figure was prepared by overlaying the E2 structures from the UbcH5b, Ubc1, UbcH8, and Ubc13 conjugate structures. The surface of UbcH5b is shown in pale blue. The ubiquitin moieties for each conjugate are shown as ribbons (dark blue is from PDB ID: 2KJH, red: 3A33, gray: 2JMI, and pink: 1FXT). The left panel is oriented with the active site at front, and the right panel is a 180° rotation.

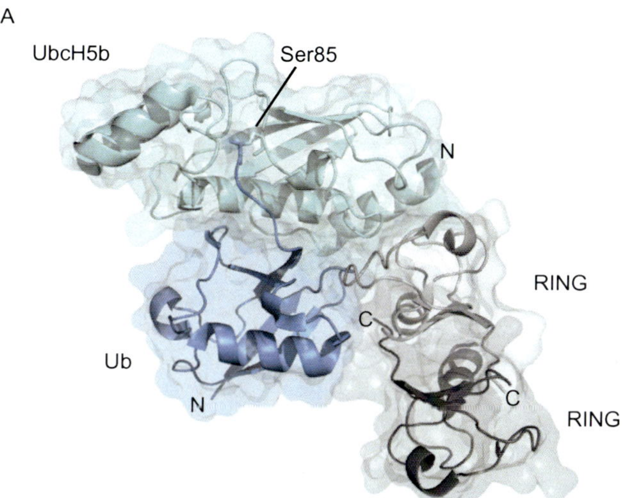

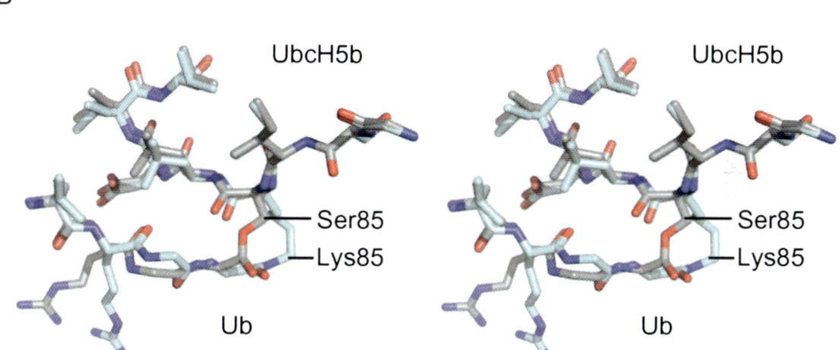

Adam J. Middleton *et al.*, Figure 10.5 Structure of RING-E2 ~ Ub conjugate complexes. (A) Structure of a dimeric RING (BIRC7) in complex with the oxyester-linked UbcH5b ~ Ub conjugate (PDB ID: 4AUQ). E2 is in light blue, ubiquitin in dark blue, and the RING dimer is shown in light and dark gray. (B) Overlay showing the oxyester and isopeptide linkages of closed-conformation conjugates in stereo view. E2 molecules were overlaid from PDB IDs 4AP4 and 4AUQ.

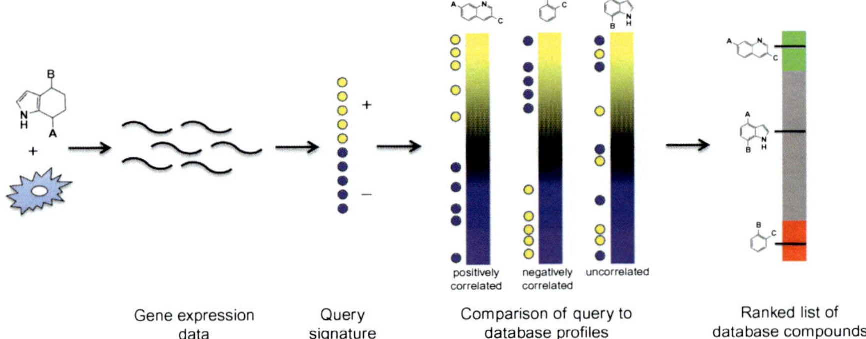

Adam J. Wolpaw and Brent R. Stockwell, Figure 11.3 Connectivity Map schematic. An outline of the process of querying the Connectivity Map is shown. Gene expression information is generated from an independent source, shown here as small-molecule treatment of cells. Those data are used to generate a unitless query signature which is then compared to the database of gene expression profiles. Small molecules in the database are ranked based on how well their profiles are correlated to the query signature.

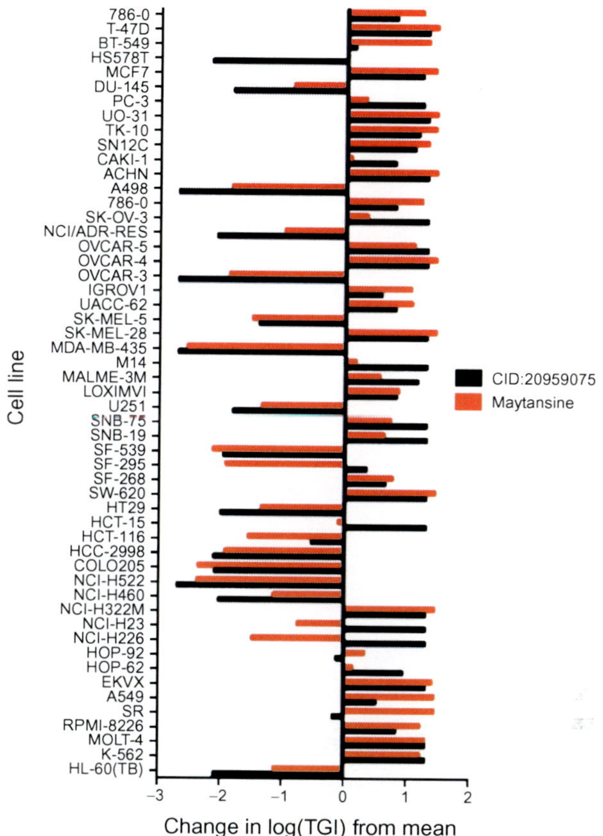

Adam J. Wolpaw and Brent R. Stockwell, Figure 11.4 NCI60 data display and COMPARE analysis. Yang and colleagues (Yang et al., 2012) identified the highly potent cytotoxic compound 20959075. Its profile across the NCI60 cell lines is shown in the black bars. The bars represent the distance of the total growth inhibition (TGI) from the mean of all cells lines tested. COMPARE analysis identified maytansine, a known microtubule destabilizer, as the most highly correlated compound in the database with a correlation of 0.808. Its profile is shown in the red bars. The authors went on to show that 20959075 is in fact a microtubule destabilizer.

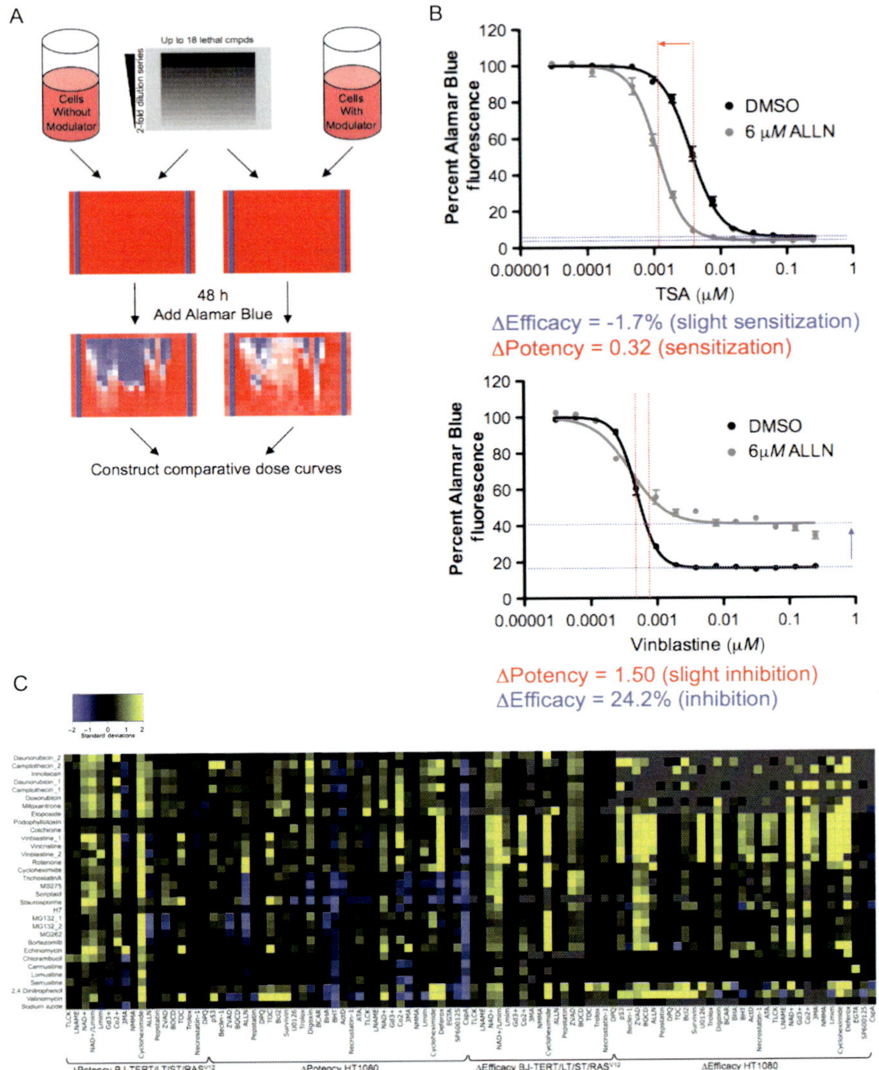

Adam J. Wolpaw and Brent R. Stockwell, Figure 11.5 Creating modulatory profiles. (A) Cells with or without modulator were seeded into 384-well plates and lethal compounds were added in a dilution series. Viability was measured after 48 h with Alamar Blue and comparative concentration–response curves were constructed from the data. (B) Two examples of comparative dose curves. These illustrate the two parameters extracted from each pair of curves, the change in potency and the change in efficacy. (C) Heat map illustrating the modulatory profiles for a number of characterized lethal compounds. Each row represents the modulatory profiles for a different compound. *Reproduced with permission from Wolpaw et al. (2011).*

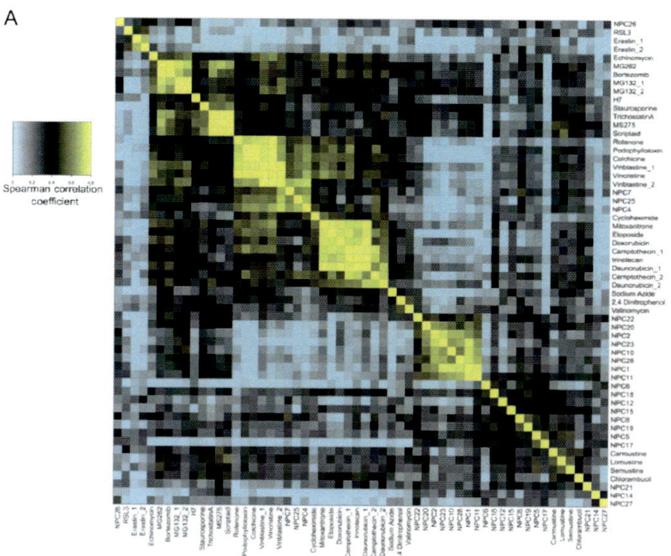
A
Spearman correlation coefficient
NPC26
RSL3
Erastin_1
Erastin_2
Echinomycin
MG262
Bortezomib
MG132_1
MG132_2
H7
Staurosporine
TrichostatinA
MS275
Scriptaid
Rotenone
Podophyllotoxin
Colchicine
Vinblastine_1
Vincristine
Vinblastine_2
NPC7
NPC25
NPC4
Cycloheximide
Mitoxantrone
Etoposide
Doxorubicin
Camptothecin_1
Irinotecan
Daunorubicin_1
Camptothecin_2
Daunorubicin_2
Sodium Azide
2,4 Dinitrophenol
Valinomycin
NPC22
NPC20
NPC2
NPC23
NPC10
NPC28
NPC1
NPC11
NPC6
NPC18
NPC12
NPC15
NPC8
NPC19
NPC5
NPC17
Carmustine
Lomustine
Semustine
Chlorambucil
NPC21
NPC14
NPC27

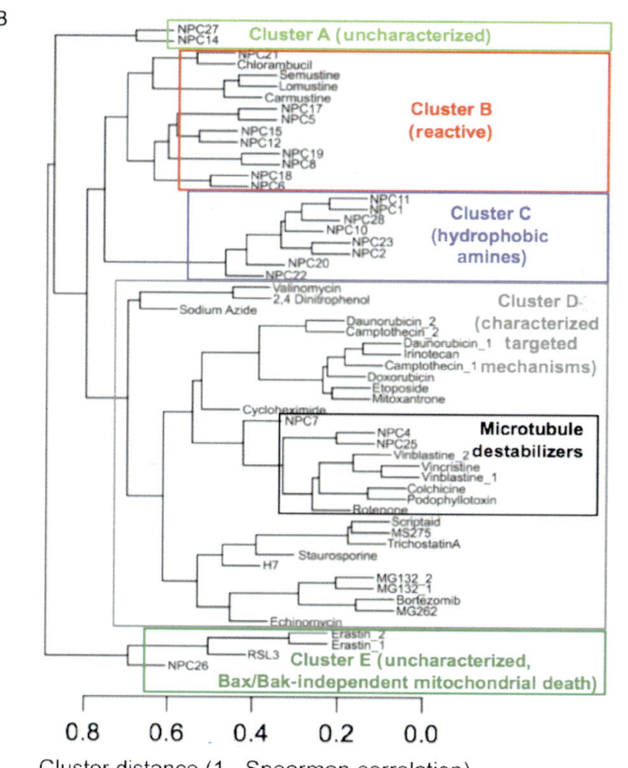
B
NPC27
NPC14
Cluster A (uncharacterized)
NPC21
Chlorambucil
Semustine
Lomustine
Carmustine
NPC17
NPC5
NPC15
NPC12
NPC19
NPC8
NPC18
NPC6
Cluster B (reactive)
NPC11
NPC1
NPC28
NPC10
NPC23
NPC2
NPC20
NPC22
Cluster C (hydrophobic amines)
Valinomycin
2,4 Dinitrophenol
Sodium Azide
Cluster D (characterized targeted mechanisms)
Daunorubicin_2
Camptothecin_2
Daunorubicin_1
Irinotecan
Camptothecin_1
Doxorubicin
Etoposide
Mitoxantrone
Cycloheximide
NPC7
Microtubule destabilizers
NPC4
NPC25
Vinblastine_2
Vincristine
Vinblastine_1
Colchicine
Podophyllotoxin
Rotenone
Scriptaid
MS275
TrichostatinA
Staurosporine
H7
MG132_2
MG132_1
Bortezomib
MG262
Echinomycin
Erastin_2
Erastin_1
RSL3
NPC26
Cluster E (uncharacterized, Bax/Bak-independent mitochondrial death)
0.8
0.6
0.4
0.2
0.0
Cluster distance (1– Spearman correlation)

Adam J. Wolpaw and Brent R. Stockwell, Figure 11.6 Comparing and clustering modulatory profiles. (A) Heat map of the similarity matrix showing the Spearman correlation between modulatory profiles of both characterized and uncharacterized lethal compounds. (B) Dendrogram derived from clustering the similarity matrix shown in (A). Five broad clusters are highlighted and lettered. In addition, microtubule destabilizers are shown in black, a cluster that includes three previously uncharacterized compounds. Other features that are not highlighted include clustering of characterized compounds according to their known mechanisms of action—alkylating agents, mitochondrial poisons, topoisomerase inhibitors, histone deacetylase inhibitors, and proteasome inhibitors. *Reproduced with permission from Wolpaw et al. (2011) and slightly altered.*

CPI Antony Rowe
Eastbourne, UK
August 05, 2014